XXI International Scientific Conference of Young Scientists and Specialists (AYSS-2017)

EPJ Web of Conferences Volume 177 (2018)

Dubna, Russia
2 – 6 October 2017

Editors:

A. Verkheev
A. Aparin
I. Bobrikov
V. Chudoba
A. Friesen

ISBN: 978-1-5108-6195-4

Printed from e-media with permission by:

Curran Associates, Inc.
57 Morehouse Lane
Red Hook, NY 12571

Some format issues inherent in the e-media version may also appear in this print version.

Printed by Curran Associates, Inc. (2018)

For additional information, please contact EDP Sciences – Web of Conferences at the address below.

EDP Sciences – Web of Conferences
17, Avenue du Hoggar
Parc d'Activité de Courtabœuf
BP 112
F-91944 Les Ulis Cedex A
France

Phone: +33 (0) 1 69 18 75 75
Fax: +33 (0) 1 69 28 84 91

contact-edps@webofconferences.org

Additional copies of this publication are available from:

Curran Associates, Inc.
57 Morehouse Lane
Red Hook, NY 12571 USA
Phone: 845-758-0400
Fax: 845-758-2633
Email: curran@proceedings.com
Web: www.proceedings.com

TABLE OF CONTENTS

The reduction of atmospheric emissions after the implementation of first Polish nuclear power plant

Maciej Cholewiński[1,*]

[1]Chair of Energy Technologies, Turbines and Modelling of Thermal and Fluid Flow Processes, Faculty of Mechanical and Power Engineering, Wrocław University of Science and Technology, Wybrzeże Wyspiańskiego 27, 50-370 Wrocław, Poland

Abstract. In this work the environmental benefits in the atmospheric emissions after the implementation of 3,000 MW nuclear power plants were assessed and presented. To determine the quantity of avoided emissions of CO_2, NO_x, SO_2 and Hg compounds, harmonised stoichiometric combustion model dedicated to solid fuel fired power plant was created. To increase the credibility of the studies, future strict emission standards (Directive 2010/75/EU, BAT documents for LCP) were included as well. In conducted studies, representative samples of 3 different Polish solid fuels were examined (by comprehensive proximate and ultimate analysis) and used in assessment. It was proven that by the replacement of thermal solid fuel power plant by nuclear unit (with annual production rate of 22.4 TWh net) up to 16.4 million tonnes of lignite, 8.9 million tonnes of hard coal or 13.1 million tonnes of solid biomass can be saved. Further, for the case of lignite, the emission, at least, of 21.29 million tonnes of CO_2 (6.9% of all Polish emission in 2015), 1,610 tonnes of dust (0.4%), 16,102 tonnes of NO_x (2.2%), 16,102 tonnes of SO_2 (2.0%) and 564 kg of mercury (5.9%) can be avoided. For selected hard coal, calculated emission savings were equal to 17.60 million tonnes of CO_2 (5.7%), 1,357 tonnes of dust (0.4%), 13,566 tonnes of NO_x (1.9%), 13,566 tonnes of SO_2 (1.7%), 271 kg of mercury (2.9%), and for biomass - equal to 20.04 million tonnes of CO_2 (6.5%), 1,471 tonnes of dust (0.4%), 14,712 tonnes of NO_x (2.0%), 14,712 tonnes of SO_2 (1.8%) and 294 kg of mercury (3.1%).

1 Introduction

Polish power sector is currently undergoing a period of fundamental economic and technological reorganisation. In order to execute upcoming international pro-environmental programmes (i.e. Minamata Convention on Mercury, new EU energy policy, European CO_2 Emission Allowances Trading System), several low-emission power units and energy storage facilities need to be connected to the Polish electrical grid within the following 10-20 years. By the replacement of coal-fired (mainly lignite) units with renewable energy sources (RES) and nuclear power plants, it will be possible to reduce the quantity of

* Corresponding author: maciej.cholewinski@pwr.edu.pl

national emissions of several pollutants substantially. Moreover, reorganised fuel structure of power sector will allow Polish authorities to adjust national emission rates (i.e. regarding CO_2 releases per every MWh of electricity) to the values proposed in new energy policy of European Union (i.e. within the EU Winter Package) [1]. What is particularly interesting in Poland, while RES have been already integrated in power grid (22.7 TWh generated in 2015 – approx. 45% from wind turbines and 50% from solid biomass boilers), nuclear power plants have not been operated in Poland yet.

To support national generation emission factors (i.e. concerning CO_2 or Hg), first Polish nuclear power plant (NPP) is going to be built in Poland in the next 12-15 years. It is planned to operate a 2-3 reactors with total capacity of 3,000 MW and to generate 22.4 TWh of electricity (approx. 14% of total generation in Poland in 2015) by 2030-2035 as well as up to 43.2 TWh from 6,000 MW of nuclear units by 2050. While first Polish NPP will be probably placed in Pomerania Voivodeship, second one will be located in eastern Poland [1,2]. Bidding process for selection of NPP technology for first power unit should start in 2018 – from the operational point of view, several generation III or III+ water cooled technologies will be taken into account, both PWR and BWR class).

As mentioned before, one of the main goals of the replacement of coal-fired power plants by NPP and RES in the reduction of atmospheric emissions. Nowadays, due to the vast consumption of hard coals (50.1% of total electricity in Poland was generated in coal-fired units in 2015) and lignite (31.5% of 165 TWh provided to the grid), power sector in Poland has become the largest emitter of CO_2 (51.5% of 310.3 million tonnes released from Poland in 2015), SO_2 (53.5% of 690.3 thous. tonnes), NO_x (28.8% of 713.8 thous. tonnes) and Hg (49.6% of 10.6 tonnes). Interestingly, only 8.8% of total quantity 317.7 thous. tonnes of dusts were introduced into the atmosphere from professional power sector [3].

In conclusion, when introduced, first Polish NPP should lead to the notably decrease of CO_2, SO_2, NO_x and Hg emissions first and seems to be a crucial part of future low-emission power market in Poland. By the replacement of coal-fired units with NPP, national consumption of hard coal and lignite in power sector may be lowered as well. To evaluate the possible both fuel and emission savings, calculation method dedicated to solid fuel combustion units was prepared.

2 Materials and methods

To assess the environmental benefits of the implementation of 3,000 MW net NPP (with total annual production of electricity equal to 22.3 TWh net), calculation tool based on stoichiometric method was adopted [4]. Air-fuel equivalence ratio was set at 1.2, average net efficiency of solid fuel power plants at 40% and annual capacity ratio at 85% (7,446 hr/yr). To simplify the calculations, the value of carbon content in slag was fixed at 3%, while in fly ash – at 4% (relatively high, i.e. to enhance the uptake of mercury on fly ash). Then, equation (1) was used to identify the quantity of CO_2, H_2O, O_2, N_2, SO_2, HCl and Hg in raw flue gas. To calculate the mass concentration at standard temperature and pressure (S_{STP}), Avogadro's hypothesis was included. Finally, all values at STP were corrected for the water vapour content of the waste gases and at a standardised O_2 content of 6% (S_{st} – reference conditions) [5].

$$\begin{aligned} &8.33\cdot10^{-2}\,a\,\mathrm{C} + 0.5\,b\,\mathrm{H_2} + 3.57\cdot10^{-2}\,c\,\mathrm{N_2} + 3.13\cdot10^{-2}\,d\,\mathrm{S} + 3.13\cdot10^{-2}\,e\,\mathrm{O_2} + \\ &+ 5.56\cdot10^{-2}\,f\mathrm{H_2O} + 4.98\cdot10^{-3}\,g\,\mathrm{Hg} + 3.13\cdot10^{-2}\,h\,\mathrm{Cl} + \lambda\,M_s\,(\mathrm{O_2} + 3.76\,\mathrm{N_2}) = \\ &= n_1\,\mathrm{CO_2} + n_2\,\mathrm{H_2O} + n_3\,\mathrm{O_2} + n_4\,\mathrm{N_2} + n_5\,\mathrm{SO_2} + n_6\,\mathrm{HCl} + n_7\,\mathrm{Hg} + n_8\,\mathrm{C(a)} + n_9\,\mathrm{C(s)}, \end{aligned} \quad (1)$$

where: *a-h* – different constituents contents (weigh fraction) in fuel (*a* – carbon, *b* – hydrogen, *c* – nitrogen, *d* – sulphur, *e* – oxygen, *f* – moisture, *g* – mercury, *h* – chlorine; see Table 1. and Table 2.), λ - air–fuel equivalence ratio, M_s – number of moles of oxygen delivered per 1 mole of fuel in stoichiometric conditions, n_i – calculated number of moles of selected component accumulated in flue gases from combustion process of 1 g of fuel (1 – carbon dioxide, 2 - water vapour, 3 – oxygen, 4 – nitrogen, 5 – sulphur dioxide, 6 - hydrogen chloride, 7 – elemental mercury, 8 – carbon in fly ash, 9 – carbon in slag).

Table 1. Ultimate analysis of selected fuels (air-dried samples) [4,6].

Fuel	C	H	N	S	O	Hg	Cl
	wt%					ppb	%
hard coal	60.05	4.07	1.36	0.46	7.83	76	0.13
lignite	55.19	4.49	0.65	1.76	17.32	545	0.07
solid biomass	50.32	5.64	1.52	0.08	31.95	38	0.30

Legend: C – carbon content, H – hydrogen content, N – nitrogen content, S – total sulphur content, O – oxygen content, Hg – mercury content, Cl – chlorine content

Table 2. Proximate analysis and calorific values of selected fuels (air-dried samples) [4,6].

Fuel	W	W_{ex}	A	V	FC	HHV	LHV
	wt%					MJ/kg	
hard coal	1.05	10.07	25.05	28.41	45.49	23.43	22.52
lignite	4.39	35.08	16.13	44.44	35.03	19.99	18.90
solid biomass	1.50	16.51	8.69	68.88	20.93	18.86	17.59

Legend: W – moisture content , W_{ex} – external moisture content, A – ash content, V - volatile matter content, FC – fixed carbon content, Q - higher heating value, LHV - lower heating value

To calculate the dust concentration in raw flue gases (at STP), empirical equation (2) were implemented.

$$S_{dust} = (10\, A^r L_{dry}\, a_u + C_{fa}\, L_{dry}^{-1} B^{-1})\, (21 - O_2^{st})\, (21\text{-}O_2)^{-1}, \tag{2}$$

where: S_{dust} - fly ash concentration in g/m^3_{st}, A^r – ash content in received fuel in %, L_{dry} – the quantity of dry flue gases per 1 kg of fuel, a_u – the ratio of the quantity of mineral matter accumulated in fly ash to total mineral matter in received fuel (fixed at 0.9 - for pulverised fuel fired boilers), C_{fa} – the stream of carbon in fly ash in kg/s, B – fuel consumption in kg/s, O_2^{st} – standardised oxygen concentration in flue gas (6% for solid fuels), O_2 – calculated oxygen concentration in raw flue gases. To determine the quantity of wet flue gas created per every 1 g of solid fuel, n_1 .. n_7 shall be summed up.

Finally, to determine the annual emissions E_i of selected pollutants (restricted by the emission standards – Table 3), equation (3) was used.

$$E_i = 22.42\, n_i\, B\, \rho_{iSTP}\, (1 - \eta_i) = 8.07 \cdot 10^{10}\, n_i\, P\, \eta_{el}^{-1}\, LHV_r^{-1}\, \rho_{iSTP}\, S_{ies}\, S_{ist}^{-1}, \tag{3}$$

where: B – annual fuel consumption in kg, ρ_{iSTP} – density of gaseous form of component at STP in kg/m^3, η_i – capture efficiency of selected pollutant, -, P – annual net electricity generation in TWh, η_{el} – total net efficiency of power plant, -, LHV_r – lower heating value of fuel (in received state), S_{ies} – emission standard for selected pollutant in $\mu g/m^3_{st}$, S_{ist} – pollutant concentration in raw flue gases in $\mu g/m^3_{st}$.

To identify the impact of NPP on emission and fuel savings, 3 different types of solid energy carriers were analysed and used within calculations. They represent 3 solid fuels from Poland: hard coal, lignite and straw pellet. The results of ultimate and proximate analysis dedicated to selected fuels were presented in Table 1. and Table 2. All lab test were performed using PN/EN ISO standards i.e. to determine the contents of 7 different elements (using atomic absorption spectroscopy and ion chromatography – LECO CHNS TruSpec, AMA 254, Dionex ICS-1100 devices), moisture, ash, volatile matter (using gravimetric methods) and to evaluate calorific values (IKA C-2000 Basic) of analysed fuels. The methodology was presented in Author's previous works [4,6].

In calculations, maximum values of already introduced emission standards were adopted – in accordance to Directive 2010/75/EU (for SO_2, NO_x, dust) and BAT Reference Document for LCP (Hg, HCl) (see Table 3.). As a result, highest possible emission and fuel savings (using presented model and chosen energy carriers) were evaluated.

Table 3. Yearly average emission standards (all values at the STP: temperature 273.15 K and pressure 101.3 kPa, no water vapour content in gas and standardised O_2 content in gas equal to 6%) according to Directive 2010/75/EU on industrial emissions (IED) and BAT Reference Document for combustion plants using solid fuels with the total rated thermal input >300 MW [5,7]

Pollutant	**Coal and lignite and other solid fuels**		**Biomass and peat**	
	BAT	IED	BAT	IED
SO_2, mg/m^3	10-180	150-200	<10-100	150-200
NO_x, mg/m^3	50-175	150-200	40-160	150-200
Dust, mg/m^3	2-12	10-20	2-10	20
Hg, µg/m^3	< 1-7	-	<1-5	-
HCl, mg/m^3	< 1-7	-	1-25	-

To investigate the concentration of NO_x in raw fumes, empirical correlations for coal, lignite and biomass were included [8]. For hard coal and lignite, NO_x concentration in raw fumes was set as a level of 800 and 700 mg/m$^3_{st}$ (pulverized fuel boilers with tangential burner configurations), respectively, for biomass – 400 mg/m$^3_{st}$ (boiler with circulating fluidised bed).

3 Results

Calculated quality and quantity of raw fumes from the combustion of selected fuels were presented in Table 4. All values regarding emission standards were corrected to reference conditions. Additionally, CO_2 net emission factors were identified.

Table 4. Calculated quality and quantity of raw flue gas (for raw samples).

Fuel	L_{wet}	L_{dry}	O_{2dry}	S_{SO2}	S_{HCl}	S_{Hgtot}	S_{dust}	W_{CO2}	
	m^3_n/kg	m^3_n/kg	%	mg/m$^3_{st}$	mg/m$^3_{st}$	µg/m$^3_{st}$	g/m$^3_{st}$	kg/GJ	kg/MWh
hard coal	7.497	6.521	3.84	1109	161.1	9.16	27.2	97.32	875.8
lignite	5.348	4.270	3.75	4655	95.2	72.07	19.2	113.72	1023.5
solid biomass	5.923	4.859	3.67	238	458.7	5.65	11.6	107.08	963.76

Legend: L_{wet} – quantity of wet flue gases per 1 kg of fuel, L_{dry} – quantity of dry flue gases per 1 kg of fuel, O_{2dry} – concentration (by volume) of oxygen, SO_2 – sulphur dioxide concentration, S_{HCl} - hydrogen chloride concentration, S_{Hgtot} – total mercury concentration (>700°C), S_{dust} – fly ash concentration, W_{CO2} – CO_2 emission factors (kg of carbon dioxide per GJ energy accumulated in fuel and kg of carbon dioxide per MWh net introduced to the electrical grid)

Assuming that consumption of hard coal in Poland in 2015 (excluding electricity, gas, steam and air conditioning supply) reached 72.3 million tonnes (51% was consumed in power sector) and lignite – 63.1 million tonnes (99% was dedicated to power industry), the adaptation of first NPP in Poland should lead to the decrease of consumption by 12% and 26% (in comparison with 2015), respectively [9]. Moreover, it would result in reducing the total annual emission of CO_2 in Poland by 5.7-6.8%, SO_2 by 2.0-2.3%, NO_x by 2.0-2.3%, dust by 0.01-0.05% and mercury by 2.6-5.3% (including new environmental emission standards). Calculated absolute quantities of emission savings for each of the fuel were presented in Table 5.

Table 5. Calculated emission and fuel savings after implementation of 3,000 MW NPP

Fuel	Fuel savings	CO_2	SO_2	NO_x	Dust	Hg
	milion tonnes		thous. tonnes		tonnes	kg
hard coal	8.9	17.6	13.57	13.57	1.36	271
lignite	16.4	21.29	16.10	16.10	1.61	564
biomass	13.1	20.04	14.71	14.71	1.47	294

4 Conclusions

The replacement of conventional, mostly coal-fired, professional power units in Poland by first nuclear power plant will result in significant emission and fuel savings. For instance, the 3,000 MW class NPP in 2015 would help to reduce the total annual emissions of CO_2 by 5.7-6.8% and mercury up to 5.3%. For SO_2, NO_x and dust emission, obtained emission savings would not exceed 3% each. Therefore, when introduced, first NPP in Poland will undoubtedly become a vital part of pro-environmental power sector and significantly decrease the emission rates of Polish economy (CO_2 emission factor for analysed hard coal stood at 876 kg/MWh_{net}, for biomass – at 964 kg/MWh_{net} and for lignite – at 1024 kg/MWh_{net}).

References

1. Ministerstwo Gospodarki, *Wnioski z analiz prognostycznych na potrzeby Polityki energetycznej Polski do 2050 roku. Załącznik 2. do Polityki energetycznej Polski do 2050 roku* (in Polish) (Warsaw, June 2015)
2. Central Statistical Office (GUS), Production Department, *Energy statistics in 2014 and 2015* (Statistical Information and Elaboration, Warsaw, October 2016)
3. The National Centre For Emissions Management (KOBiZE), *Krajowy bilans emisji SO_2, NO_x, CO, NH_3, NMLZO, pyłów, metali ciężkich i TZO za lata 2014 – 2015 w układzie klasyfikacji SNAP. Raport syntetyczny* (in Polish) (Warsaw, February 2017)
4. M. Cholewiński, *Zeszyty Energetyczne* **4**, 73-87 (2017)
5. Official Journal of the European Union, *Directive 2010/75/EU of the European Parliament and of the Council of 24 November 2010 on industrial emissions (integrated pollution prevention and control)*
6. M. Cholewiński, *Zeszyty Energetyczne* **2**, 65-81 (2015)
7. Joint Research Centre, Institute for Prospective Technological Studies, Sustainable Production and Consumption Unit in European IPPC Bureau, *Best Available Techniques (BAT) Reference Document for Large Combustion Plants* (Final Draft, June 2016)
8. W. Kordylewski, *Spalanie i paliwa* (in Polish) (Oficyna Wydawnicza PWr, Wrocław, 2008)
9. Central Statistical Office (GUS), Production Department, *Zużycie Paliw i Nośników Energii w 2015 r.* (in Polish) (Statistical Information and Elaboration, Warsaw, December 2016)

Solvothermal synthesis of Au@Fe_3O_4 nanoparticles for antibacterial applications

Zhazgul Kelgenbaeva[1,*], *Zhypargul* Abdullaeva[2], and *Bektemir* Murzubraimov[1]

[1]Institute of Chemistry and Phytotechnology, National Academy of Sciences of the Kyrgyz Republic, 720071 Bishkek, Kyrgyzstan
[2]Department of Material Sciences and Engineering, Kumamoto University, 860-8555 Kumamoto, Japan

Abstract. We present Au@Fe_3O_4 nanoparticles obtained from Fe nanoparticles and $HAuCl_4$ using a simple solvothermal method. Trisodium citrate ($C_6H_5Na_3O_7$*$2H_2O$) served as a reducing agent for Au. X-ray diffraction analysis, electronic microscopes and energy-dispersive X-ray spectroscopy revealed cubic structure, elemental composition (Au, Fe and O) and spherical shape of nanoparticles. Antibacterial activity of the sample was tested against *E. coli* bacteria and obtained results were discussed.

1 Introduction

Bimetallic nanoparticles have recently attracted much attention, because of their unique catalytic, electronic, optical, magnetic and other novel dual (depending on metals) properties and wide range of possible applications [1]. In particular, binary 3d-5d nanoparticles formed by transition metals such as Fe or Co together with 5d noble metals such as Au or Pt allow the possibility to tune the magnetic properties based on an in-depth knowledge of their geometrical and magnetic behaviour [2-3]. The bimetallic nanoparticles are generally divided into two types in the structure: the alloyed and the core/shell structured bimetallic nanoparticles [4].

Concerning the synthesis of Au@Fe_3O_4 nanoparticles, Seino et al. [5] synthesized magnetic carriers consisting of nano iron oxide and gold core-shell structure in an aqueous solution by using γ–ray irradiation. Lyon et al. [6] prepared core-shell structures with gold coating layers on the surface of either Fe_2O_3 or partially oxidized Fe_3O_4 via iteration hydroxylamine seeding. Yu et al. [7] also synthesized the so-called dumbbell-like Au-Fe_3O_4 bifunctional nanoparticles with interesting novel properties.

Among existed techniques [8-9], solvothermal synthesis can be one of the simplest and environmentally friendly ways for preparation of bimetallic nanomaterials [10]. The method is based on chemical reaction between initial substances and solvent mixture, and being heated at temperatures from 100 to 1000 °C, and pressure starting from 1 MPa. Solvothermal reaction takes place inside the stainless steel autoclave, covered by stainless steel lid, and fixed tightly by screws. Experimental conditions of this method can be varied

* Corresponding author: jaza-86@mail.ru

and allow to synthesize different types of materials. In this work, Au@Fe_3O_4 bimetallic nanoparticles were synthesized from the mixture of Fe nanoparticles with $HAuCl_4$ solution using solvothermal reaction and their properties were studied. For biomedical applications, synthesized Au@Fe_3O_4 nanoparticles were tested against Escherichia coli bacteria. *E. coli*, a gram-negative bacterium can enter human intestines and cause urinary tract infection, cholecystitis, or septicaemia [11].

2 Experimental

Experimental setup presented in Figure 1a was used for Au@Fe_3O_4 nano-sample preparation. Fe nanoparticles with an average diameter of 5 nm, synthesized by pulsed plasma in liquid method [12], were used as Fe sources. Teflon cup (total volume of 60 mL) was filled with Fe nanoparticles, 1.0 mM tetracloroauric acid ($HAuCl_4$) as Au source, Trisodium Citrate Dehydrate (TCD: $C_6H_5Na_3O_7{*}2H_2O$) as a reducing agent for Au nanoparticles and deionized water. Experimental conditions were as following the temperature T = 150 °C, pressure P =1.8 MPa and reaction time t = 2 hours. After the reaction, the temperature of autoclave cooled to room temperature and the liquid was taken out of the autoclave. The colour of the liquid changed from transparent to bright brown, sample was separated from the liquid, dried and then characterized. The starting materials, experimental conditions and process of the nanoparticles' formation are schematically represented in Figure 1b.

The sample was characterized by X-ray diffraction (XRD) performed on a Rigaku RINT-2500 HV diffractometer (Cu- Kα radiation, λ= 0.15406 nm; voltage of 40 kV, and current of 200 mA). Surface view and elemental composition analysis were examined by JEOL JSM-7600F Field-Emission Scanning Electron Microscope (FE-SEM) equipped with an Energy-Dispersive X-ray Spectroscopy (EDS) operated at 200 keV. High-resolution transmission electronic microscopy (HR-TEM) with a Philips Tecnai F20S-Twin instrument was used to determine the size and morphology of the nanoparticles at 200 keV. For analysis the particles dispersed in ethanol were drop cast onto a carbon-film-coated copper grid, followed by evaporation and drying.

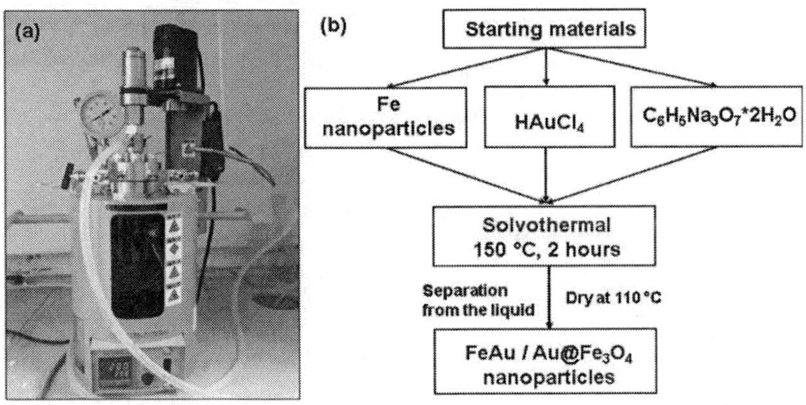

Fig. 1. a) Experimental setup of solvothermal method; b) Au@Fe_3O_4 nanoparticles preparation process by solvothermal synthesis

 https://doi.org/10.1051/epjconf/201817701002

3 Results and Discussion

XRD is a rapid analytical technique primarily used for phase identification of a crystalline material and can provide information on unit cell dimensions. Figure 2 shows an XRD pattern of the Au-Fe bimetallic nano-sample. The diffration peaks at (2Θ) 38.180, 44. 658, 64. 542, 77. 536 and 82. 300 were attributed to gold, which can be indexed to (111), (200), (220), (311) and (222) lattice planes of gold in cubic phase Fm-3m (JCPDS card no 04-0784). The peaks at (2Θ) 30.167, 35. 336, 36.136, 43.255, 57.382, 62. 595, 65.00, 89. 617 were identified as magnetite ($Fe+2Fe_2+3O_4$) with cubic spinel (Fd-3m) structure (JCPDS card no 19-0629), indicating the presence of both Fe_3O_4 (magnetite) and Au (gold) in the sample.

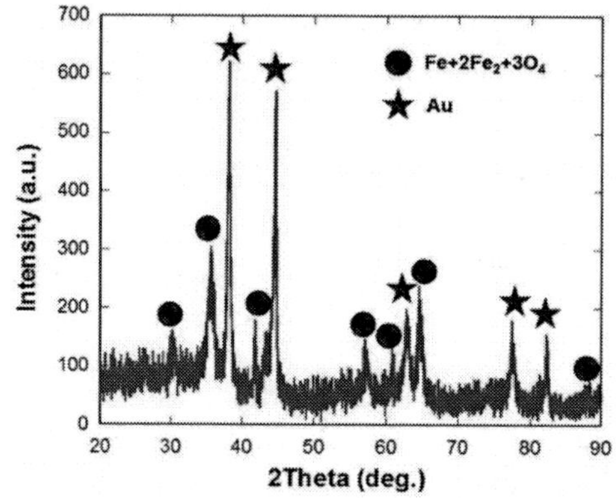

Fig. 2. X-Ray diffraction pattern of $Au@Fe_3O_4$ nanoparticles synthesized by solvothermal method

Further, the evidence for the presence of Au and Fe_3O_4 in the sample was proved by EDS, as shown in Figure 3 (left), where the elements gold (Au), iron (Fe), and oxygen (O) have been detected. The composition of the magnetite phase (Fe_3O_4) is 43 at% of iron and 57 at% of oxygen. In addition, peaks belonging to sodium (Na), chlorine (Cl) and carbon (C) were detected due to the reducing agent (TCD) and sample holding carbon tape.

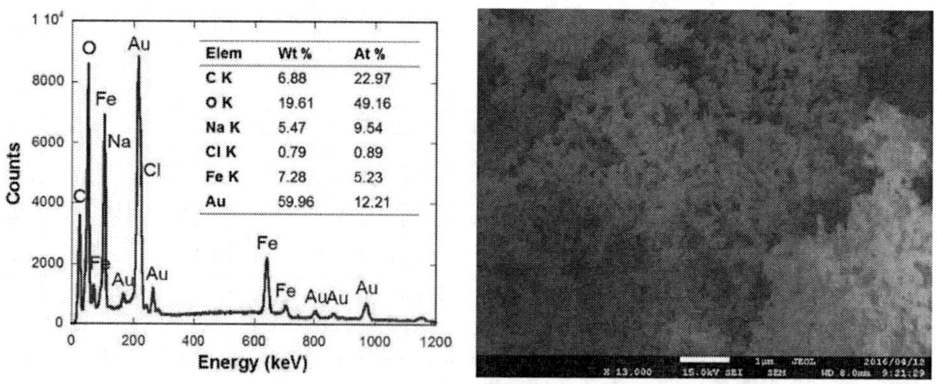

Elem	Wt %	At %
C K	6.88	22.97
O K	19.61	49.16
Na K	5.47	9.54
Cl K	0.79	0.89
Fe K	7.28	5.23
Au	59.96	12.21

Fig. 3. Elemental composition and microphotography of $Au@Fe_3O_4$ nanoparticles synthesized by solvo-thermal synthesis

Surface view of the $Au@Fe_3O_4$ nanoparticles taken by FE-SEM is presented on microphotography on the right (Figure 3). Image was captured at resolution of ×13000, and the agglomerates are clearly displayed. Because of the high surface energy of the nanoparticles and chemical bonds between the particle's surfaces, their tendency to form agglomerates is quite strong. Magnetic dipole forces also potentially contribute to inter-particulate forces [13]. As it can be seen from Figure 4 the size of obtained nanoparticles ranged between 10-13 nm.

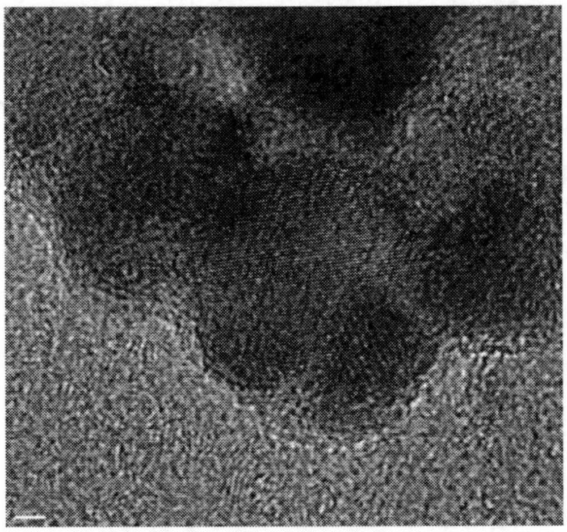

Fig. 4. HR-TEM image of several $Au@Fe_3O_4$ nanoparticles synthesized by solvothermal synthesis

Bacterial infections were one of the major causes of mortality in the nineteenth century. Nowadays, antimicrobial effects are intensively studied due to an enormously increasing bacterial resistance against excessively and repeatedly used classical antibiotics. Magnetic iron oxide nanoparticles are of particular interest as antibacterial agents, as they can be prepared with extremely high surface areas and unusual crystalline morphologies with a high number of edges and corners, and other potentially reactive sites [14]. Also, inorganic antibacterial agents such as metal and metal oxides are advantageous compared to organic compounds due to their stability. Ismail et al. studied that iron oxide nanoparticles with antibacterial properties and superparamagnetic nature are very prominent in biomedicine [15]. Antibacterial activity of the $Au@Fe_3O_4$ sample prepared by solvothermal synthesis was tested against *E coli* bacteria. Inhibition effect of the nanoparticles against the target microorganism was evaluated using 3M Petrifilm *E coli* / Coliform Count plates and inhibition time was 24 hours. Number of colonies were counted using the technique called TNTC (too numerous to count) and the results are represented in Table 1.

Table 1. Antibacterial activity of $Au@Fe_3O_4$ nanoparticles synthesized by solvothermal method

Nano particles	1 % concentration			0.1 % concentration			Control
	Probe I	Probe II	Probe III	Probe I	Probe II	Probe III	
$Au@Fe_3O_4$	101	102	97	200	207	199	350 counts
	Average: 100 counts			Average: 202 counts			

Following results were obtained after an average *E. coli* bacteria colonies calculation: as for control sheet there are 350 counts ; at 0.1 % concentration of the nanoparticles in medium - 202 counts and at 1.0 % concentration of the nanoparticles in media the value was 100. As it can be seen from obtained results, the bacteria count number decreased with increase of the nanoparticles concentration in medium. This means that $Au@Fe_3O_4$ nanoparticles synthesized by solvothermal synthesis can inhibit bacteria growth indicating on their good antibacterial activity.

4 Conclusions

Magnetic $Au@Fe_3O_4$ nanoparticles were synthesized from Fe nanoparticles and $HAuCl_4$, using TCD as reducing agent by solvothermal synthesis method. XRD and EDS analysis indicated the presence of Fe_3O_4 and Au phases in the sample. The $Au@Fe_3O_4$ nanoparticles have potent antibacterial activity against E coli bacteria and the inhibitory effect increases with increasing the nanoparticles concentration. Obtained results highlighte possible biomedical applications of $Au@Fe_3O_4$ nanoparticles as biocompatible components of antibiotics as well as medicaments.

This research was funded by start-up research acceleration fund (Funding number: 535-11535016) of the Kumamoto University) and by the International Innovative center of Nanotechnology, CIS (Funding number: 038/110).

References

1. O N. Toshima and T. Yonezawa, New J Chem., **22**, 1179-1201 (1998).
2. G. Rollman, M. E. Gruner, A. Hucht, R. Meyer, P. Entel, M. L. Tiago and J. R. Chelikowsky. Phys. Rev. Lett. **99,** 083402 (2007).
3. R. Singh and P. Kroll, Phys. Rev. **B 78**, 245404 (2008).
4. A. Zaleska-Medynska, M. Marchelek, M. Diak and E. Grabowska, Adv in Colloid and Interface Scie., **229,** 80-107 (2016).
5. S. Seino, T. Kinoshita, T. Nakagawa, T. Kojima, R. Taniguchi, S. Okuda and T. Yamamoto, J Nanopart Res., **10**, 1071-76 (2008).
6. J. L. Lyon, D. A. Fleming, M. B. Stone, P. Shiffer and M. E. Williams, Nano Let., **4,** 719-23 (2004).
7. H. Yu, M. Chen, P. M. Rice, S. X. Wang, R. L. White and S. Sun, Nano Let., **5**, 379-82 (2005).
8. H. Liu, P. Hou, W. Zhang, Y. K. Kim and J. Wu. Nanotechnology, **33,** 335602 (2010).
9. I. C. Chiang and D. H. Chen, Adv Functional Mater., **17,** 1311-16 (2007).

10. H. Chen, S. M. Yu, D. Shin and J. Yoo, Nanoscale Res Lett., **5**, 217-223 (2010).
11. N. Allocati, M. Masulli, M. F. Alexeyev, C. D. Ilio, Int J Environ Res Public Health, **10,** 6235-6254 (2013).
12. Z. Kelgenbaeva, E. Omurzak, S Takebe, S. Sulaimakulova, Z. Abdullaeva, C Iwamoto and T. Mashimo, J Nanopart Res., **16,** 2603 (2014).
13. R. Bandyopadhyaya, W. Rong, S. K. Friedlander, Chem. Mater., **16**, 3147-3154 (2004).
14. V. Velmurugan, G. Arunachalam, V. Ravichandran, Archives of Appl. Sci. Res., **2**, 147-150 (2010).
15. R. A. Ismail, G. M. Sulaiman, S. A. Abdulrahman, T. R. Marzoog, Materials Sci Eng.:C **53**, 286-297 (2015).

Influence of media with different acidity on structure of FeNi nanotubes

Alena Shumskaya[1,*], *Egor* Kaniukov[1], *Maksim* Kutuzau[1], *Victoria* Bundyukova[1], *Dinara* Tulebayeva[2,3], *Artem* Kozlovskiy[2,3], *Daryn* Borgekov[2,3], *Inesh* Kenzhina[2,3] and *Maxim* Zdorovets[2,3]

[1]Scientific and Practical Materials Research Centre of the National Academy of Sciences of Belarus, 220072 Minsk, Belarus
[2]L.N. Guilyov Eurasian National University, Engineering laboratory, 010008 Astana, Kazakhstan
[3]The Institute of Nuclear Physics, Astana branch, 010008 Astana, Kazakhstan

Abstract. A detailed analysis of the structure features of FeNi nanotubes exposed at environment with different acidity is carried out. It is demonstrated that the exposure of the nanostructures in the environment with high acidity causes the structure deformation, leading to sharply increasing of the presents of oxide phases and partial amorphization of nanotubes walls that determined the rate of FeNi nanotubes destruction. It was established that the evolution of the crystal structure parameters concerned with appearance of oxide phases and with formation of disorder regions as a result of oxidation processes.

1 Introduction

Nowadays, precise attention is paid to nanostructured materials, which in the future will take their place in practically all fields of science and technology: from magnetic recording devices to sensors and carriers of drugs [1-3]. An alloy of iron and nickel is one of the unique materials for practical applications, due to its physical-chemical properties [4-7] . The composition of alloy strongly influences on such properties [8].

One of the essential criteria for evaluating the potential application of magnetic nanotubes is their reactivity as well as rate of oxidation and destruction in environment with different value of acidity (pH). Oxidation process lead to evolution of the crystal structure parameters, Fe : Ni ratio, presence of amorphous phase and the material destruction. These processes can produce changes in target magnetic and electrical properties, which are critical for micro- and nanoelectronic applications.

The information about the changes occurring in FeNi nanostructures during holding in environments with different acidity is fragmentary. In this paper, the dynamics of FeNi nanotube degradation in environment with different acidity is considered. An analysis of the change in structural properties is carried out. Morphology and structural properties were

* Corresponding author: lunka7@mail.ru

investigated using scanning electron microscopy, energy dispersive and X-ray diffraction analysis.

2 Experimental part

The synthesis of FeNi nanotubes was conducted in the pores of the polyethylene terephthalate (PET) ion-track membranes with a thickness of 12 μm, with pore density 4×10^7 cm^{-2} and diameters of 400 nm. The cathode during deposition was a 10 nm thick gold layer, which was deposited by magnetron sputtering in a vacuum. With this thickness of the Au layer, the pores remained open, and the growth of the nanostructures began at the points of contact of gold with the electrolyte.

Electrochemical deposition was carried out at a potential of 1.75 V using an electrolyte: $FeSO_4\times6H_2O$ (20 g/l), $NiSO_4\times6H_2O$ (100 g/l), H_3BO_3 (45 g/l), $C_6H_8O_6$ (1.5 g/l) at temperature of 25 °C, the pH of the electrolyte solution was 3. The deposition process was monitored chronoamperometrically using an Agilent 34410A multimeter.

Three aqueous solutions were chosen to study the reactivity with different values of the hydrogen index, ranging from pH=1 (strongly acid environment) to pH=7 (neutral environment). The most common chemical reagent used to reduce the high pH of aqueous solutions is hydrochloric acid. A gradual increase in the concentration of hydrogen ions in solutions was achieved due to the use of slightly concentrated (0.01 M/l) hydrochloric acid. Value of pH was monitored with pH meter (Hanna Instruments HI2210-02).

The nanotubes were taken out from the polymer template by dissolving it in an alkaline solution for conduction of experiment. 9 M/l solution of sodium hydroxide was used to dissolve the film. Pure nanotubes were placed in prepared aqueous solutions with different pH values. The time range was 1, 5, 10 and 20 days. At the end of each time interval, changes in the structure and morphology of nanotubes were investigated.

Characterization of structural features was conducted by scanning electron microscopy method (SEM, Hitachi TM3030), energy dispersive analysis (EDA, Bruker XFlash MIN SVE), X-ray diffraction analysis (XRD, Bruker D8 ADVANCE) using Cu K_α radiation.

3 Results and discussion

Typical SEM image of FeNi nanotubes obtained by the template synthesis are shown in Figure 1.

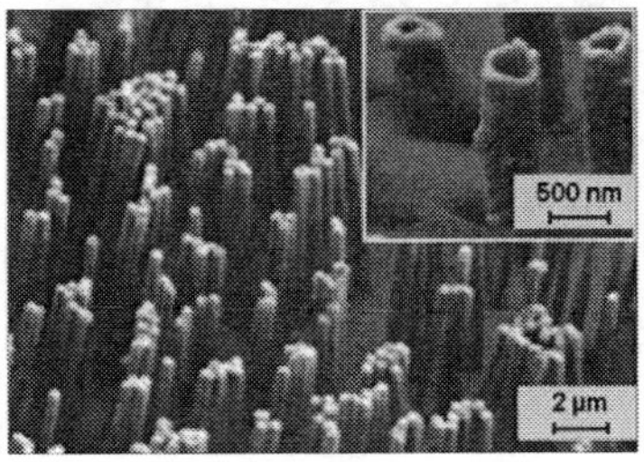

Fig. 1. SEM images of pristine FeNi nanotubes (nanotubes taken out from the PET templates).

From SEM microphotographs it is seen that the nanotubes are isotropic and cylindrical, with a length of 12 μm and an outer diameter of 400 nm, which corresponds to the pore size of the initial template. The effect of environment acidity on the crystal structure of FeNi nanotubes was studied using EDA and XRD. Pristine FeNi nanotubes composition was Fe : Ni = 20 : 80 (permalloy).

Based on the EDA the composition of FeNi nanotubes after exposure in media with different pH was provided (Table 1).

Table 1. Composition of the FeNi nanotubes, bases on EDA, %.

Time, days	pH=1			pH=5			pH=7		
	Fe	Ni	O	Fe	Ni	O	Fe	Ni	O
0	21	79	0	21	79	0	21	79	0
5	18	77	5	20	77	3	21	79	0
10	15	75	10	18	75	7	20	78	3
15	13	71	16	16	73	11	18	75	7
20	10	68	22	14	70	16	15	73	12

To understand the changes occurring in the structure of nanotubes, XRD analysis was performed. Based on the analysis of the change in intensity and shape of XRD peaks (Figure 2), as well it is possible to assume of structure amorphization and partial nanotubes degradation in acidic environment.

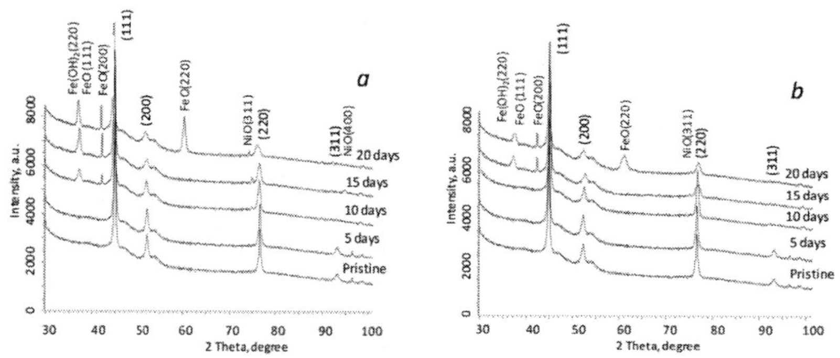

Fig. 2. XRD patterns of FeNi nanotubes exposed in media with various acidity: a) pH=1, b) pH=5.

On the diffractograms for aggressive environment, there are iron hydroxide ($Fe(OH)_2$) and oxide (FeO) for the 10th day, and for environment with pH=5 for the 15th day. Also, there are nickel oxide (NiO) phase with Miller indices (311) and (400). At the same time, as the exposition time in solutions increases, a decrease in the intensity of the peaks characteristic for the FeNi phase is observed, while the intensity of peaks of $Fe(OH)_2$ and FeO phases increases. Based on the analysis of the shape of the peaks, the contributions of various phases to the crystal structure of the nanotubes were obtained during the oxidation process. The results of the assessment of contributions are presented in Table 2.

Table 2. Phase state of the FeNi nanotubes, bases on analysis of XRD patterns, %.

Time, days	pH 1				pH 5				pH 7			
	FeNi	FeO	$Fe(OH)_2$	NiO	FeNi	FeO	$Fe(OH)_2$	NiO	FeNi	FeO	$Fe(OH)_2$	NiO

0	100	-	-	-	100	-	-	-	100	-	-	-
5	100	-	-	-	100	-	-	-	100	-	-	-
10	91	6	2	1	100		-	-	100	-	-	-
15	86	9	3	2	94	3	2	1	100	-	-	-
20	77	15	5	3	88	7	3	2	100	-	-	-

As can be seen from the presented data, the formation of hydroxide phase $Fe(OH)_2$, oxide phase FeO begins on the 10th day, when the oxygen content in the structure of nanotubes exceeds 10%. At the same time, for nanotubes in neutral environment, the presence of oxide phases is not detected. However, the appearance of oxygen is observed by EDA (Table 1) as well. Probably the oxide phase is formed only on the nanotubes surface.
The crystallinity degree was estimated by decreasing the integrated intensity of diffraction peaks automatically with using of software BrukerAXSDIFFRAC.EVAv.4.2. X-ray diffraction patterns were processed by the required number of symmetric pseudo-Voigt functions, it made possible to determine the FWHM of the registered lines, which is necessary for further calculations. The degree of crystallinity, which characterizes the fraction of regularly packed molecules with respect to the number of disordered molecules, i.e. the ratio of crystalline (S_{cr}) and amorphous (S_{am}) phases in polymer (relative degree of crystallinity),%, was determined on the basis of equation:

$$\text{Crystallinity} = \frac{S_{cr}}{S_{cr} + S_{am}} * 100\%$$

Reduction of the integrated intensity of the diffraction peaks, observed in the analysis of diffractogramms, corresponds to a decrease in the degree of crystallinity of the study samples.

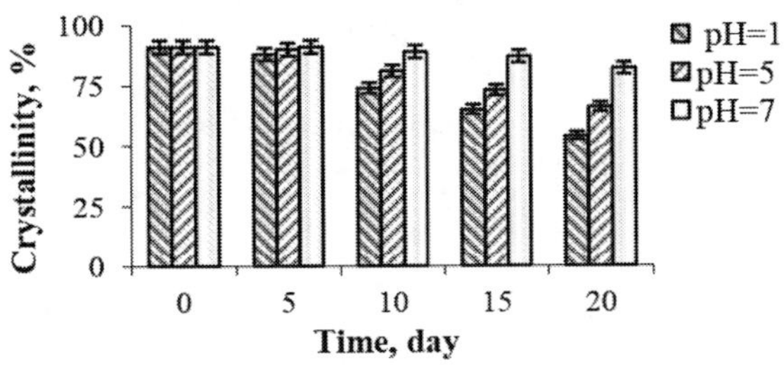

Fig. 3. Diagram of the changes in the FeNi nanotubes crystallinity degree exposed in media with various acidity.

From the presented data, for an environment with pH=1, a decrease in the degree of crystallinity from 91% to 65% for the pristine sample is observed at the 15th day and 54% for the 20th day, and for an environment with pH=5, a decrease in the degree of crystallinity to 73% for the 15th day and 66% on the 20th day. It was found that partial destruction of the structure due to the formation of oxide compounds on the surface of nanotubes and subsequent corrosion processes can cause structural failure, leads to a high degree of structure amorphization and decreases the degree of crystallinity below 70%. In turn, for samples in an environment with pH=7, an insignificant decrease in the degree of crystallinity is observed. The results, obtained by XRD analysis are in a good correlation with EDA. For nanotubes, exposed in media with pH=1, the outgrowths on surface were detected. According to the results of EDA, a high concentration of iron oxides was observed in the outgrowths, which indicates the oxide structure of the growths. On the 15th and 20th day, an increase in the number of outgrowths is observed that cause amorphization of nanotubes surface, which is confirmed by the results of these changes in the degree of crystallinity. For environment with pH=5 and pH=7 and according to the EDA, the appearance of oxygen in the structure is observed on the 5th day for pH=5 and for the 10th

day for pH=7. However, according to XRD, oxide compounds in the crystal structure appear on the 15th day for the environment with pH=5. The presence of small impurities of oxygen on the 5th day is due to the oxidation of the surface layer of nanotubes.

4 Conclusion

The synthesis and study of the physical and chemical properties of nanotubes based on the Fe : Ni alloy (permalloy) synthesized in the pores of PET membranes were carried out. Investigation of the dependence of structural and morphological parameters of nanotubes allowed to evaluate the degree of crystallinity, and also to establish the dynamics of crystal structure amorphization in environment with different pH. On the basis of the analysis of the structure change dynamics, an increasing of oxides phases in the structure with an increasing of nanotubes exposition time in solutions with different acidities was shown. For example, when nanotubes were kept in environment with pH = 1 for the 10th day and in environment with pH 5 for the 15th day, the iron oxide (FeO) and iron hydroxide ($Fe(OH)_2$), as well as nickel oxide (NiO) were established. Most of the oxides were concentrated in the near-surface layers and were localized in the form of outgrowths mainly on the surface of the amorphized oxide phase. Conversely, when samples were held in environment with pH 7, even at the maximum exposure time, only a slight decrease in the degree of crystallinity was observed. The nickel and iron oxide compounds were not found. It should be mentioned that understanding of the degree of nanotubes degradation in different environments could make it possible to significantly clarify the terms of FeNi nanotubes applicability without any special processing, such as covering with protection coats.

1. Z. Wang, G. Shimon, X. Liu, C. V. Thompson, C.A. Ross, W.K. Choi, A.O. Adeyeye, J. Appl. Phys. **113**(2013).
2. V. Haehnel, S. Fähler, P. Schaaf, M. Miglierini, C. Mickel, L. Schultz, H. Schlörb, Acta Mater. **58** (2010).
3. S.J. Son, J. Reichel, B. He, M. Schuchman, X. Bai, S.B. Lee, Abstr. Pap. Am. Chem. Soc. **230** (2005).
4. B. Kalska-Szostko, U. Wykowska, D. Satuła, Nukleonika **60** (2015).
5. M.A. Kashi, A. Ramazani, V. Asgari, E.J.- Khamse, J. Nanostruct. **3** (2013).
6. A.K. Singh, K. Mandal, J. Nanosci. Nanotechnol. **14** (2014).
7. A.E. Shumskaya, E.Y. Kaniukov, A.L. Kozlovskiy, D.I. Shlimas, M. V Zdorovets, M.A. Ibragimova, V.S. Rusakov, K.K. Kadyrzhanov, Prog. Electromagn. Res. C **75** (2017).
8. M.A. Kashi, A. Ramazani, S. Doudafkan, A.S. Esmaeily, Appl. Phys. A Mater. Sci. Process. **102** (2011).
9. A.L. Kozlovskiy, D.I. Shlimas, A.E. Shumskaya, E.Y. Kaniukov. Phys Met Metallogr. **118** 174-179 (2017).

New experimental research stand SVICKA neutron field analysis using neutron activation detector technique

Jan Varmuza[1,*], *Karel* Katovsky[1] *Miroslav* Zeman[1], *Ondrej* Stastny[1], Ivan Haysak[2] and *Robert* Holomb[1]

[1] Brno University of Technology, FEEC, Technicka 12, 616 00, Brno, Czech Republic
[2] Uzhhorod National University, Narodna Square 3, 88000, Uzhhorod, Ukraine

Abstract. Knowledge of neutron energy spectra is very important because neutrons with various energies have a different material impact or a biological tissue impact. This paper presents basic results of the neutron flux distribution inside the new experimental research stand SVICKA which is located at Brno University of Technology in Brno, Czech Republic. The experiment also focused on the investigation of the sandwich biological shielding quality that protects staff against radiation effects. The set of indium activation detectors was used to the investigation of neutron flux distribution. The results of the measurement provide basic information about the neutron flux distribution inside all irradiation channels and no damage or cracks are present in the experimental research stand biological shielding.

1 Introduction

In 2016 a new americium-beryllium neutron source was acquired at Department of electrical power engineering at Brno university of Technology. The main goal of the purchase was to provide better conditions for the basic research and the neutron energy spectrum analysis using the neutron activation detectors and measurement. The new measuring stand was designed for this purpose. The stand contains several irradiation channels and allows long term storing of the neutron source at the same time. The basic knowledge of the neutron flux distribution is required because of the further experiments and R&D. The paper focuses on the measurement of neutron activation detectors and comparison of detector's response inside investigated irradiation channels of the irradiation stand. The basic characterization of thermal neutron spectra is performed using a cadmium spectral cover. The cadmium cover absorbs thermal neutrons in neutron spectra and calculated spectral index brings information about the ration between slow and fast plus epithermal neutrons.

* Corresponding author: varmuza@feec.vutbr.cz

2 Stand description

The basic element of SVICKA (Stand with Various Irradiation Channels and Key Access) stand is composed of a high density polyethylene (HDPE) barrel full of paraffin. The second iron cask has the diameter 570 mm. The icon cask is located inside the HDPE cask in its center.

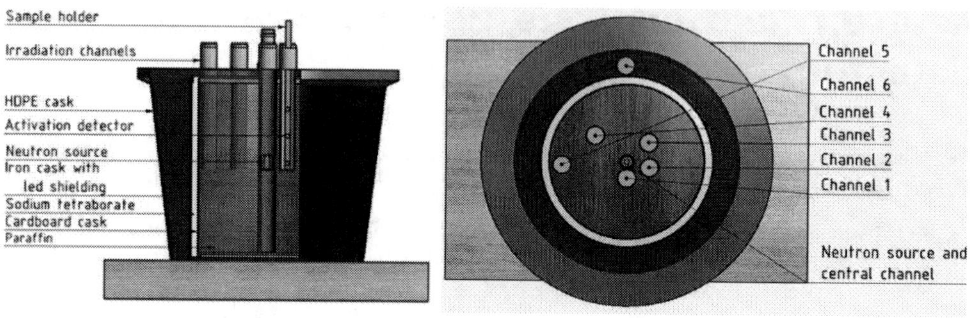

Fig. 1. The experimental irradiation stand.

The iron cask is covered with the 4 mm led layer that provides the basic gamma radiation shielding function. The next cardboard cask (the diameter 520 mm) is in the center of iron cask. The gap between iron and cardboard casks is filled with a sodium tetraborate. The gab is 25 mm wide and the same thickness also has the sodium tetraborate layer on the top and on the bottom of iron cask. The sodium tetraborate provides neutron biological shielding function for slow neutrons.

There are 6 polyvinylchloride (PVC) tubes inside the cardboard cask that is also filled with the paraffin. These irradiation channels have 63 mm in the diameter and can be closed by removable plugs. The end of the plug is also filled with the sodium tetraborate because of the slow neutron leakage. The central channel is 50 mm in the diameter. There is the next removable tube with the diameter 36 mm containing the neutron source in the central channel. The removable tube also contains the removable plug full of paraffin and the end of the plug is filled with the sodium tetraborate.

The upper part of the stand also is covered with the 4 mm thick led layer. This layer provides gamma radiation shielding as well as a protection of paraffin and a working space. The central channel ends in the middle of the HDPE cask height. The channel 1 is located just next to the central channel and goes through the entire stand up to the cardboard cask bottom. Other channels end at the same vertical position like the neutron source. The channel 6 is located behind the sodium tetraborate layer and led shielding because of the low doses irradiation. Distances between central and irradiation channels are 0, 2, 4, 8, 16 cm. The exception is the channel 6. Its distance is 26.5 cm from the central channel because of inner shielding.

 https://doi.org/10.1051/epjconf/201817701004

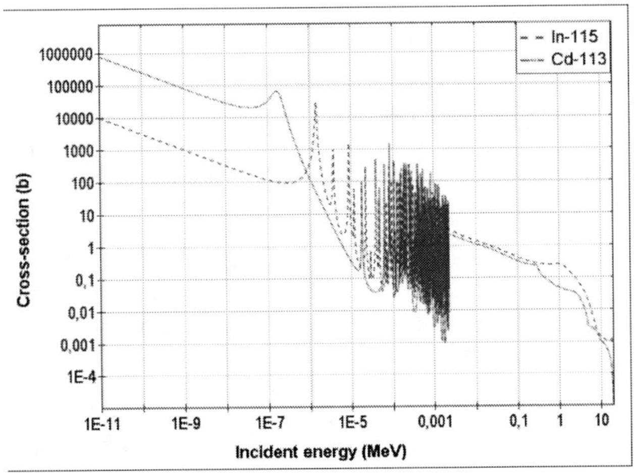

Fig. 2. Neutron cross section comparison [3].

3 Experiment

The neutron activation detectors and analysis were used for the basic analysis of irradiation channels neutron field and neutron energy spectra. Indium activation detectors with the diameter 16 mm and thickness 0.5 mm were used for irradiation in positions 0, 10 and 20 cm above the neutron source. Then the second irradiation was carried out with the cadmium spectral cover in 0 cm vertical position. The fig. 2 shows the neutron capture cross section of Cd-113 and In-115.

The spectral index provides information about the ratio between slow and fast plus epithermal neutrons in the neutron spectrum. The spectral index can be calculated using measurement with the cadmium filter. The cadmium edge is approximately 0.5 eV which is the value where the cross section falls before the local extreme [1]. That means neutrons with lower energy are strongly absorbed by the Cd-113. That means the response of the detector is caused by fast and epithermal neutrons while the detector was hidden inside the cover because all slow neutrons were captured in the cadmium cover. That is the reason why the comparison of the reaction rates with and without spectral cover provides the rough information about amount of slow neutrons in the spectra. All activation detectors were measured using the high purity germanium detector (HPGe).

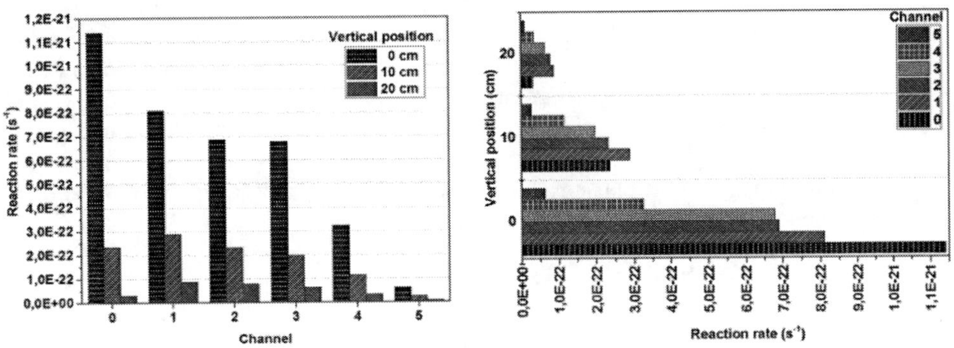

Fig. 3. Neutron flux distribution in irradiation channels and different height levels

4 Evaluation

Reaction rates and spectral indices were calculated based on the indium detector measurement inside all channels and all height positions except of the channel 6. The reaction rates and its distribution are shown in fig. 3. The left part of fig. 3 shows the dependence of the reaction rate distribution inside irradiation channels and the distance from the neutron source.

The reaction rate of the detector depends mostly on the neutron flux, neutron energy spectra and material of the detector. The neutron flux is decreasing while the distance is increasing. When the neutron energy spectrum is not changing with increasing distance the reaction rate will also decrease. This can be seen in channels 3, 4 and 5. When the neutron energy spectrum is significantly changing between channels than the detector response is strongly affected by the neutron capture cross section. It can be seen in channels 2 and 3 where the decrease of the reaction rate is not significant but the difference in spectral index is significant. Reaction rates are almost the same in channels 2 and 3. That is related to diffuse length which is required for neutron slow down. There is just 2 cm of paraffin between central channel and channel 2. The paraffin diffuse length is longer than the distance between the neutron source and the irradiation channel [2]. That means the neutron spectrum is not fully thermalized. This can be seen in fig. 4 where reaction rates and spectral indices are calculated according formulas (1) and (2). In the fig. 4 the reason of the reaction rate increase is visible. The distance between channels 2 and 3 is greater than the diffuse length therefore the amount of thermal neutrons is higher than fast neutrons. These thermal neurons are absorbed in Cd filter and spectral index is higher. The reaction rate is relative to one target nucleus of irradiated material and one incident particle.

$$R_R = S(\mathrm{E}_\gamma)\cdot\frac{1}{N_0}\cdot\frac{1}{t_{live}}\cdot\frac{1}{1-e^{-\lambda\cdot t_a}}\cdot\frac{1}{e^{-\lambda\cdot t_c}}\cdot\frac{\lambda\cdot t_{real}}{1-e^{-\lambda\cdot t_{real}}}\cdot\frac{1}{I_\gamma(E_\gamma)}\cdot\frac{1}{\varepsilon^{\gamma}_{FEP}(E_\gamma)} \tag{1}$$

$$spectral\ index = \frac{R_R}{R_{R,Cd}} \tag{2}$$

Where R_R is the reaction rate without cover, S the peak area, E_γ the gamma line energy, N_0 the number of detector nuclei (calculated for each detector separately) , t_{live} the live time of measurement, t_a the irradiation time, λ the decay constant, t_d the cooling time between irradiation end and measurement start, t_{real} the real time of measurement, I_γ the gamma line intensity, $\varepsilon^{\gamma}_{FEP}$ the detector efficiency and $R_{R,Cd}$ is the reaction rate with Cd cover.

The lower reaction rate is in the channel 5 because of the sodium tetraborate shielding that absorbs part of reflected neutrons. The fig. 3 also shows that the reaction rates are lower in the central channel than reaction rates in the channel 1 in vertical positions 10 and 20 cm above the neutron source. This is because of the activation detector mounting position. Activation detectors were located in the center of the central channel in the vertical position. This position caused that the self-shielding effect was present and total reaction rate was lower.

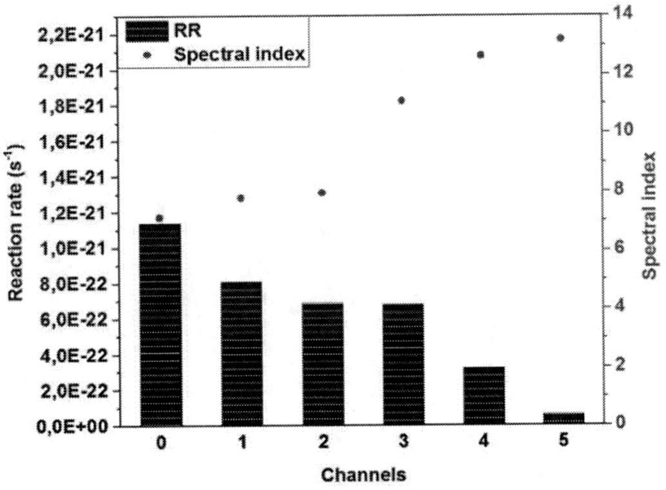

Fig. 4. Irradiation channel and spectral index comparison

The neutron energy characterization of carried out for the position 0 cm above the neutron source because of the highest detector response in all channels. The fig. 4 shows the change of ration between slow and fast + epithermal neutrons. The spectral index increases almost linearly. The lower value is in the last channel because part of slow neutrons is captured in shielding barrier. The drop signalizes lower presence of slow neutron in the channel 2. A cavity could be responsible for harder neutron energy spectra between the neutron source and the channel 2. The other option is the relationship to the absorption cross section. The presence of the cavity has not very a high probability between the neutron source and the irradiation channel 2 therefore further investigation is required. The investigation will focus on the slighter changes of the moderator between the source and the activation detector.

5 Conclusion

The reaction rates were measured using neutron activation method and the analysis of the neutron flux distribution was done. The significant change of neutron energy spectra is shown in the fig. 4 in irradiation channels 2 and 3. The change of the reaction rate is very low even the distance between channels 2 and 3 is two times higher. This information is also supported by the spectral index calculation. The spectral index is almost 30% higher in the channel 3 than in the channel 2. Based on the measured data the significant decrease of the reaction rate is shown in the fig. 3 and fig. 4 in channel 5 because the sodium tetraborate shielding is present near this channel. The irradiation channels 2 and 3 are suitable for the further spectral filter experiment because the significant change of the neutron energy spectrum was observed.

This research work has been carried out in the Centre for Research and Utilization of Renewable Energy (CVVOZE). Authors gratefully acknowledge financial support from the Ministry of Education, Youth and Sports of the Czech Republic under NPU I programme (project No. LO1210), the Technology Agency of the Czech Republic (project No. TE01020455) and Joint Institute for Nuclear Research (project No. JINR2017_3a3_259_16).

References

1. G.F.Knoll, *Radiation detection and measurement*: revue littéraire mensuelle. 4th ed. Hoboken, N.J.: John Wiley, 2010. ISBN 04-701-3148-9.
2. Y. San, *Investigation of Neutron Diffusion Length in Wax Medium*, Universities Research Journal 2011, Vol. 4, No. 4, [online] 2017 [cit. 2017-08-22] Available form: http://umoar.mu.edu.mm/bitstream/handle/123456789/192/URJ-Vol4-No4%28% 201D %20%20Heat%20Eq%29.pdf?sequence=1&isAllowed=y
3. Nuclear Energy Agency. *Janis 4: Java-based Nuclear Data Information System* [online] 2017. [cit. 2017-01-13]. Available form: http://www.oecd-nea.org/janis/

Delithiated states of layered cathode materials: doping and dispersion interaction effects on the structure

Roman Eremin[1,*], *Pavel* Zolotarev[1] , and *Ivan* Bobrikov[2,3]

[1]Samara Center for Theoretical Materials Science, Samara University, Samara, Russia
[2]Frank Laboratory of Neutron Physics, Joint Institute for Nuclear Research, Dubna, Russia
[3]Saratov State University, Saratov, Russia

Abstract. Here we present results of density functional theory (DFT) study of delithiated structures of layered $LiNiO_2$ (LNO, $Li_{12}Ni_{12}O_{24}$ model) cathode material and its doped analogue $LiNi_{0.833}Co_{0.083}Al_{0.083}O_2$ ($N_{10}C_1A_1$, $Li_{12}Ni_{10}CoAlO_{24}$ model). The paper is aimed at independent elucidation of doping and dispersion interaction effects on the structural stability of cathode materials studied. For this purpose, the LNO and $N_{10}C_1A_1$ configurational spaces consisting of 87 and 4512 crystallographically independent configurations (obtained starting from 2×2×1 supercell of R-3m structure of LNO) are optimized within a number of DFT models. Based on a comparison of the calculated dependencies for the lattice parameters with the results of *in situ* neutron diffraction experiments, the most pronounced effect of cathode material stabilization is due to the dispersion interaction. In turn, the doping effect is found to affect cathode structure behavior at the latest stages of delithiation only.

1 Introduction

At present, rechargeable Li-ion batteries have become an important component of modern electronic devices (portable electronics, variety of vehicle, medical equipment, *etc.*). In an effort to improve their performance both theoreticians and experimentalists investigate structure and properties of electrode materials as well as the ways to modify/improve the working characteristics [1–4]. Among the theoretical approaches, a number of computer simulation techniques, such as density functional theory (DFT) calculations, are applied for predicting properties of the electrode materials [5].

In order to perform a comprehensive modeling, it is necessary to take into account disordering of the electrode structure. Moreover, different types of corrections are needed to improve descriptive power of a DFT model, e.g. the dispersion interactions for layered cathode materials [6]. There are two types of approaches to handle the disordering: modeling of a limited number of random configurations [7] or employing fast screening

* Corresponding author: eremin_roman@inbox.ru

methods for all possible structure realization [8]. The issues of structural stability, ion migration hindrance and voltage dependence *versus* deintercalation of the mobile ions can be efficiently simulated as well as cation/anion doping effect on the characteristics [8, 9]. Modern experimental techniques allow tracking structural changes of the electrodes during charge/discharge processes. For the commercial $LiNi_{0.8}Co_{0.15}Al_{0.05}O_2$ (NCA) cathode studies were carried out by means of *in situ* x-ray [10] and neutron [11, 12] diffraction.

Previously, the full $LiNiO_2$ (LNO) and NCA configurational spaces were set using topological approach and studied by means of the DFT calculations and machine learning algorithms [13]. In the scope of the current research, the topological approach is applied for sampling of the configurational space of the $LiNi_{0.833}Co_{0.083}Al_{0.083}O_2$ ($N_{10}C_1A_1$) cathode. The work is aimed at the computational study of the effect of structural stabilization of the LNO-based cathodes doping and dispersion interactions. The approach is based on the comparison of results of computer modeling and *in situ* neutron diffraction data [12].

2 Methods

The step-by-step methodology of sampling of the configurational spaces for $N_{10}C_1A_1$ was used as described in [13] for $N_9C_2A_1$ using the ToposPro package [14]. For the current study, change of the composition (from $N_9C_2A_1$ to $N_{10}C_1A_1$) is reasonable from the computational complexity point of view. *In situ* neutron diffraction experiments were carried out during a charge-discharge process using the Real Time Diffractometer and the High Resolution Fourier Diffractometer (IBR-2, FLNP, JINR). The procedure of collection of neutron diffraction patterns and structure refinement was described [12] in details.

The obtained configurational spaces were optimized within the DFT relaxation using the projector-augmented wave method and Perdew-Burke-Ernzerhof (PBE) exchange-correction functional within the generalized gradient approach (GGA) as implemented in the *Vienna Ab Initio Simulation Package* (VASP) [15]. At the next stage of analysis, the DFT-D3 method [16] was applied as implemented in VASP (noticed hereafter by abbreviation 'vdW' stands for 'van der Waals' forces). The further model refinement was carried out in the spin-restricted mode of DFT. The spin-polarized (SP) models were considered with non-magnetic (NM), ferromagnetic (FM), and two antiferromagnetic (AFM) initial guesses (magnetic moment of 2 μ_B was associated with each Ni atom). Final improvement of the SP models was the Hubbard approach for descriptions of the localized electron states (*d* orbitals) of transition metal (Ni), U=6.0 eV [17].

The DFT modeling allowed us to perform averaging of the mentioned structural parameters of the cathode material over the configurational space [18]:

$$y(x) = \Sigma_x y_i \omega_i \exp(- E_i / kT) / \Sigma_x \omega_i \exp(- E_i / kT), \quad (1)$$

where $y(x)$ is the mean value of a lattice parameter at a certain delithiation level (in Å), x; ω_i, y_i (in Å), E_i (in eV/atom) have meanings of the statistical weight of a certain configuration, the current parameter value and the normalized free energy value of ith configuration, respectively; the Boltzmann constant $k = 8.617 \cdot 10^{-5}$ eV/K and T is the temperature (in K). The summation in formula (1) is performed over the whole set of studied configurations corresponding to a certain delithiation level, x ($Li_{1-x}MeO_2$, $x = 0 \div 1$).

3 Results

3.1 Configuration space sampling

The LNO configurational space consisting of 87 entries with respect to Li content and its distribution in the model cell was obtained during full delithiation [13]. For each independent Li disposition the substitution of two Ni atoms (over twelve possible atomic positions in the sublattice) by dopants (one Al and one Co atoms) leads to totally 132 possible configurations. Thus, 11484 configurations were obtained as the reducible configurational space of $N_{10}C_1A_1$. After that, 4512 configurations (instead of 20760 ones in case of $N_9C_2A_1$ model of NCA [13]) were found to be structurally independent at the topological selection stage. The counts of the obtained configurations of LNO and $N_{10}C_1A_1$ are listed in Table 1 for each delithiation level, x. The statistical weights of all independent configurations were collected for the subsequent averaging procedure.

Table 1. Statistics of the independent configurations *versus* delithiation, x, for LNO and $N_{10}C_1A_1$.

x	**0**	**1/12**	**1/6**	**1/4**	**1/3**	**5/12**	**1/2**	**7/12**	**2/3**	**3/4**	**5/6**	**11/12**	**1**
LNO	1	1	3	6	11	13	17	13	11	6	3	1	1
$N_{10}C_1A_1$	3	21	88	258	547	849	980	849	547	258	88	21	3

3.2 DFT-based geometry optimization

The statistically averaged dependencies ($T \to \infty$) of a and c lattice parameters of the LNO crystal structure are shown in Fig. 1. Obviously, prediction of the GGA-PBE model on the c parameter strongly deviates in comparison to the other DFT models that take into account dispersion interactions and results in structural stabilization (non-monotonic behavior of c parameter). Thus the obtained results are in a good agreement with the fact that dispersion correction becomes important at high deintercalation levels [6]. The most important observation at the current stage of analysis is the fact that relative variances of the a and c lattice parameter values that were calculated using the dispersion corrected models (+vdW, +vdW+SP and +vdW+SP+U) do not exceed 1.8% and 1.4%, respectively.

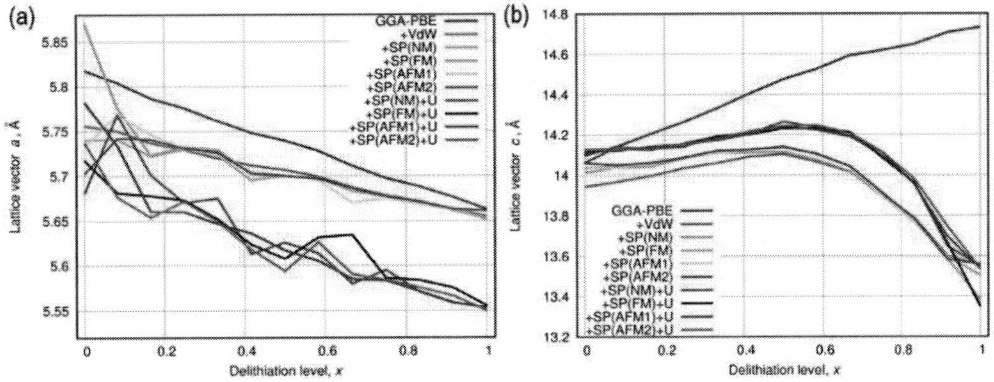

Fig. 1. The averaged dependencies of the LNO (a) a and (b) c lattice parameter values evaluated by means of the DFT calculations using formula (1) at the $T \to \infty$ limit.

For this reason, only the GGA-PBE together with its dispersion corrected (+vdW) models were used for the modeling of the $N_{10}C_1A_1$ configurational space. For the GGA-PBE model, the relaxed structures, their energies and statistical weights are used for calculation by formula (1) of the average behavior of lattice parameters shown in Fig. 2a, where they are compared to the quantities observed from the *in situ* neutron diffraction experiments [12].

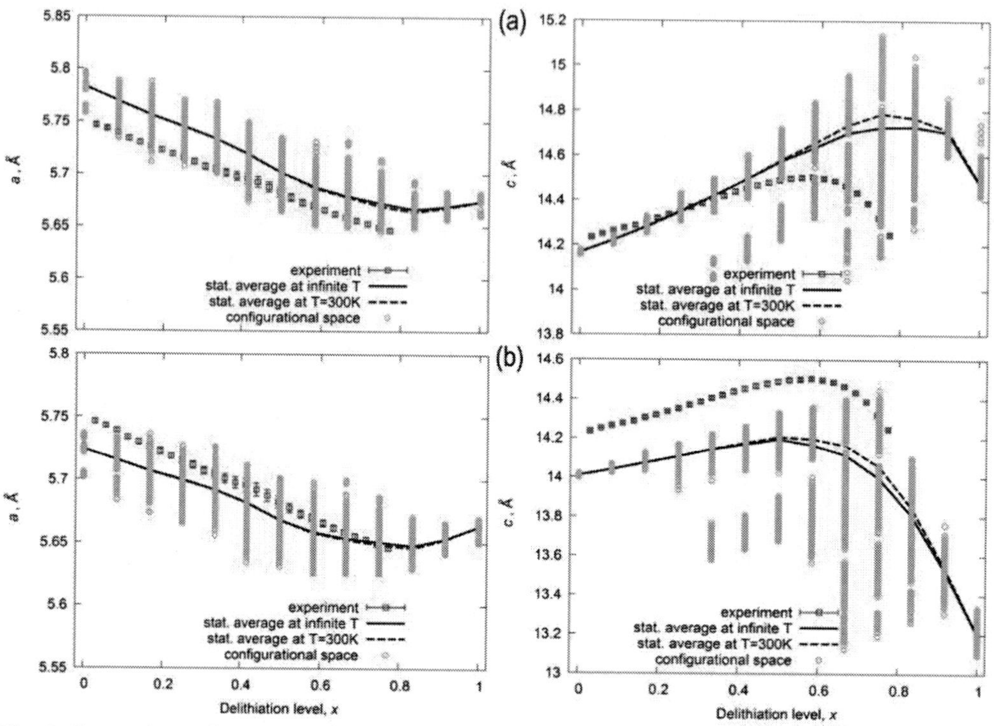

Fig. 2. Comparison of *in situ* neutron diffraction data for *a* (left) and *c* (right) lattice parameters of the NCA structure (squares) with the dependencies computed for the configurational space relaxed within the (a) GGA-PBE and (b) GGA-PBE-vdW models of $N_{10}C_1A_1$ (circles). An averaging is performed by formula (1) with (at temperature of 300 K) and without accounts for energy balance of the configurational space.

The *a* lattice parameter decreases at the initial stage of delithiation and increases after $x \approx 0.8$. This observation is in a good agreement with x-ray diffraction data for deep-charged states of the NCA cathode [10]. Even a more pronounced effect is observed for the mainly increasing dependency of *c* parameter, which starts to decrease at the latter stage of delithiation ($x = 0.75 \div 0.80$). According to the model applied, the described structural stabilization of the cathode can be only associated with doping.

The dependencies of the *a* and *c* lattice parameter values computed by means of the GGA-PBE-vdW model are depicted in Fig. 2b and are much better consistent with the *in situ* neutron diffraction results. Based on the results of modeling of the LNO configurational space (Fig. 1), further improvement of the DFT model (by applying of the +SP and +U corrections) is expected to reduce the deviations for *c* lattice parameter. On the other hand, the corresponding deviation of *a* values should become higher. For these reasons, the subsequent extension of the DFT models was not considered.

For the computed *c* parameter dependencies the averaging at the finite temperature (T=300 K) and pure statistic averaging ($T \rightarrow \infty$) deviate in the *x* range of 0.6 to 0.8 (*cf* Fig. 2 *a* and *b*). This fact is caused by the presence of the configurations with two fully empty Li layers in the structure [13]. The *a* parameter behavior does not depend on the averaging method for the both models studied.

4 Conclusions and remarks

For the cathode material studied the doping effect results in non-monotonic behavior of the structural parameters (*a* and *c*) at the latest stages of delithiation. In contrast, based on the results of LNO and $N_{10}C_1A_1$ modeling, the dispersion interactions significantly stabilize cathode material structure (only *c* parameter) starting at $x \approx 0.5$. The proposed approach of the configurational space sampling seems to be promising tool for the computational investigation of the cathode materials. The development of the modeling approaches allows either refining of the experimental data or even predicting of the properties of perspective materials, avoiding costly and long-term synthesis, post-synthetic treatment and subsequent experimental studying. Possibility of a complement use of the full-configurational-spaces approaches with machine learning (regression) algorithms can significantly reduce a computational complexity of cathode materials investigations and, consequently, increase a predictive power of the modeling methods.

We are obliged to computational facilities of the Samara University complexes, namely, 'Sergey Korolev' and 'Zeolite' (Samara Center for Theoretical Materials Science). The computational studies were supported by the Russian Federation Government (projects No.14.B25.31.0005 and No.3.6588.2017/9.10). The work on the neutron diffraction experiments was supported by Russian Science Foundation (project No. 15-13-10006).

References

1. V. L. Chevrier, L. Liu *et al. J. Electrochem. Soc.* **161,** A783–A791 (2014).
2. M. N. Obrovac, V. L. Chevrier *Chem. Rev.* **114,** 11444–11502 (2014).
3. B. Xu, D. Qian et al. *Mater. Sci. Eng. R Reports* **73,** 51–65 (2012).
4. E. V. Antipov, N. R. Khasanova, S. S. Fedotov *IUCrJ* **2,** 85–94 (2015).
5. M. D. Bhatt, C. O'Dwyer *Phys. Chem. Chem. Phys.* **17,** 4799–4844 (2015).
6. M. Aykol, S. Kim, C. Wolverton *J. Phys. Chem. C* **119,** 19053–19058 (2015).
7. Z. Yang, Z. Zhang *et al. J. Solid State Chem.* **244,** 52–60 (2016).
8. M. Dixit, M. Kosa *et al. Phys. Chem. Chem. Phys.* **18,** 6799–6812 (2016).
9. F. Kong, C. Liang *et al. Chem. Mater.* **28,** 6942–6952 (2016).
10. R. Robert, C. Bunzli *et al. Chem. Mater.* **27,** 526–536 (2015).
11. W. K. Pang, M. Alam *et al. J. Mater. Res.* **30,** 373–380 (2014).
12. I.A. Bobrikov, N.Yu. Samoylova *et al. J. Power Sources* **372,** 74–81 (2017).
13. R.A. Eremin, P.N. Zolotarev *et al. J. Phys. Chem. C* **121**, 28293–28305 (2017).
14. V.A. Blatov, A.P. Shevchenko, D.M. Proserpio *Cryst. Growth Des.* **14,** 3576–3586 (2014).
15. G. Kresse, J. Furthmüller *Phys. Rev. B* **54,** 11169–11186 (1996).
16. S. Grimme, J. Antony *et al. J. Chem. Phys.* **132,** 154104 (2010).
17. A. Jain, G. Hautier *et al.. Phys. Rev. B* **84,** 045115 (2011).
18. R. Grau-Crespo, S. Hamad *et al. J. Phys. Condens. Matter* **19,** 256201 (2007).

EPJ Web of Conferences **177**, 02002 (2018) https://doi.org/10.1051/epjconf/201817702002
AYSS-2017

Measurements of the gamma-quanta angular distributions emitted from neutron inelastic scattering on ^{28}Si

N.A. Fedorov[1,2,⋆], *D.N.* Grozdanov[1,3,⋆⋆], *V.M.* Bystritskiy[1], *Yu.N.* Kopach[1], *I.N.* Ruskov[1,3], *V.R.* Skoy[1], *T.Yu.* Tretyakova[4], *N.I.* Zamyatin[1], *D.* Wang[1,5], *F.A.* Aliev[1,6], *C.* Hramco[1,7], *A.* Gandhi[8], *A.* Kumar[8], *S.* Dabylova[1], *E.P.* Bogolubov[9], and *Yu.N.* Barmakov[9]

[1]*Joint Institute for Nuclear Research, Dubna, Russia*
[2]*Lomonosov Moscow State University, Moscow, Russia*
[3]*Institute for Nuclear Research and Nuclear Energy of the Bulgarian Academy of Sciences, Sofia, Bulgaria*
[4]*Skobeltsyn Institute of Nuclear Physics, Lomonosov Moscow State University, Moscow, Russia*
[5]*Xi'an Jiao Tong University, Xi'an, China*
[6]*Institute of Geology and Geophysics of Azerbaijan National Academy of Sciences, Baku, Azerbaijan*
[7]*Institute of Chemistry of the Academy of Sciences of Moldova, Chisinau, Republic of Moldova*
[8]*Banaras Hindu University, Varanasi, India*
[9]*All-Russia Research Institute of Automatics (VNIIA), Moscow, Russia*

Abstract. The characteristic gamma radiation from the interaction of 14.1 MeV neutrons with a natural silicon sample is investigated with Tagged Neutron Method (TNM). The anisotropy of gamma-ray emission of 1.779 MeV was measured at 11 azimuth angles with a step of ∠15°. The present results are in good agreement with some recent experimental data.

1 Introduction

Natural silicon (Si) consists of three stable isotope ^{28}Si, ^{29}Si, and ^{30}Si, abundances of which are 92.2%, 4.7%, and 3.1%, respectively. Si and Si-based products have important scientific and industrial applications, for example, it is used in radiation detectors, computers and electronics (semiconductor integrated circuits), energy production (solar cells). Therefore, accurate knowledge of the behaviour of natural silicon materials in 14.1 MeV neutron field is very important for the nuclear electronics development.

From fundamental science viewpoint the chain of Si isotopes has attracted increased attention in recent years. New experimental possibilities for studying unstable nuclei with neutron excess made it possible to trace the changes in the shell structure with the neutron asymmetry growth up to the extremely neutron-rich double-magic ^{42}Si nucleus [1, 2], and to obtain the first experimental evidence of the formation of proton density bubble in the ^{34}Si [3]. New experimental results are a challenge to theoretical considerations and make it necessary to consider the chain of Si isotopes in general, including model calculations and experimental data for stable isotopes ^{28,29,30}Si [4].

⋆e-mail: na.fedorov@physics.msu.ru
⋆⋆e-mail: dgrozdanov@mail.ru

2 The Tagged Neutron Method

The method is based on the fact that the neutron and the α-particle formed in the reaction:

$$d + t \rightarrow \alpha(3.5\text{MeV}) + n(14.1\text{MeV}) \tag{1}$$

in the laboratory frame are scattered in almost opposite directions. Therefore, knowing the direction of emission of α-particle, it is possible to recover the direction of the neutron, i.e. to "tag" it. In practice, the "tagging" of a neutron is done by using a position-sensitive (pixelated) alpha detector embedded in the neutron generator.

Registration of α-particles permits to determine the intensity of tagged neutron flux and realize α-γ (detectors) coincidence scheme. The tagged neutrons irradiate the target and induce inelastic scattering reactions $(n, n'\gamma)$, resulting in the photon emission. The γ-quanta are detected with coincident α particle impulses. It has been shown [5] that using the $(\alpha - \gamma)$ coincidences helps reduce the signal-to-noise ratio by more than 200 times and significantly increase the accuracy of the experiment.

3 The experimental setup

For the investigation of neutron-induced reactions at En = 14.1 MeV, at the Joint Institute for Nuclear Research (JINR), Frank Laboratory of Neutron Physics (FLNP), the experimental setup called TANGRA (Tagged Neutrons & Gamma Rays) was created [6]. It consists of an ING-27 neutron generator, operating in continuous mode, in which the Deuteron and Triton ions are accelerate to energies of ~80-100 keV and focused on a Ti-target, forming this way a self-made Tritium enriched target. The maximum 'tagged' neutron flux from the reaction 1 in 4π-geometry, produced by the generator, is $\sim 5*10^7$n/s. The α-particles with energy of 3.5 MeV are registered with the built-in the generator 64-pixel Si-detector located at a distance of 100 mm from the TiT-target. The size of a pixel is 6×6 mm. Registration of the gamma rays is carried out by 22 scintillation detectors based on NaI(Tl)-crystal. The scheme of the TANGRA experimental setup is shown in Fig. 1.

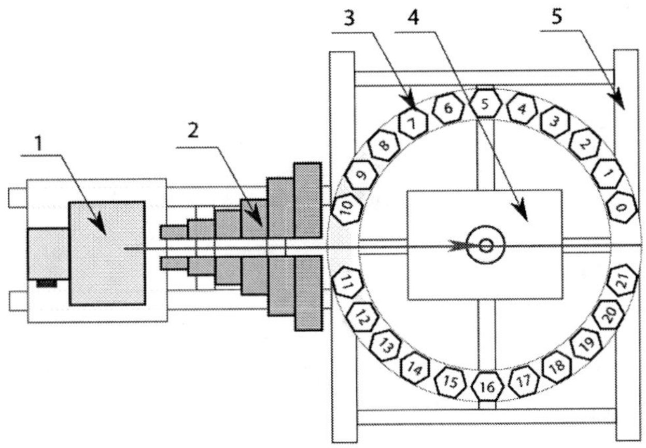

Figure 1. Scheme of TANGRA-setup: 1 – neutron generator ING-27, 2 – collapsible shielding-collimator made of iron (Fe), 3 – array of 22 NaI(Tl) gamma spectrometers, 4 – table with investigated samples, the 5 – frame. The direction of the neutron beam through the collimator and sample is shown by an arrow

Gamma-ray detectors are spaced by the angle of ~15°. To protect the NaI(Tl) gamma-detectors from ING-27 direct radiation, we used a compact shielding-collimator made of iron plates with a total thickness of ~40 cm. Separation of background events is done by the method of time-of-flight (ToF), using the known energy of the incident neutron (14.1 MeV), the distance from the n-generator's tritium target to the irradiated (Carbon, Oxygen) sample position, and that from the sample to the γ-ray detectors. Therefore, for further processing we selected the events lying within a narrow time interval, the beginning of which is given by α-particle of reaction 1. This allows to make an efficient separation of γ-rays from the neutrons, impinged the γ-detector, by the ToF-method. For the collection and prior analysis of the experimental data we used a PC with two ADCs ADCM-16 [7].

As targets in the experiments on inelastic scattering of fast neutrons with light nuclei, we used blocks (10×10×5 cm) from pure substances or plastic containers of the same size, filled with powder from element oxides.

4 Data processing

The signals from α- and γ- detectors of TANGRA-setup are digitized with ADCM and recorded on PC's hard-disk, after that there are off-line analyzed by plotting the time- and amplitude- spectra of the events caused by gamma-rays and neutrons separated by ToF-method, as shown in Fig. 2.

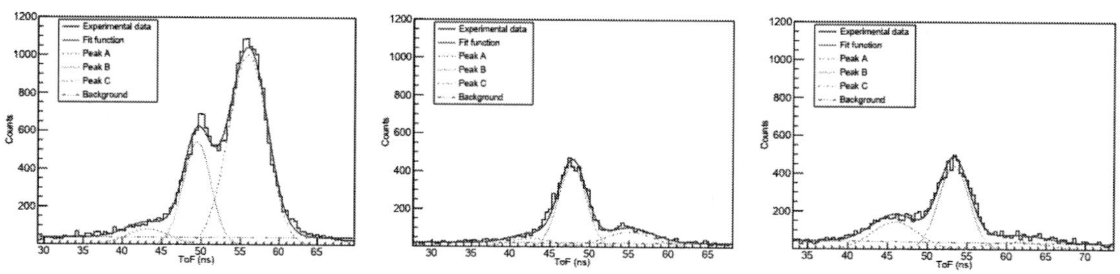

a) b) c)

Figure 2. (Color online) TOF-spectra and their approximation with Gaussian-functions for detector on ∠15° (a), ∠90° (b) and ∠165° (c). In each graph, Peak A (green dashed line) - is from neutron interactions with shielding-collimator, Peak B (blue dotted line) - is from neutron interactions with sample, Peak C (magenta dash-dotted line) - is from direct beam neutron interactions with detector. The total fit function is presented by solid red line.

Further, using the energy calibration of the gamma detectors, we obtained the energy spectrum of events that fall into the time-window corresponding to gamma-rays (Peak B in Fig. 2).

In this experimental setup configuration we used a single pixel of alpha-detector. This allowed us to reduce the number of background events significantly, to simplify and speed-up the data analysis, but in the same time, this method significantly increased the time required to get enough statistics. Registering γ-spectra, we obtain information about the number of events corresponding to the emission of gamma-rays during the transition of the nucleus from a certain excited state to a lower energy state. Typically, one counts only events in the gamma-ray full-energy absorption peak (FEAP) or in single annihilation gamma-quantum escape peak at a lower energy (FEAP minus 0.511 MeV).

5 Results

In inelastic scattering of neutrons with 14.1 MeV energy on ^{28}Si nuclei occurs excitation of a large number of states that can decay by emission of a gamma-ray, but in our experiment we succeed to get a 'correct' angular distribution of gamma-radiation only for the transition from the first excited state ($E^* = 1.779$ MeV, $J^\pi = 2^+$) to the ground state.

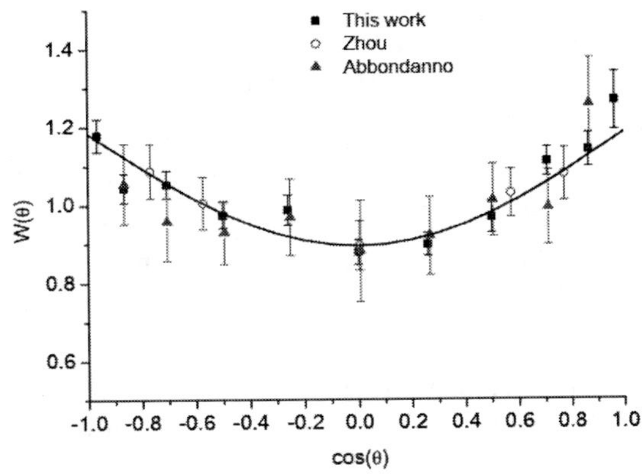

Figure 3. The experimental angular distribution of gamma-rays with $E_\gamma = 1.779$ MeV from the inelastic scattering of 14.1 MeV neutrons on ^{28}Si obtained with TANGRA-setup (square) together with the data from [8] (circle) and [9] (triangle). The black line is Legendre approximation of our data

The angular dependent parameter of the anisotropy of the irradiation of the γ-rays from the inelastic scattering of neutrons $W(\theta)$ which is defined as the ratio of the number of events recorded by the detector located at an angle θ to the number of events recorded by a detector at the angle of 90°, can be well described by the following expression:

$$W(\theta) = A \sum_{i=0}^{2J} a_i P_i(\cos(\theta)) \tag{2}$$

where A-is a normalization constant, chosen so that $a_0 = 0$, a_i are the coefficients of the Legendre polynomials, and J – is the multi-polarity of the γ-transition. The summation index i takes only even values.

Fig. 3 presents the angular dependence of the parameter of anisotropy of the γ-quanta emission from the reaction $^{28}\mathrm{Si}(n, n')^{28}\mathrm{Si}^*$ was obtained from the analysis of the reported experimental data in comparison with the previous experimental results from [8, 9]. The Legendre polynomial coefficients are in agreement with those obtained from the experimental data of the other authors, within the frame of the achieved experimental uncertainties, as shown in the table 1. It is interesting to note the fact that the contribution of the components of the 4th degree Legendre polynomials is insignificant. To explain more research is needed.

Table 1. Legendre coefficients in $W(\theta)$ (2) for gamma rays from inelastic scattering of 14 MeV neutrons.

Experiment	a_2	a_4
14.9 MeV (Zhou [8])	0.21 ± 0.02	-
14.2 MeV (Abbondanno [9])	0.20 ± 0.09	0.11 ± 0.14
14.1 MeV (This work)	0.19 ± 0.02	0.02 ± 0.03

6 Conclusion

- The inelastic scattering of neutrons with an energy of 14.1 MeV on natural silicon sample was investigated by using the method of tagged neutrons. The angular distribution of the γ-rays with energy of 1.779 MeV from $^{28}Si(n, n')^{28}Si^*$ was obtained.
- Within the limits of statistical errors, the calculated values of Legendre coefficient for the angular anisotropy is found to be in good agreement with the results published earlier [8, 9].
- Unlike the results for the angular distribution of gamma-rays from $^{12}C(n, n'\gamma)^{12}C$ with 14.1 MeV neutrons [10], it was confirmed that the contribution of the components of the 4th degree Legendre polynomials is insignificant for gamma-radiation from the first excited state $E^*(2^+) = 1.779$ MeV in ^{28}Si.

It is planned to do some model calculations of the angular distribution of gamma rays from the inelastic neutron scattering of 14.1 MeV on ^{28}Si nuclei in order to reveal the possible mechanism of this reaction.

References

[1] S. Takeuchi, M. Matsushita, N. Aoi *et al.*, Phys. Rev. Lett **109**, 182501 (2012)
[2] J. Fridmann, I. Wiedenhöver, A. Gade *et al.*, Nature **435**, 922 (2005)
[3] A. Mutschler, A. Lemasson, O. Sorlin *et al.*, Nature Physics **13**, 152 (2017)
[4] M.L. Markova, T.Yu. Tretyakova, N.A. Fedorov, Phys. Atom. Nucl. **80**(2017) (in press)
[5] V.M. Bystritsky, N.I. Zamyatin, E.V. Zubarev *et al.*,Phys. of Part. and Nucl. Lett., **10**, 442 (2013).
[6] I.N. Ruskov, Yu.N. Kopatch, V.M. Bystritsky *et al.*, EPJ Web of Conferences **146**, 03024 (2017)
[7] ADCM – An universal Digital Pulse Processing system for nuclear physics experiments: http://afi.jinr.ru/ADCM16-LTC.
[8] H. Zhou, F. Deng, Q. Zhao *et al.*, Phys.Rev. C **82**, 047602 (2010); H. Zhou, F. Deng, W. Cheng *et al.*, Nucl. Inst. and Meth. A **648**, 192 (2011)
[9] U. Abbondanno, R. Giacomich, M. Lagonegro *et al.*, J. Nucl. Energy. **27** 227 (1973)
[10] V.M. Bystritsky , D.N. Grozdanov, A.O. Zontikov *et al.*, Phys. Part. Nucl. Lett. **13** 504 (2016)

EPJ Web of Conferences **177**, 02003 (2018) https://doi.org/10.1051/epjconf/201817702003
AYSS-2017

Optimization of Shielding- Collimator Parameters for ING-27 Neutron Generator Using MCNP5

Aya Hamdy Hegazy [1,*], *V. R. Skoy* [2], *and K. Hossny* [1]
[1] Nuclear and Radiation Engineering department, Faculty of Engineering, Alexandria University, Alexandria Egypt.
[2] Joint Institute for Nuclear Research (JINR), Joliot Curie St., 6, 141980 Dubna, Russian Federation.

Abstract. Neutron generators are now used in various fields. They produce only fast neutrons; D-D neutron generator produces 2.45 MeV neutrons and D-T produces 14.1 MeV neutrons. In order to optimize shielding-collimator parameters to achieve higher neutron flux at the investigated sample (The signal) with lower neutron and gamma rays flux at the area of the detectors, design iterations are widely used. This work was applied to ROMASHA setup, TANGRA project, FLNP, Joint Institute for Nuclear Research. The studied parameters were; (1) shielding-collimator material, (2) Distance between the shielding-collimator assembly first plate and center of the neutron beam, and (3) thickness of collimator sheets. MCNP5 was used to simulate ROMASHA setup after it was validated on the experimental results of irradiation of Carbon-12 sample for one hour to detect its 4.44 MeV characteristic gamma line. The ratio between the signal and total neutron flux that enters each detector was calculated and plotted, concluding that the optimum shielding-collimator assembly is Tungsten of 5 cm thickness for each plate, and a distance of 2.3 cm. Also, the ratio between the signal and total gamma rays flux was calculated and plotted for each detector, leading to the previous conclusion but the distance was 1 cm.

1. Introduction

Neutron generators are now used in various fields and their applications are enormously increasing. They produce only fast neutrons; the D-D neutron generator produces 2.45 MeV neutrons and the D-T produces 14.1 MeV neutrons. One of the problems of neutron generators is optimization of the parameters of neutron-gamma shielding-collimator assembly [1]. The purpose of the optimization is to reach an acceptable value of the ratio between neutron flux in the investigated sample (the signal) and neutron flux in the area of the detectors (the background) also to reach as low as possible gamma rays from inelastic scattering of fast neutrons with the shield. Various simulations and experiments are needed for the design of the shielding-collimator assembly.

E.S.Konobeevsky et al, have optimized the collimator and shielding of the NG 430 neutron generator at the Institute for Nuclear Research, Russian Academy using the MCNP5 and

* Corresponding author: Aya_Hamdy_Anwar@yahoo.com

https://doi.org/10.1051/epjconf/201817702003

SHIELD transport codes [2]. D.L. Chichester et al, have optimized the biological shield for D-T neutron generator. The modeling used in these simulations was performed using the MCNP 4C Monte Carlo code [3]. Bergaoui et al. have optimized a neutron radiography system for non-destructive testing that uses 2.45 MeV D-D neutron generator using Monte Carlo N-Particle Transport Code (MCNPX 2.7.0) [4].

This work was applied on ING 27 D-T neutron generator, TANGRA (TAgged Neutron and Gamma RAys) for studying neutron-induced nuclear reactions, ROMASHA setup, Frank laboratory of neutron physics (FLNP), Joint Institute for Nuclear Research (JINR) [5]. The studied parameters were; (1) the shielding-collimator material; three materials were studied; Iron, Tungsten, and Tungsten alloy (90% Tungsten, 7% Nickel, and 3% Iron), (2) the distance between the shielding-collimator assembly's first plate and the center of the neutron beam; the distance was varied from 0.1 cm to 5.6 cm with 56 bins, while the other distances were set to be functions of the studied distance, and (3) the thickness of shielding-collimator sheets; two thicknesses for each plate were studied 4 cm and 5 cm. The criterion upon which the different collimator parameters were evaluated was based on the ratio between the flux read by the detector and the flux reaching the sample for both neutrons and gammas, such ratio was defined as the Q-value.

1.1 TANGRA Setup

TANGRA-setup consists of a portable neutron generator (ING-27) for producing a continuous beam of 14.1 MeV neutrons, a compact neutron-gamma shielding-collimator assembly and an array of 22 hexagonal NaI (Tl), 10 BGO cylindrical detectors for gamma-ray and neutron spectroscopy in variable configurations. The current configuration utilizes 10 cylindrical BGO gamma-ray detectors (ROMASHA setup), figure 1. While the previous configuration was 22 hexagonal NaI (TI) gamma-ray detectors (ROMASHKA setup) [6]. A 32 channel multi-parametric digital data acquisition system (DAQ) is used for analog signal processing.

Figure 1. ROMASHA Setup

2. Methodology

An MCNP5 model was developed for the ROMASHA setup assuming the neutron source to be a point source and the 10 BGO detectors to be cylinders placed at the specified locations in space. Also, the structural material of the experiment has been neglected as shown in figure 2. The structural material was manufactured of stainless steel which is well known to be transparent to neutrons, so they will not affect the readings of the detectors. The experiment upon which the model was validated, was the irradiation of a parallelepiped Carbon-12 sample for one hour to detect the Carbon-12 4.44 MeV characteristic gamma line. The 4.44

MeV peak output from the MCNP5 was fitted to a Gaussian distribution and compared to the experimental output of the 10 detectors (channels), peak energy counts for both experimental results and MCNP5 results were compared as shown in table 1.

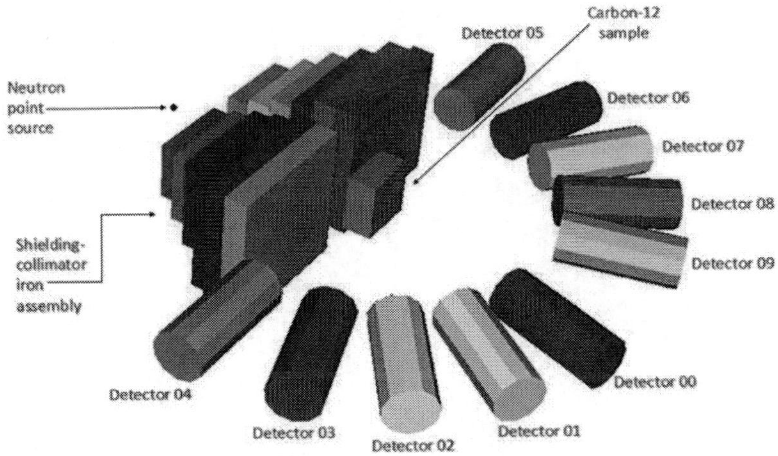

Figure 2. ROMASHA Setup MCNP5 Model

Table 1. Validation Results of the ROMASHA Setup MCNP5 Model

Detector Number	Peak Counts of Experimental Results	Peak Counts of MCNP5 Results	MCNP5 Relative Error	Difference Between Experimental and MCNP5 Peak Counts
00	537	548.21	3.63E-02	11.21
01	486	527.32	3.76E-02	41.32
02	387	329.09	4.19E-02	57.91
03	285	287.77	4.71E-02	2.76
04	250	298.69	4.80E-02	48.69
05	232	245.24	5.00E-02	13.24
06	270	237.23	4.89E-02	32.77
07	338	334.69	4.33E-02	3.31
08	423	404.19	3.83E-02	18.81
09	431	420.66	3.61E-02	10.34

Another model was developed based on using the Artificial Neural Networks (ANNs) to predict the MCNP results in order to be used later in generating the data required for the optimization process. The algorithm upon which the developed package is working is illustrated in figure 3.

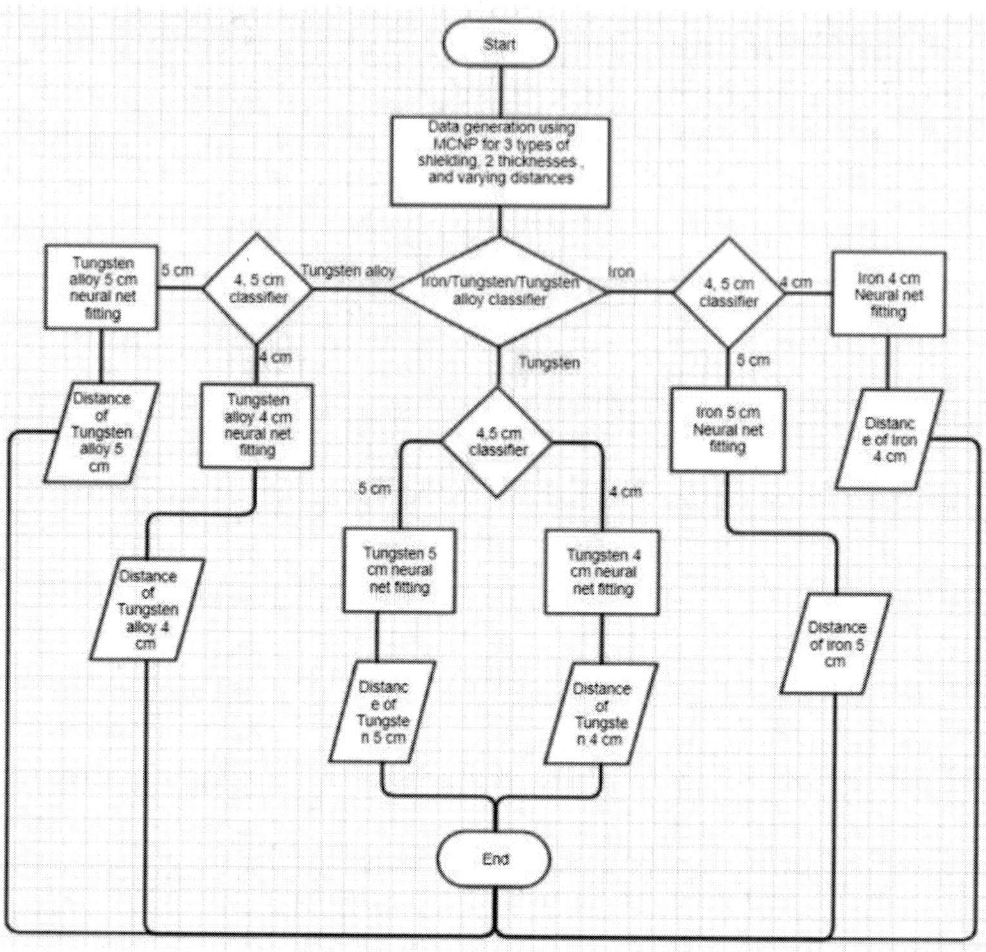

Figure 3. The methodology of Development of the ANNs based Model.

3. Results and Discussion

3.1. Neutrons Q-value Calculations

The main reason for calculating the Q-value is to optimize the different collimator parameters in order to achieve the highest Q-value in order to protect the BGO detectors from the radiation damage that can be caused by the neutrons reaching the detector. The neutrons Q-value was calculated for all the proposed materials and the two proposed thicknesses of the collimator with variation in the distance of the first collimator plate from the center of the neutron beam for each BGO detector. Symmetry has been observed between the first five channels and the second five channels, due to the symmetry in their spatial distribution as shown in figure 1. The simple fluctuations shown in figure 4 are due to the variation of the distance mentioned later which affects the differential scattering cross-section of both; the collimator and the sample. In order to simplify the process of analyzing the results, the symmetry mentioned previously was used to compare the different parameters for half the number of channels. It was observed that a five cm thick tungsten collimator gives the highest Q-value for five of the studied channels, and the highest Q-value is at these four channels existed at a distance of the first plate from the center of the neutron beam ranging from zero to 3 cm as shown in figure 4. It was also observed that at a distance ranging from zero to 3 cm the curves have almost identical behavior except that each channel has a higher curve than the previous one. Therefore the distance is chosen for the highest point in the lowest curve, which was found to be 2.3 cm.

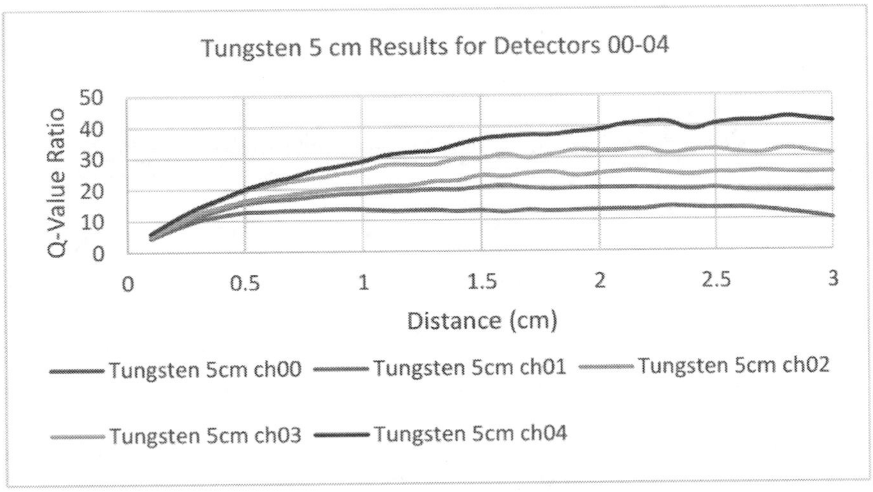

Figure 4. Neutron Q-Value versus the Distance between the Shielding Plates for Tungsten Thickness 5 cm Results for Detectors 00-04

3.2. Gamma Q-value Calculations

Like the neutrons Q-value, the gamma Q-value was calculated for all the channels but for a different reason which is to minimize the gamma rays emitting from the inelastic scattering, radiative capture, and gamma Compton scattering of the collimator material to the neutrons and gammas in order to minimize the background radiation reaching the detectors in order to prevent the overlapping of the collimator gamma peaks with the sample peaks in order to facilitate the process of the neutron spectroscopy. The fluctuations shown in figure 5 are due to the variation of the distance between the plates which affects the differential Compton scattering cross-section of both; the collimator and the sample. As in neutron Q-value, it was observed that the highest of the Q-values are achieved when the collimator is Tungsten and its sheets are of 5 cm thick, and the operating distance ranges from 0 to 3 cm. Therefore, the operating distance will be chosen based on the highest Q-value for the lowest channel curve as shown in figure 5, which was found to be 0.7 cm. Since the 0.7 cm distance is not practical, therefore another distance was chosen based on the same methodology which was found to be 1 cm.

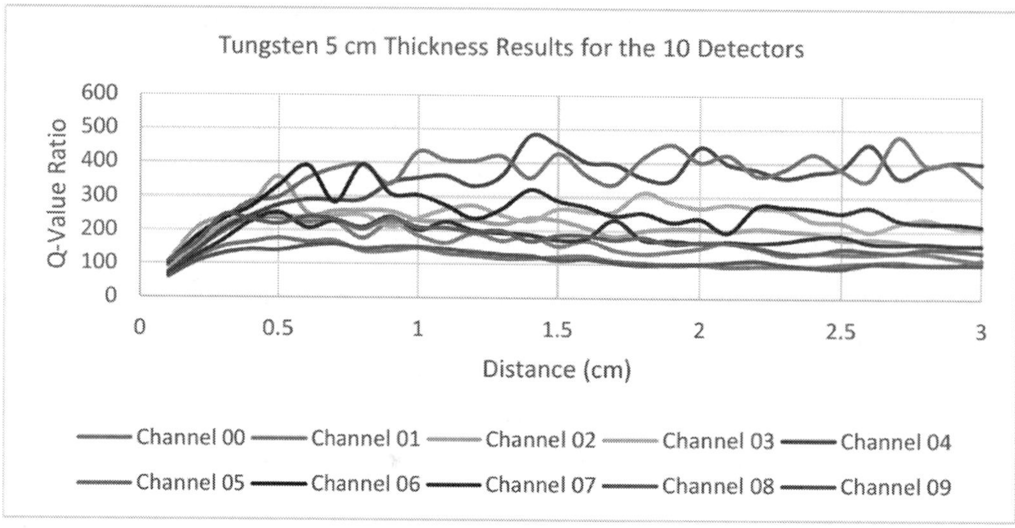

Figure 5. Gamma Q-Value versus the Distance between the Shielding Plates for Tungsten Thickness 5 cm Results for the 10 BGO Detectors

4. Conclusion

From the above discussion, it was concluded that the highest Q-Value for the neutrons was that of the collimator made of tungsten of 5 cm thickness and with a distance of 2.3 cm. While that of photons was found to be the same except for the distance of 1 cm. Therefore, the distance to be chosen is the one for neutrons as the higher priority is for protecting the detectors array from neutron radiation damage. One of the problems faced in the optimization process is the huge runtime required for the data generation for such optimization, Future work will include the optimization of the same parameters using the validated Artificial Neural Networks (ANNs) based model described in the methodology.

5. References

1. "Neutron Generators for Analytical Purposes," IAEA, Vienna, 2012.
2. E. S. Konobeevsky, L. N. Latysheva, N. M. Sobolevsky, and R. D. Ili," Optimizing the Collimator/Shielding Configuration of the NG-430 Neutron Generator", Bulletin of the Russian Academy of Sciences. Physics, 2011, **Vol. 75**, No. 4, pp. 449–453, 2011.
3. D.L. Chichester and B.W. Blackburn, "Radiation fields from neutron generators shielded with different materials", Nuclear Instruments and Methods in Physics Research B 261 (2007) 845–849.
4. K. Bergaoui, N. Reguigui, C. K. Gary, M. A. Pinstripe and J. T. Cremer, "Design, testing and optimization of a neutron radiography system based on a Deuterium-Deuterium (D–D) neutron generator," Journal of Radioanalytical and Nuclear Chemistry, 2013.
5. I.N. Ruskov, Yu.N. Kopatch, V.M. Bystritsky et al., TANGRA-Setup for the Investigation of Nuclear Fission induced by 14.1 MeV neutrons. *Physics Procedia*, 64 (2015) 163-170.
6. Skoy V.R., Kopatch Yu.N., Ruskov I., A versatile multi-detector gamma-ray spectrometry system for investigation of neutron induced reactions. *21st International Seminar on Interaction of Neutrons with Nuclei,* ISINN-21, 20-25 May 2013, Alushta, Ukraine, pp. 242-248.

Possible Experiment for the Demonstration of Neutron Waves Interaction with Spatially Oscillating Potential

Mădălina Mihaela Miloi [1,2,*], *Semyon* Goryunov [2], and *German* Kulin[2]

[1]Faculty of Physics, University of Bucharest, Bucharest, Romania
[2]Joint Institute for Nuclear Research, Frank Laboratory of Neutron Physics, Dubna, Russia

Abstract. A wide range of problems in neutron optics is well described by a theory based on application of the effective potential model. It was assumed that the concept of the effective potential in neutron optics have a limited region of validity and ceases to be correct in the case of the giant acceleration of a matter. To test this hypothesis a new Ultra Cold neutron experiment for the observation neutron interaction with potential structure oscillating in space was proposed. The report is focused on the model calculations of the topography of sample surface that oscillate in space. These calculations are necessary to find an optimal parameters and geometry of the planned experiment.

1 Introduction

The beginning on the study of the interaction of neutrons with a moving matter was laid by the work [1]. Authors drew attention to the fact that the translational motion of the sample should lead to a change in the phase of the transmitted wave and it can take place only due to the movement of boundaries of matter. Later in experiments [2,3] the existence of an effect was observed when the sample moved along the direction of the neutron pulse and its absence when the sample moved parallel to its surface (moves inside stationary boundaries) [4].

The case of arbitrary motion of the sample remained for a long time practically unexplored. The question of the passage of neutrons through a material layer moving with acceleration was theoretically considered in [5-7] and it was concluded that energy of neutrons after passing it should be changed. The phenomenon of the change in the energy of neutrons passing through an accelerating matter was observed in the experiment with ultra-cold neutrons (UCN) [8, 9].

In [10,11], suggestions were made about a possible violation of the dispersion law in an accelerating substance. The critical acceleration at which a significant deviation from dispersion law can be expected was estimated in [12]. This critical acceleration is proportional to the energy of neutrons and for UCN an estimate of the critical acceleration gives 4×10^7 cm/sec^2 [12]. This acceleration is achievable at laboratory experiments. The

* Corresponding author: madalina.mihaela.miloi@gmail.com

experimental approach for the observation of interaction of UCN with matter moving with giant acceleration was proposed in [13].

2 The possible realization of the experiment

In work [13] was proposed to compare the results of UCN passing through the NIF oscillating in space, with the results of the calculation in assumption that the potential model is valid for this system. It is planned to carry out measurements, for relatively low acceleration, which is lower than critical value, and for the acceleration of matter significantly exceeding the critical one.

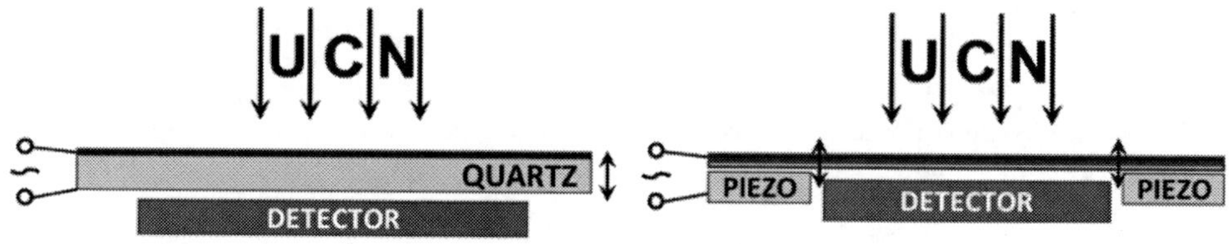

Fig. 1. Experimental scheme [13]. On the left side: a piezo-driver combined with NIF wafer, high frequency regime. On the right side: low frequency regime with a thick piezo.

In high frequency regime it is proposed to depose NIF directly to the quartz plate which serves as resonant piezo driver [13] (see fig. 1, left). The thickness of the plate must be equal or multiple to the half of ultrasonic wavelength. In the case of low frequency about 100 kHz it is not possible to use the same configuration as for high frequency. In this case quartz plate should be ten times thicker that will lead to serious losses of UCN due to absorption in it. As possible solution, it was proposed [13] to use NIF deposed to the thin wafer which is fixed at the annular Piezo driver (see fig. 1, right).

3 Equations and mathematics

It is important that the surface of the sample is oscillated at frequencies from several kHz to several MHz as a whole, together with its boundaries or without significant deformation. The aim of this work was to find optimal geometric parameters of system at which the surface deformation would remain within 10%. The difficulty lies in the fact that the thickness and material of the wafer have to be such that it remains sufficiently transparent for UCN.

Different configurations for the wafer were studied. All the simulations were done in ANSYS software. In simulations, the oscillation is represented by a sinusoidal time-dependent displacement that is applied to the boundaries of the wafer. The rest of the disk is free to move following the inertial laws. Hence, a relative deformation could be defined in percents as

$$D_{rel}=(A_{max}-A)/A*100 \quad (1)$$

where A *is* the displacement amplitude characterizing the forcing load. A_{max} the maximum deformation on axis that is identical to the direction perpendicular to the plane of the disk that could appear because of the inertia of the disk, under the acceleration. The amplitude of oscillations was fixed and was about 14 μm.

At figure 2 (left) results of simulations for plain disk of 12mm in diameter for different thickness are shown. The simulations that were done, the range of frequencies was varied from 1 kHz to 0.1 MHz. As we can see, at this diameter of free area it is not possible to achieve 100 kHz at reasonable values of relative deformation. At 2 mm of sample thick it is possible to reach 50 kHz only, but transparency of the sample for UCN in this case will be very small. Optimal parameters for relative deformation up to 100 kHz reachable for the diameter of the free area of 5mm and thickness of 0.6mm only (figure 2, right), but this geometry is impossible to use in the experiment due to very small size of the free area.

It was understood that the deformation is affected by two parameters. The first is inertia of the freely moving region inside the oscillating annular surface and the second is elastic bonds at the boundary of the free region. Moreover, the inertia of the free region is proportional to the mass and the elastic coupling is proportional to the cross-sectional area at the boundary of the free region and these two parameters act against each other.

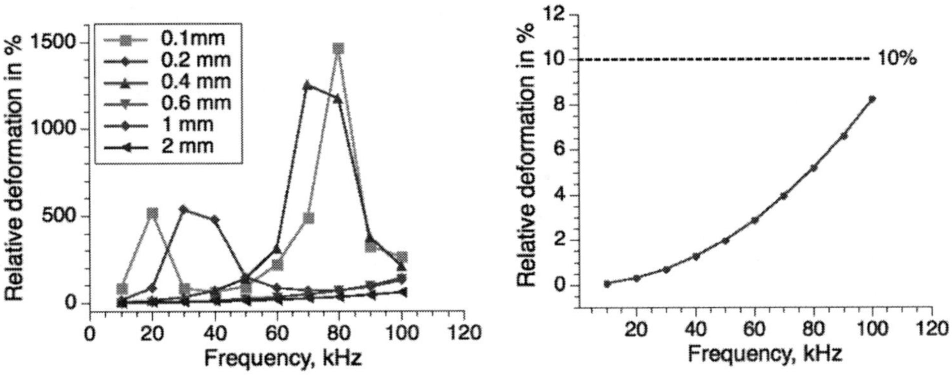

Fig. 2. The relative deformation dependence on frequency. On the left: for different thicknesses of the plain disk of 12mm in diameter. On the right: for the plain disk of 5mm in diameter and thickness 0.6 mm.

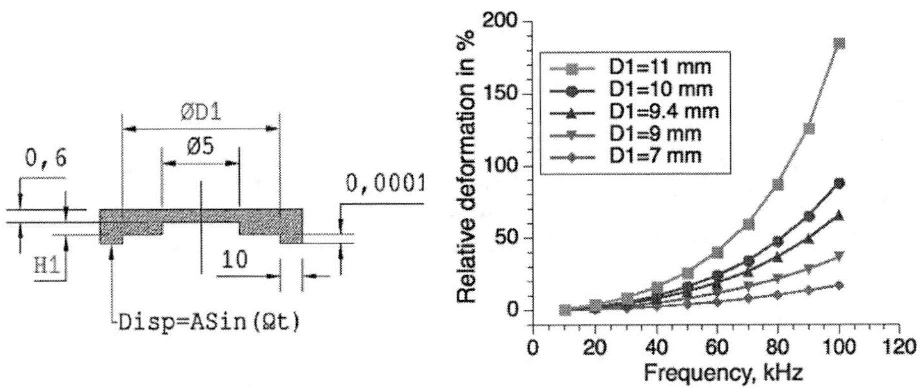

Fig. 3. The geometry of the disk with step-wise decreasing of the thickness. Dependence of the deformation on the frequency for different diameters D1 at a thickness of H1 = 0.6 mm.

Also, attempts were done to use step-wise decreasing of the thickness from the boundary of the free region to the center of the disk (figure 3) and to use the disk with inner part like a honeycomb grating. It allowed to reduce the mass of the free region without decreasing the cross section at the boundary of the free and oscillating region. Unfortunately, these attempts did not lead to any profit in comparing with plain disk geometry.

4 Conclusions

A wide range of problems in neutron optics is well described by a theory based on application of the effective potential model. It was supposed that the concept of the effective potential in neutron optics have a limited region of validity and ceases to be correct in the case of the giant acceleration of a matter.

To test this hypothesis a new Ultra Cold neutron experiment for the observation neutron interaction with potential structure oscillating in space was proposed. For the implementation of the experiment, it is important that the surface of the sample is oscillated at frequencies from several kHz to several MHz as a whole, together with its boundaries or without significant deformation.

The model calculations of the topography of sample surface that oscillate in space were shown that, at frequencies of the order of 100 kHz a relative deformation about 10% is impossible to be obtained for large size of the sample free area. It means that experimental approach itself should be seriously reconsidered.

Authors would like to thank A. I. Frank for permanent support and valuable suggestions.

References

1. M. A. Horn and A. Zeilinger, *Clarendon Press, Oxford*, 350 (1979)
2. A. G. Klein, G. I.Opat, A. Cimmino, et al., Phys. Rev. Lett. **46**, 1551 (1981)
3. U. Bonse and A. Rumpf, Phys.Rev. Lett. **56**, 2411 (1986)
4. M. Arif, H. Kaiser, S. A. Werner, et al., Phys. Rev. A **31**, 1203 (1985)
5. V. I. Mikerov, *PhD thesis The P.N. Lebedev Physical Institute, Moscow*, 105 (1977)
6. F. V. Kowalski, Phys. Lett. A **182**, 335 (1993)
7. V. G.Nosov, A. I. Frank, Phys. At. Nucl. **61**, 613 (1998)
8. Frank A. I., Geltenbort P., Kulin G. V., Kustov D. V., Nosov V. G. and Strepetov A. N., JETP Letters **84**, 363 (2006)
9. A. I. Frank, P. Geltenbort, P. Geltenbort, D. V. Kustov, G. V. Kulin, A. N. Strepetov, JETP Letters, **93**:7 (2011)
10. A. I. Frank, P. Geltenbort, G. V. Kulin, D. V. Kustov, V. G. Nosov and A. N. Strepetov, Phys. At. Nucl. **71**, 1656 (2008)
11. Frank A. I., Phys. Usp. **52**, 397 (2009)
12. Frank A. I., JETP Lett. **100** (10), 613 (2015)
13. A.I. Frank, D.V. Kustov, G.V. Kulin, S.V. Goryunov, D.V. Roshchupkin, D.V. Irzhak, J. of Phys. Conf. Series **746**(1), 012054(2016)
14. I. V. Bondarenko, V. I. Bodnarchuk, S. N. Balashov et al., Physics of Atomic Nuclei **62**, 721 (1999)

EPJ Web of Conferences **177**, 02005 (2018) https://doi.org/10.1051/epjconf/201817702005
AYSS-2017

Topology-based description of the NCA cathode configurational space and an approach of its effective reduction

Pavel Zolotarev*, *Roman* Eremin

Samara Center for Theoretical Materials Science, Samara University, Samara, Russia

Abstract. Modification of existing solid electrolyte and cathode materials is a topic of interest for theoreticians and experimentalists. In particular, it requires elucidation of the influence of dopants on the characteristics of the studying materials. For the reason of high complexity of the configurational space of doped/deintercalated systems, application of the computer modeling approaches is hindered, despite significant advances of computational facilities in last decades. In this study, we propose a scheme, which allows to reduce a set of structures of a modeled configurational space for the subsequent study by means of the time-consuming quantum chemistry methods. Application of the proposed approach is exemplified through the study of the configurational space of the commercial $LiNi_{0.8}Co_{0.15}Al_{0.05}O_2$ (NCA) cathode material approximant.

1 Introduction

Nowadays, the Li-ion batteries (LIBs) are of great importance in production of electronic devices, energy storage systems, electric vehicle manufacturing and many other industry realms [1,2]. In LIBs, the properties of cathode material are largely responsible for the most relevant battery characteristics (energy density, output voltage, cycling stability, operation safety, *etc*). The improvement in the LIBs' performance can be achieved either by utilization of new cathode compounds or by the improvement of the existing cathode materials. The latter can be done through the doping the pristine lattices in the existing cathode materials. As an example, such modification route resulted in the whole family of layered cathode materials derived from the lithium nickel oxide $LiNiO_2$ (LNO) by the substitution of Ni atoms by Mn, Co, Al, *etc*. These are NMC (Mn, Co), NCA (Co, Al), LNMO (Mn) materials with different amount of dopants that are already used in LIB manufacturing. Understanding of the structural changes caused by the dopant atoms is paramount in the pursuit for new cathode materials with advanced characteristics. Modeling of the structure and physical properties of crystals is currently affordable due to the development of different quantum-mechanical algorithms, among which density functional theory (DFT) approach is the most popular due to its high transferability. Recently, DFT calculations were employed in studying the ion transport processes, structural changes upon

* Corresponding author: pavelzolotarev@sctms.ru

delithiation and doping effect on the characteristics of the cathode materials [3–6]. However, direct application of the modeling for a comprehensive study of the cathode materials features is still a complicated task because of high complexity of a disordered system' configurational space. This obstacle is manifested in the necessity to carry out the quantum-mechanical calculation of many thousands (up to millions) configurations belonging to the configurational space of a modeled crystal structure [7,8]. The purpose of the current study is to propose and to verify the scheme that will give an opportunity to perform quantum-mechanical calculation for reduced subsets of the complete configurational spaces of a modeled crystal structure. Application of the proposed scheme is exemplified through the study of the NCA cathode material configurational space that was explored using DFT modeling in our previous study [8].

2 Methods

2.1 Sampling of the NCA configurational space, DFT calculations and structural descriptors

The configurational space used in this study and the results of the DFT modeling are taken from the previous study [8]. In total 20760 independent doped configurations were generated from the 2x2x1 supercell of the parent LNO compound (Fig. 1). The composition of the generated NCA configurations corresponds to $Li_{1-\beta}Ni_{0.75}Co_{0.167}Al_{0.083}O_2$, which is quite close to the composition $LiNi_{0.8}Co_{0.15}Al_{0.05}O_2$ of the commercial NCA material.

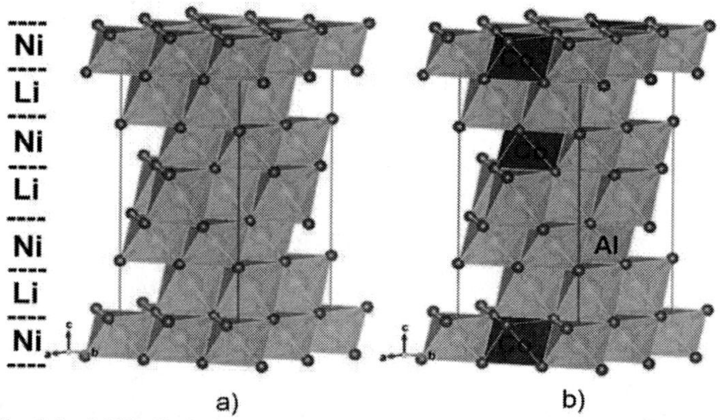

Fig. 1. Unit cell of the LNO (2x2x1 supercell) (a) and one of the configurations from the generated NCA configurational space (b). Nickel-oxide and lithium-oxide layers are designated.

In order to build a regression model, we applied a set of structural descriptors [8] for entries of the NCA configurational space that serve as independent variables of the model. The pertinent types of descriptors calculated by means of the ToposPro [9] package are: (1) Li distribution within the structure. This set characterizes the Li atoms positioning in each Li layer (*LiO_net_2D*) and mutual arrangement of the Li layers with respect to each other (*LiO_net_3D*); (2) dopants positions characterized by the topology of the Ni sublattice (*Ni_sublattice)*; (3) disposition of the dopants described by the topology of the Al-Co-O sublattice (*AlCo_sublatice)* and mutual arrangement of the dopants *(Co_around_Al, Co_around_Co*); (4) mutual arrangement of Li and dopants (*Al_close, Al_far, Al_all, Co_close, Co_far, Co_all*) determined as the numbers of the AlO_6/CoO_6 octahedra around each LiO_6 octahedra having common edge (*_close*), common vertex (*_far*) or both edge and vertex (*_all*).

2.2 Ridge regression algorithm

The ridge regression as implemented in the *scikit-learn* library [10] was employed to evaluate the energy per modeled cell of NCA configurations and structural characteristics such as unit cell parameter *c* or Ni atomic charges. For each delithiation, an individual regression model was built using the abovementioned structural descriptors as regressors. The size of the training set was varied: 20%, 50% and 80% of the configurations at each delithiation level were used. The performance of the regression models built in the prediction of the aforementioned target variables was assessed by evaluation of the mean absolute error (MAE) and determination coefficient (R^2) estimators.

3 Results and discussion

3.1 Prediction of NCA configurations energy

Expectedly, the quality of prediction of configuration energies depends on the number of configurations that were used during regression models training. The plots of the estimators of regression models are shown in Figure 2 (a, b). As can be seen from the dependencies, the predictions of the model trained on 50% and 80% sets of configurations for all delithiations differ negligibly. Further reduction of the training set down to 20% leads to subsequent decline in the prediction quality, which is clearly demonstrated by the changes of the R^2 values. Another point to be stressed is that the errors in prediction of the high-energy structures are generally larger that is visible in the plots of the $E_{predicted}$ *vs.* E_{DFT} dependencies in Figure 2 (c, d).

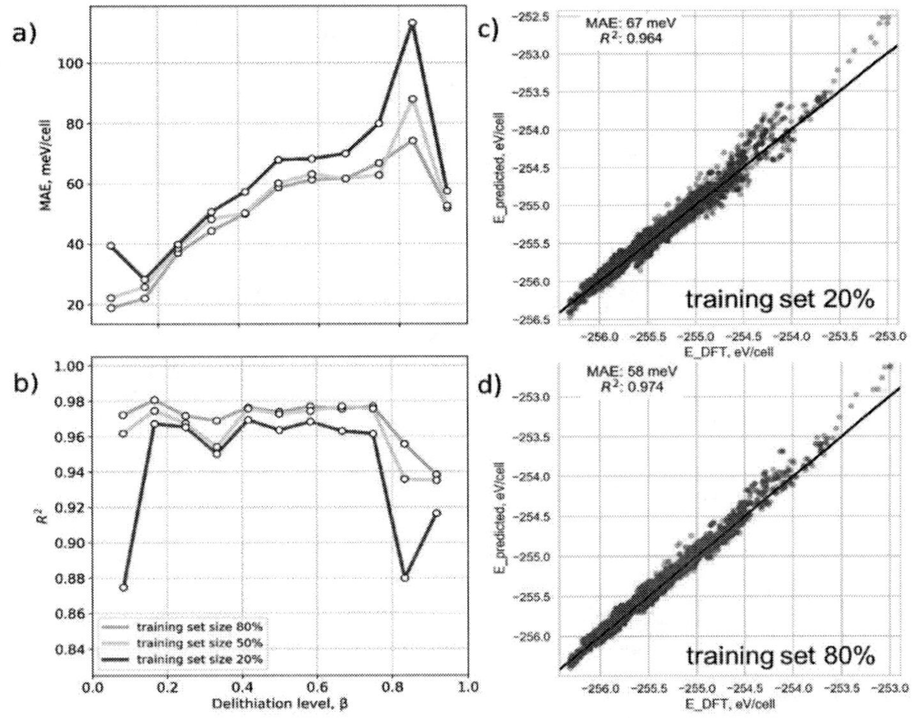

Fig. 2. (a, b) Estimators of the quality of the configuration energy (eV/cell) prediction made by the ridge regression models trained on 20% (red), 50% (yellow) and 80% (green) of configurations; (c, d) scatter plots $E_{predicted}$ *vs.* E_{DFT} for the set of configurations that correspond to the delithiation β=0.5 obtained for different training set sizes.

Deviation of the points from the line corresponding to equality of $E_{predicted}$ and E_{DFT} values becomes more pronounced with the increase of the configuration energy. Moreover, for the configurations of the highest energy the difference $E_{predicted}$ - E_{DFT} is always negative. This can be attributed to the deficiencies of the used set of structural descriptors. Possibly, some additional descriptors are needed for the high-energy structures that are able to grasp their structural peculiarities.

3.2 Prediction of other NCA configurations parameters

Evaluation of the structural characteristics of NCA configurations is also possible with moderate level of accuracy. The results of the unit cell parameter *c* predictions are presented in the Figure 3a. One have to note that, again, the prediction errors of the unit cell parameter *c* are larger for the structures with higher energies of formation, which have smaller *c* lattice parameter values due to the nonuniform distribution of Li ions manifested by the presence of the fully vacant Li-O layers. On the contrary, estimations of the Ni mean Bader charges (Figure 3b) show the absence of such dependence.

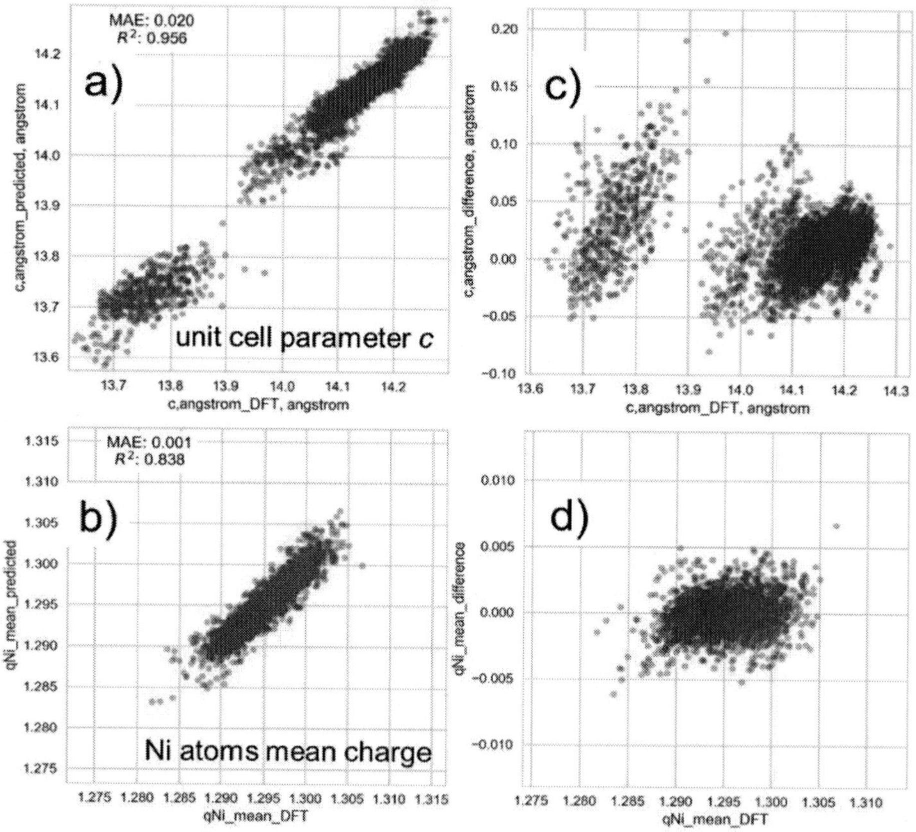

Fig. 3. Scatter plots of $c_{predicted}$ *vs.* c_{DFT} (a) and $\bar{q}(Ni)_{predicted}$ *vs.* $\bar{q}(Ni)_{DFT}$ (b) for a set of configurations that correspond to the delithiation β=0.5 obtained for the training set size 50%. The 'mean' charges $\bar{q}$ were obtained by the averaging the Bader charges over all Ni atoms in a given configuration. Scatter plot of the differences $c_{predicted}$ - c_{DFT} (c) and $\bar{q}(Ni)_{predicted}$ - $\bar{q}(Ni)_{DFT}$ (d) for the set of configurations.

3.3 Reduction of the modeled set

In the previous sections we have shown that the ridge regression models are able to predict variety of characteristics of NCA configurations fairly well. Besides, we have demonstrated that the quality of parameter predictions depends on the size of the training set used to obtain the model, and it declines progressively as the training set shrinks. Taking that into account, it is possible to carry out the DFT calculations only for subsets of NCA configurations and then build regression model to predict the characteristics of a certain NCA configuration such as formation energy, *c* axis length, partial atomic charges, *etc*. The overall scheme proposed for the reduction of the DFT calculations set can be summarized in the block diagram depicted in Figure 4.

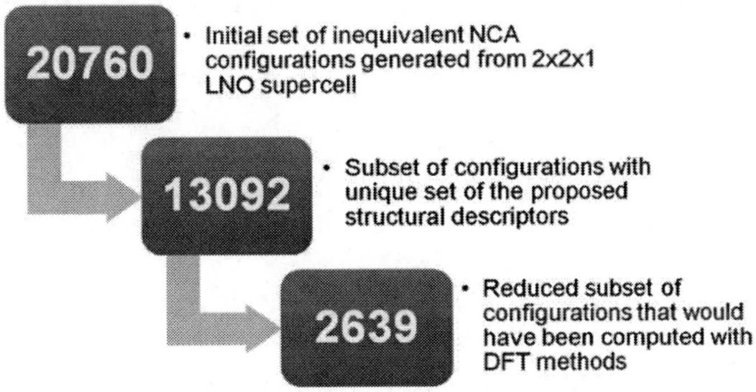

Fig. 4. Block diagram of the scheme for the reduction of the NCA configurations set. The smaller set of 2639 structures is used in the building of a regression model and corresponds to the use of 20% of the configurations at each delithiation in the training set.

4 Conclusions

The machine-learning techniques have already been applied for similar tasks that require processing of large sets of structural information on Li superionic conductors [11] or single- and binary-component solids [12]. The scheme presented in this study could be of particular interest when modeling large configurational spaces. Almost 8-fold reduction of the DFT calculation set (2639 *vs.* 20760 structures) can be achieved at the expense of moderate errors in the forecast of a configuration characteristics. Thus, the presented combination of the topological analysis, DFT calculations and machine learning algorithms seems to be a promising tool for the comprehensive investigations of the existing materials for electrochemistry applications and forecasting new ones.

We are obliged to computational facilities of the Samara University computing complexes ‘Sergey Korolev’ and ‘Zeolite’ (SCTMS). The computational studies were supported by the Russian Federation Government (projects No.14.B25.31.0005 and No.3.6588.2017/9.10).

References

1. E. V. Antipov, N. R. Khasanova, S. S Fedotov, IUCrJ, **2**, 85 (2015)
2. B. Xu, D. Qian, Z. Wang, Y. S. Meng, Mater. Sci. Eng. R Reports, **73**, 51 (2012)
3. M. S. Islam, C. A. Fisher, J. Chem. Soc. Rev., **43**, 185 (2014)
4. J. Kang, H. Chung, C. Doh, B. Kang, B. Han, J. Power Sources, **293**, 11 (2015)
5. C. James, Y. Wu, B. W. Sheldon, Y. Qi, Solid State Ionics, **289**, 87 (2016)
6. Y. Yamauchi, H. Nakai, J. Electrochem. Soc., **160**, A1364 (2013)
7. M. Dixit et al., Phys. Chem. Chem. Phys., **18**, 6799 (2016)
8. R.A. Eremin, P.N. Zolotarev et al., J. Phys. Chem. C, **121**, 28293 (2017)
9. V. A. Blatov, A. P. Shevchenko, D. M. Proserpio, Cryst. Growth Des., **14**, 3576 (2014)
10. F. Pedregosa et al., J. Mach. Learn. Res., **12**, (2011)
11. K. Fujimura et al., Adv. Energy Mater., **3**, 980 (2013)
12. A. Seko, T. Maekawa, K. Tsuda, I. Tanaka, Phys. Rev. B, **89**, 1 (2014)

EPJ Web of Conferences 177, 03001 (2018) https://doi.org/10.1051/epjconf/201817703001
AYSS-2017

First radioactive beams at ACCULINNA-2 facility and first proposed experiment

A.A. Bezbakh[1,2,3,⋆], W. Beekman[4], V. Chudoba[1,3], A.S. Fomichev[1,5], M.S. Golovkov[1,5], A.V. Gorshkov[1,2], L.V. Grigorenko[1,6,7], G. Kaminski[1,8], S.A. Krupko[1,2], M. Mentel[1,10], E.Yu. Nikolskii[1,6], Yu.L. Parfenova[1,9], P. Plucinski[1,10], S.I. Sidorchuk[1], R.S. Slepnev[1], P.G. Sharov[1,2], G.M. Ter-Akopian[1,5], and B. Zalewski[1,8]

[1]Flerov Laboratory of Nuclear Reactions, JINR, Dubna RU-141980, Russia;
[2]SSC RF ITEP of NRC "Kurchatov Institute", Moscow RU-117218, Russia;
[3]Institute of Physics, Silesian University in Opava, Bezrucovo nam. 13, Opava 74601, Czech Republic;
[4]SIGMAPHI, F-56000 Vannes, France;
[5]Dubna State University, Dubna RU-141982, Russia;
[6]National Research Center "Kurchatov Institute", Kurchatov sq. 1, RU-123182 Moscow, Russia;
[7]National Research Nuclear University "MEPhI", Kashirskoye shosse 31, 115409 Moscow, Russia;
[8]Heavy Ion Laboratory, University of Warsaw, Pasteura 5a, 02-093 Warsaw, Poland;
[9]Skobeltsyn Institute of Nuclear Physics, Moscow State University, Moscow RU-119991, Russia;
[10]AGH University of Science and Technology, Faculty of Physics and Applied Computer Science, 30-059 Kraków, Poland;

Abstract. New fragment separator ACCULINNA-2 was installed at the primary beam line of the U-400M cyclotron in 2016. Recently, first radioactive ion beams were obtained. The design parameters of new facility were experimentally confirmed. Intensity, purity and transverse profile of several secondary beams at the final focal plane were studied. The intensities obtained for the secondary beams of ^{14}B, ^{12}Be, ^{9,11}Li, ^{6,8}He in the fragmentation reaction ^{15}N (49.7 AMeV) + Be (2 mm) are in average 15 times higher in comparison to the ones produced at its forerunner ACCULINNA separator. The ACCULINNA-2 separator will become a backbone facility at the FLNR for the research in the field of light exotic nuclei in the vicinity of the nuclear drip lines. The planned first experiment, aimed for the observation of the ^{7}H nucleus at ACCULINNA-2, is outlined.

1 Present status of ACCULINNA-2

Since yearly 1990's, intense radioactive ion beams (RIBs) play a crucial role in experiments aimed for study of exotic nuclei lying near and beyond the nucleon drip lines. The fragmentation of incident projectile is proven to be a powerful tool for the RIB production [1]. This technique is employed in a wide range of beam energies (30-1000 AMeV) with the subsequent separation of fragmentation products in-flight. As compared to the known ISOL method used for the production of secondary beams, the in-flight method has two important advantages: (i) typical flight time of fragment from the production target to the final focus of the separator is rather short($\sim$0.5 μs for energies of several tens of

⋆e-mail: bezbakh@jinr.ru

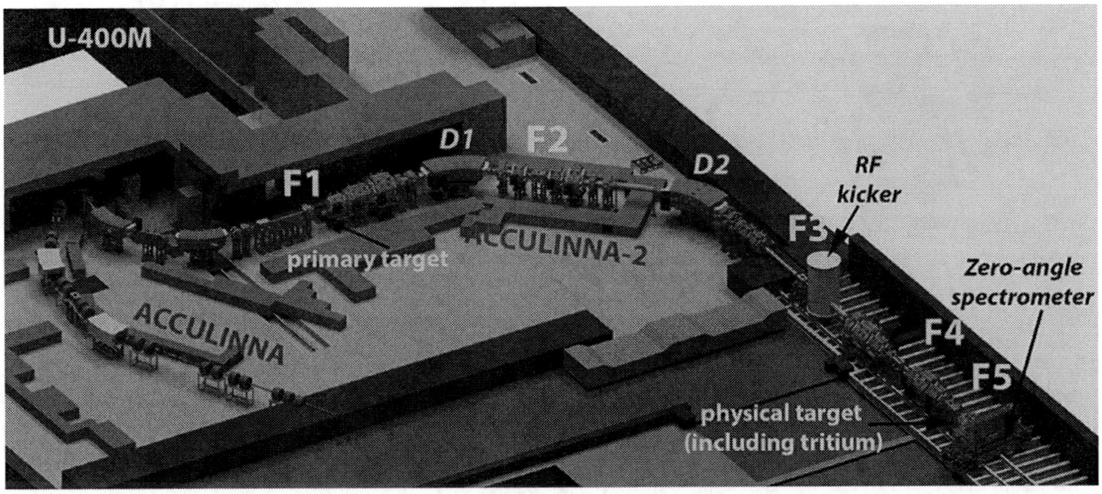

Figure 1. Layout of ACCULINNA-2 fragment separator at U-400M cyclotron. The separator focal planes are indicated by F1-F5, respectively. The main separator part (F1-F3), planed installation of the RF-deflector at F3, and the zero-angle spectrometer at F5 are presented, respectively.

AMeV), which means that there is only a smal limitation on the half-life of RIBs;) (ii) fragmentation cross section is large enough to produce with rather high intensities nuclei far from the valley of stability. The ACCULINNA facility was commissioned in 1996 at Flerov Laboratory of Nuclear Reactions, JINR Dubna [2], and upgraded in 2003. High intensity (up to 10 pμA) primary beams of ^{7}Li, ^{11}B, ^{13}C, ^{15}N and ^{18}O with energies between 32 and 50 AMeV were delivered by the U-400M cyclotron to the separator production target. For example, intensity of $6{\times}10^{5}$ s^{-1} was reached for the ^{6}He beam with energy of 25 AMeV obtained with a ^{11}B primary beam at current of 2 pμA hitting the berrylium (350 mg/cm^{2}) production target. In 2016, new fragment separator ACCULINNA-2 (see Fig. 1) has been installed at another, extended beam line of the U-400M cyclotron. The separator and a part of the primary beam line were designed and built in collaboration with SIGMAPHI company [3]. It is a achromatic separator consisting of two 45° dipoles (D1-D2), 14 quadrupoles (Q1-Q14), 5 multipoles (M1-M5) and 2 steering magnets (ST1-ST2). Projectile fragments are produced at focal plane F1 where thick rotating RIB-production target is installed. Secondary ions are separated in flight by using the combination of magnetic analysis and energy losses in achromatic wedge degrader located at the intermediate dispersive focal plane F2. In 2018 a vertically deflecting radio-frequency deflector is planned to be installed at F3 focal plane. The RF-deflector will significantly purify the proton-rich secondary beams. Technical launch of the ACCULINNA-2 facility was carried out in December 2015. The primary beam of ^{32}S with energy 51.5 AMeV was transported to the F2 focal plane, and the beam characteristics were examined using a couple of Farraday cups (located at the U-400M output and at F2) and a luminophore located at F2. The beam transmission from U-400M to F1 and from F1 to F2 was found to be more than 90% and 100%, respectively. These values, together with the size and the beam profile observed in F2, were in accordance with the simulations and fully met the requirements of the project [4]. All commissioning works related to power-supply, water-cooling of the magnetic elements, ventilation and fire-safety were performed during 2016. The first radioactive ion beams were produced at ACCULINNA-2 in March 2017. The primary beam of ^{15}N with energy of 49.7 AMeV was delivered to the F1 focal plane where the 2 mm thick beryllium

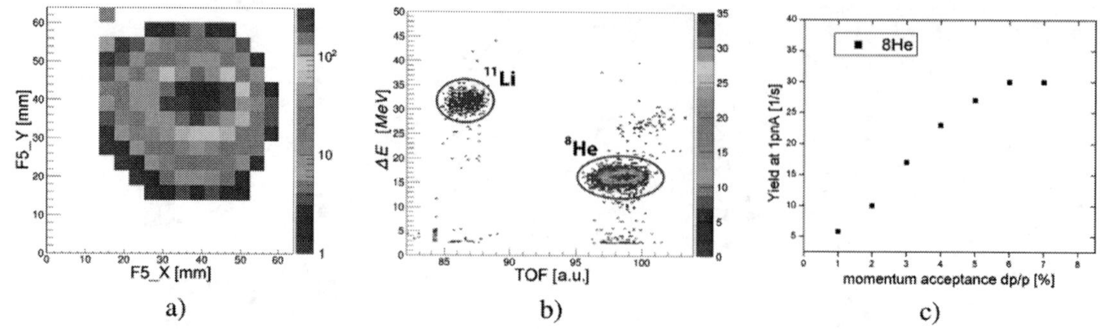

Figure 2. a) Beam spot in the F5 plane obtained by a double-side strip Si detector. The separator elements were tuned for the delivery of ^{8}He to F5. The multipole magnets were swithched off; b) ID plot ΔE vs TOF obtained on the 12.3 m base between F3 - F5 with ACCULINNA-2 tuned on the ^{8}He; c) Dependence of the ^{8}He yield on the momentum acceptance of ACCULINNA-2 measured in the focal plane F5. The measurements were performed with 1 mm beryllium wedge and dp/p=$\pm$2% at F2, Brho(D2)=2.67Tm

production target was installed. The secondary beam was cleaned by means of wedge degrader, made of natural beryllium with thickness of 1 mm, located at F2. The products of fragmentation were registered by the system of RIB diagnostics consisting of a time-of-flight (TOF) and energy losses silicon detectors. The TOF array consisted of a couple of plastic scintillators (BC404) housed in the round-form frames. The thickness of each scintillator was 250 μm, with diameter of sensitive area of 60 mm. Each scintillator was coupled with four R7600 photomultiplier tubes. The TOF detectors were located at the F3 and F5 focal planes. The distance between both detectors was 12.3 m and time resolution of TOF system better than 500 ps was established. The 300 μm thick double-sided strip silicon detector (Micron-Semiconductor BB7) with sensitive area 64$\times$64 mm^2 and with 32 strips on each side was located behind the second TOF detector at F5. This detector served for the measurement of energy losses and beam profile at the final focal plane.

Typical two-dimensional plots presenting the beam spot measured in the F5 final focal plane and the ΔE-TOF identification pattern obtained when the separator was tuned for the ^{8}He secondary beam are shown in Fig. 2a and Fig. 2b, respectively. Operating with the magnetic multipoles switched off ACCULINNA-2 provides the RIB spots in F5 with typical X/Y dimensions of 28 / 23 mm as shown in Fig. 2a, while the beam size on the production target was 7 mm in diameter. The switched on multipoles are important for RIBs with $Z > 3$ to compensate 2-th order aberrations at F2.

Another characteristic for the RIBs production is the dependence of its intensity on momentum acceptance set in the wedge position, i.e. in the F2 focal plane. Such dependence shown for the ^{8}He in Fig. 2c means that the maximum momentum acceptance of ACCULINNA-2 is about $\pm$3 %.

The basic parametrs of RIB such as intensity and purity, obtained for a different group of RIBs delivered to the final focal plane F5 of the ACCULINNA-2 separator are presented in Table 1. Intensity of the primary beam was limited to a few pnA in recent measurements because of presence of unprotected from radiation area around the production target (F1) and beam dump (F2). Radiation shielding will be fully completed in 2018 and then the high intensity secondary beams from the U-400M cyclotron (up to 3 μA for ^{15}N case) will be available. It is worthy to remark that the obtained counting rates (normalized on full primary beam intensity) were in average higher by factor of 15 in comparison with the counting rate of the same RIBs supplied by the ACCULINNA separator.

RIB	Energy (AMeV)	Intensity (s^{-1})	Purity (%)
^{14}B	37.7	120	65
^{12}Be	39.4	150	92
^{11}Li	37.0	4	67
^{9}Li	33.1	1100	50
^{8}He	35.8	25	89
^{6}He	31.5	2700	46

Table 1. Production rates and purity values for several light neutron-rich RIBs obtained in focal plane F5 when a 49.7 AMeV primary beam ^{15}N, having intensity 1 pnA, bombarded a 2 mm beryllium target installed in the F1 plane. RIB's intensities and purities were obtained with 1 mm beryllium wedge and $\Delta p/p = \pm 2$ % at F2.

2. First day experiments at ACCULINNA-2

The choice of the first experiment at ACCULINNA-2 strongly depends on the current status of radiation protection. The first experiment scheduled for March 2018 assumes the study of elastic scattering of ^{6}He beam on the deuterium target. Our objective is to obtain parameters of the optical potential for this system. In the second half of 2018, radioactive beams with intensities of up to 10^5 particles per second will be available. This will allow us to perform the study of the ^{7}H nucleus produced in the ^{2}H(^{8}He,^{3}He)^{7}H reaction. Theory predictions find it conceivable that the ^{7}H exotic nucleus, having not only the extraordinary mass-to-charge ratio, could be tested as being the probable candidate for the 4n radioactive decay. The possibility that a narrow low-lying resonance of ^{7}H can be observed experimentally was suggested by theoreticians long ago [5]. Now, when available are intensive beams of the well bound nucleus ^{8}He, having cluster structure consisting of α-core enclosed by 4n halo (skin), assumption that the ^{7}H(t+4n) system might be somehow achieved appears to be realistic.

One of the first attempts to observe the 4n radioactive decay of ^{7}H set an upper limit of 1 ns for its lifetime [6]. Searches for the ^{7}H ground-state resonance populated in the reactions ^{1}H(^{8}He,2p)^{7}H, [7], ^{11}Be(π^-,p^{3}He)^{7}H, [8], ^{2}H(^{8}He,^{3}He)^{7}H, [9, 10] and ^{12}C(^{8}He,^{13}N), [11] were undertaken. The most consistent and sensitive experiment was carried out with the ^{8}He beam at energy 42 AMeV [10]. Despite the poor resolution made available in [10] for the searched resonance state, a peculiarity observed at ~2 MeV in the measured missing-mass spectrum might be understood as indication for some low-energy state (or states) of ^{7}H. A cross section of ~30 μb·sr^{-1} was estimated for the low-lying ^{7}H resonance populated in the ^{2}H(^{8}He,^{3}He)^{7}H reaction.

Meanwhile, the authors of theory work [12] demonstrated recently that in the case of simultaneous emission of two or four neutrons (so-called "true" 2n or 4n decays), the extremely long lifetimes are possible even for a fairly large decay energies. The arrangement of experiments aimed for the search for the predicted 4n-emitters calls for taking serious steps: new methods with a significant improvement in energy resolution and higher beam intensities to achieve large statistics are needed. Among the nearest candidates for the most exotic 4-neutron radioactive decays (^{7}H, ^{18}Be, ^{28}O) the ^{7}H system is within realistic reach for the new ACCULINNA-2 facility, and we can join hunting for its discovery.

The reaction ^{2}H(^{8}He,^{3}He)^{7}H is to be measured at ACCULINNA-2. Considering the large negative Q-value inherent to this reaction the beam energy of about 35 AMeV is optimal to reach the maximum cross sections. The ^{8}He beam intensity, obtainable at ACCULINNA-2, is high enough to keep luminosity at a level of ~2×10^{26} cm^{-1}s^{-1} having the deuterium target thin enough to achieve resolution of ~500 keV in the measured missing-mass spectrum.

3. Conclusion

The new fragment separator ACCULINNA-2 was commissioned at FLNR JINR and the first radioactive beams were obtained. The basic ion-optical characteristics of ACCULINNA-2 described in Letter of Intent [4] have been confirmed experimentally for a number of RIBs. The obtained RIB intensities are by about 15 times larger in comparison with the old ACCULINNA facility. The study of ^{6}He+^{2}H scattering reaction and the search for the ^{7}H nucleus are planned to be performed at this facility in the near future.

Acknowledgements

This work was partly supported by RSF 17-12-01367 grant, MEYS Projects (Czech Republic) LTT17003 and LM2015049. The authors express their deep gratitude to FLNR director S.N. Dmitriev and to scientific leader Yu.Ts. Oganessian for the support provided to this work.

References

[1] J.P. Dufour et al., Nucl. Instr. Methods, A248 (1986) 267.
[2] A.M. Rodin et al., Nucl. Instr. Methods, B126 (1997) 236.
[3] W. Beekman et al., Proc. 13th Intern. Conf. on Heavy Ion Accelerator Technology, p.33, www.nishina.riken.jp/hiat2015.
[4] A.S. Fomichev et al., JINR Communication E13-2008-168, JINR, 2008. http://aculina.jinr.ru/acc-2.php.
[5] Ya.B. Zel'dovich, Sov. Phys. JETP 11, 812 (1960); Zh. Eksp. Teor. Fiz. 38 1123 (1960).
[6] M.S. Golovkov et al., Phys. Lett. B 588, 163 (2004).
[7] A.A. Korsheninnikov et al., Phys. Rev. Lett. 90, 082501 (2003).
[8] Yu.B. Gurov et al., Phys. Part. Nucl. 40, 558 (2009).
[9] G.M. Ter-Akopian et al., Eur. Phys. J. ST 150 61 (2007).
[10] E.Yu. Nikolskii et al., Phys. Rev. C 81, 064606 (2010).
[11] M. Caamaño et al., Phys. Rev. C 78, 044001 (2008).
[12] L.V. Grigorenko et al., Phys. Rev. C 84, 021303(R) (2011).

Band head spin for triaxial super-deformed bands in ^{165,167}Lu

Poonam Jain[1,*] and *Samit K.* Mandal[2]

[1]*Amity Institute of Nuclear Science and Technology (AINST), Amity University, U.P., INDIA*
[2]*Department of Physics and Astrophysics, University of Delhi, New Delhi-110007, INDIA*

Abstract. We use VMI model for the prediction of band head spin of Triaxial Super-Deformed (TSD) rotational bands. The calculated and observed transition energies are agreed well when an accurate band head spin (I_0) is predicted. The results are in good agreement with the experimentally known values of spin and transition energies. In the present paper, we have reported the band head spin of TSD bands for Lu isotope. This method brings comprehensive interpretation for spin assignment of TSD bands which could help in designing future experiments for these bands. Thus, we have reported the band head spin value of 5 TSD rotational band of Lu isotope.

1 Introduction

The study of Triaxial Super-Deformed nuclei has become an exciting field in the nuclear structure. The nuclei which has large triaxial deformation parameter γ and large quadrupole deformation ϵ_2 known as Triaxial Super-Deformed (TSD) nuclei [1, 2].

Until now, enough data for axial superdeformed nuclei are accumulated but the data for TSD nuclei are quite few like $^{163}Lu,^{164}Lu,^{165}Lu,^{167}Lu,^{171}Ta,^{86}Zr$ etc. were identified experimentally. The prediction of TSD nuclei in $A \sim 160$ region has been reported in Ref. [2]. However, Bohr and Mottelson predicted the existence of the nuclear wobbling motion in rapidly rotating triaxially deformed systems [3]. The TSD nuclei has the configuration with $\epsilon_2 \sim 0.4$ and $|\gamma| \sim 20^0$ has been well known in a Lu isotope since long [4]. For the first time, an excited TSD band was reported in ^{163}Lu [5].

Various empirical or semi-empirical models have been proposed for the spin assignments in SD bands. Unfortunately, very few experiments could perfectly assign spins of SD bands. The theoretical modeling is only technique to calculate the spin values. Several efforts for assigning spins of SD states have been proposed in $A \sim 190$ mass region.

Further some of the methods were proposed to assign band-head spins and transition energies as a function of spin or rotational frequency ω. However, the spin assignments in these studies were not in good agreement with the experimental values. Uma et al. [6] have proposed a modified Variable Moment of Inertia (VMI) model to predict the band head spin of the superdeformed bands. In this paper a modified Variable Moment of Inertia (VMI) model is used to assign band-head spin of TSD bands directly from experimentally observed transition energy ratio which requires less computation time [7, 8]. The extension of the VMI model to an asymmetric (γ_0) rotator can only be done by an

*e-mail: poonam.jn1@gmail.com

additional assumption. The method given is reasonable, since it reduces to the usual VMI model for ($\gamma = 0$). But the results are not changed appreciably with respect to the procedure given [9]. In this model, transition energy ratio (R) is a function of two parameters that is restoring force constant (C) and moment of inertia ($\mathcal{J}_0$). Both values are computed by a best-fitting method incorporating experimentally known transition energies with spin values and verified by r.m.s. deviation. It help to assign band-head spin correctly.

Thus, we organize the paper as : A brief description of theoretical and experimental efforts of superdeformed and triaxial superdeformed rotational band in nuclei is presented in the paper. In Sec 2, we briefly describe the formalism for assigning band-head spin of TSD rotational bands using modified VMI model. The results of these models are presented and compared in Sec 3 for the TSD bands. Finally Sec 4, contains a conclusion and brief remarks.

2 A brief description of the modified VMI model

Here we use modified VMI model to explain the relation of rotational energy with angular momentum (I) for each level. It is basically the sum of the potential energy term ($\mathcal{J}_I - \mathcal{J}_0$) and rotational energy term $[(\frac{\hbar^2}{2\mathcal{J}_I}).I(I+1)]$. However, modified VMI model is a two-parameter formula which characterizes each nucleus by band-head moment of inertia ($\mathcal{J}_0$) and restoring force constant (C). The parameters depend on the difference in moment of inertia ($\mathcal{J}_I$) from that of the ground-state moment of inertia ($\mathcal{J}_0$). The energy level ($I_0 = 0$) of rotational band is represented as:

$$E_I(\mathcal{J}) = \frac{1}{2}C(\mathcal{J}_I - \mathcal{J}_0)^2 + \frac{1}{2}\left[\frac{I(I+1)}{\mathcal{J}_I}\right] . \tag{1}$$

Here we assume $\hbar = 1$. While the band head energy level ($I_0 \neq 0$) of rotational band is given as:

$$E_I = E_0 + \frac{1}{2\mathcal{J}_I}[I(I+1) - I_0(I_0+1)] + \frac{1}{2}C(\mathcal{J}_I - \mathcal{J}_0)^2 , \tag{2}$$

where E_0 is the band-head energy of the rotational band. The variable moment of inertia $\mathcal{J}_I$ is determined by the equilibrium condition.

$$\frac{\partial E(\mathcal{J}_I)}{\partial \mathcal{J}_I} = 0 . \tag{3}$$

This is further equivalent to the cubic equation,

$$\mathcal{J}_I^3 - \mathcal{J}_I^2\mathcal{J}_0 - ([I(I+1) - I_0(I_0+1)]/2C) = 0 . \tag{4}$$

The cubic equation has one real root for any finite positive value of $\mathcal{J}_0$ and C. Finally the transition energy for TSD bands is defined as [6]:

$$E_{\gamma(I\longrightarrow I-2)} = \frac{1}{2\mathcal{J}_0}[I(I+1) - (I-2)(I-1)] + \frac{1}{8C\mathcal{J}_0^4}\left([I(I+1)]^2 - [(I-2)(I-1)]^2\right) . \tag{5}$$

In this equation, the parameters ($\mathcal{J}_0$) and (C) are determined by fitting the experimentally known transition energies using the Best-Fit Method (BFM). The band head spin is determined in terms of the observed transition energy ratio. The transition energy ratio (R) can be expressed as:

$$R = \frac{E_\gamma(I+2 \longrightarrow I)}{E_\gamma(I \longrightarrow I-2)} . \tag{6}$$

Hence the band head spin I_0 is determined from R.

The root mean square deviation (*r.m.s.*) of calculated transition energies for different I_0 values are calculated and shown in figures. The r.m.s. value is least for band-head spin value of a band. If I_0 value shifts away by ± 1 from the accurate value, a rapid shift in r.m.s. deviation (χ) can be seen. The rms deviation is taken as [6],

$$\chi = \left[\frac{1}{n} \Sigma \left| \frac{E_\gamma^{cal}(I_i) - E_\gamma^{exp}(I_i)}{E_\gamma^{exp}(I_i)} \right|^2 \right]^{\frac{1}{2}} . \tag{7}$$

Where n is the total number of fitted transitions involved. Lowest spin I_o is predicted from the least value of r.m.s., as I_o is known all the other spin values of the TSD band levels can be determined. It is clear that the fitting procedure is quite simple and straightforward [6]. Therefore, the experimentally observed band head spin for TSD bands are well justified with the I_0 predicted from VMI equations.

TSD Band	Stiffness Constant $C \times 10^7$ (keV^3)	Band Head Moment of inertia $(\mathcal{J}_0)$ (keV^{-1})	Present assigned (VMI) I_0
^{165}Lu (b1)	4.69	0.0718	12.5
^{165}Lu (b3)	2.34	0.065	20.5
^{167}Lu (b1)	1.63	0.0598	12.5
^{167}Lu (b2)	5.62	0.0687	17.5
^{167}Lu (b3)	2.32	0.0651	13.5

Table 1. The two parameters stiffness constant (C) and band head moment of inertia ($\mathcal{J}_0$) used in the fitting along with predicted band head spin (I_0) using modified VMI model are given.

It can be observed that a correct spin value of the lowest TSD level, I_o, is assigned when the calculated transition energy is almost same with the observed transition energy. However, if I_o is shifted from the correct value, even by ± 1 the r.m.s. deviation will no longer be least i.e. there is a increase in the r.m.s. deviation with the shift in band head spin. The band head spins of all the five TSD bands have been successfully predicted by this approach. Therefore, the V-shaped χ-I_0 plot gives a reasonable assignment of I_0.

3 Results

In this section, we predicted the band head spin of Lu isotopes using modified VMI model. The fitting is suceedingly done for $^{165}Lu(b1, b3)$ and $^{167}Lu(b1, b2, b3)$ isotopes, where $b1$, $b2$ and $b3$ refer to band 1, band 2 and band 3 respectively. We present our results in Table 1 and 2. The experimental data are taken from Ref. [10].

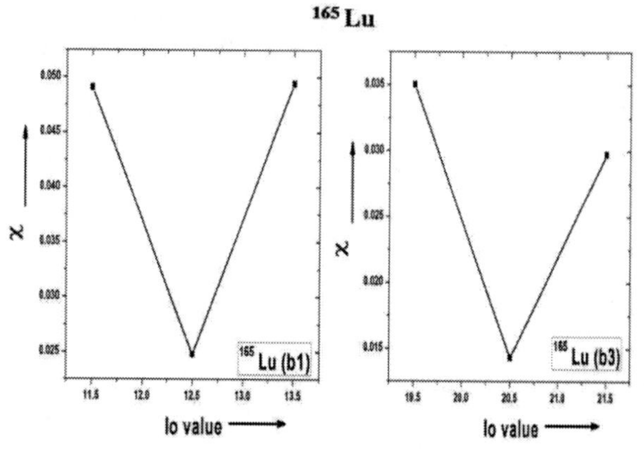

Figure 1. The r.m.s. deviation plots showing the minimum value for band head spin for ^{165}Lu

The two parameters, band head moment of inertia $\mathcal{J}_0$ and restoring force constant (C) are determined by fitting experimentally known transition energies using Best Fit Method (BFM) and the cubic equation for Lu isotopes are given in Table 1.

Using the parameters adopted in Table 1, we calculated the transition energies and with the help of calculated transition energies we are able to predict the band head spin of selected cases. The obtained gamma energies are in good agreement with the experimentally known values of energies.

TSD Band	$E_\gamma(I_0 \rightarrow I_0\text{-}2)$ Exp. (keV)	$E_\gamma(I_0 \rightarrow I_0\text{-}2)$ VMI (keV)	Band Head (I_0) Exp.	Assigned (I_0) from VMI	r.m.s. deviation (χ)
^{165}Lu (b1)	385	392	12.5	12.5	0.0248
^{165}Lu (b3)	661	651.4	20.5	20.5	0.0143
^{167}Lu (b1)	471	455	12.5	12.5	0.063
^{167}Lu (b2)	547	547.6	17.5	17.5	0.0073
^{167}Lu (b3)	457	452.7	13.5	13.5	0.0238

Table 2. Experimental transition energies, calculated transition energies using modified VMI model along with experimental and predicted band head spin are given for 5 TSD bands for Lu

We present our results in Table 2 that gives the experimental transition energy E_γ, calculated transition energy, experimental band head spin and predicted band head spin using VMI equation. It is also noticed that r.m.s. deviation is minimum for these bands which is shown in Figure 1 and 2 for ^{165}Lu and ^{167}Lu respectively. The predicted band head spin is compared with experimentally known spin values which provide a significant validity of this approach for TSD bands also. The results are encouraging. Overall, the obtained spin values are well justified by the r.m.s. deviation.

167Lu

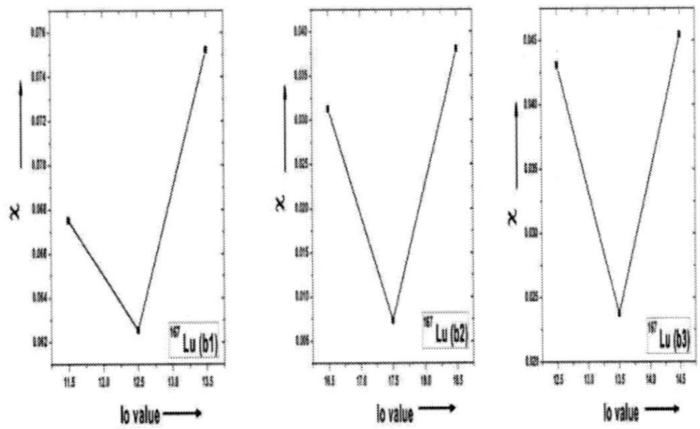

Figure 2. The r.m.s. deviation plots showing the minimum value for band head spin for ^{167}Lu

4 Conclusion

We have predicted the band head spin of 5 TSD bands in Lu isotope. Our good results for the spins and transition energies governs that the fitting parameters (C) and ($\mathcal{J}_0$) obtained here for these TSD bands are acceptable and the simple model employed here is appropriate to use for TSD bands. More profound analysis of TSD bands will be published in future.

References

[1] H. Schnack-Petersen et al., Nucl. Phys. A **594**, 175 (1995).
[2] D. G. Sarantites et al., Phys. Rev. C **57**, (1998) R1.
[3] A. Bohr and B. R. Mottelson, Nuclear Structure Benjamin, Vol. **II**, (1975).
[4] W. Schmitz, et al., Phys. Lett. B **303**, 230 (1993).
[5] S. W. Odegard, et al., Phys. Rev. Lett. **86**, 5866 (2001).
[6] V. S. Uma et al., Pram. J. Phys. **86**, No. 1, 185 (2016).
[7] M. A. J. Mariscotti et al., Phys. Rev. **178** (4), 1864 (1969).
[8] A. Goel and A. K. Jain, Mod. Phys. Lett. A **29**, 2403 (1990).
[9] H. Toki and A. Faessler, Nucl. Phys. A **253**, 231 (1975).
[10] The ENSDF and XUNDL database available on BNL website at [*http : //www.nndc.bnl.gov*].

NeuRad detector prototype pulse shape study

I. Muzalevsky[1,3,4,⋆], V. Chudoba[1,3,⋆⋆], S. Belogurov[1], O. Kiselev[5], A. Bezbakh[1,3], A. Fomichev[1], S. Krupko[1], R. Slepnev[1], D. Kostyleva[5], A. Gorshkov[1], E. Ovcharenko[2], and V. Schetinin[2] for the EXPERT/Super-FRS Experiment Collaboration

[1]Flerov Laboratory of Nuclear Reactions, JINR, 141980 Dubna, Russia;
[2]Laboratory of Information Technologies, JINR, 141980 Dubna, Russia;
[3]Institute of Physics, Silesian University in Opava, 74601 Opava, Czech Republic;
[4]Dubna State University, 141982 Dubna, Russia;
[5]GSI Helmholtzzentrum für Schwerionenforschung, 64291 Darmstadt, Germany

Abstract. The EXPERT setup located at the Super-FRS facility, the part of the FAIR complex in Darmstadt, Germany, is intended for investigation of properties of light exotic nuclei. One of its modules, the high granularity neutron detector NeuRad assembled from a large number of the scintillating fiber is intended for registration of neutrons emitted by investigated nuclei in low-energy decays. Feasibility of the detector strongly depends on its timing properties defined by the spatial distribution of ionization, light propagation inside the fibers, light emission kinetics and transition time jitter in the multi-anode photomultiplier tube. The first attempt of understanding the pulse formation in the prototype of the NeuRad detector by comparing experimental results and Monte Carlo (MC) simulations is reported in this paper.

1 Introduction

Properties of exotic nuclei represent one of the most important fields in modern nuclear physics. Such uncommon nuclei are characterized by a large excess of neutrons or protons and are located far from the nuclear stability. As far as the binding energy decreases, one may observe the transition from the discrete spectra to the nuclear resonances with many overlapping states and, as a consequence, such unique phenomena as neutron halo, soft mode of dipole excitation and others could be observed. Moreover, new decay channels, including many-body, are becoming open [1].

One of the best facilities for production of beams of unstable exotic nuclei will be the Super-FRS fragment separator at FAIR (Facility for Antiproton and Ion Research) [2]. The project EXPERT (EXotic Particle Emission and Radioactivity by Tracking [3]) dedicated to study of properties of exotic nuclei is a part of scientific program of the Super-FRS Experiment Collaboration. The EXPERT setup will be situated in intermediate focal planes of the Super-FRS beam line. The setup consists of six modules [4]: (i) The secondary target for

⋆e-mail: muzalevsky@jinr.ru
⋆⋆e-mail: chudoba@jinr.ru

production of the nuclei of interest (ii) The radiation-hard silicon strip detector for Time-of-Flight measurements and triggering. (iii) The microstrip silicon (μSi) tracking detectors. (iv) The gamma-ray and light particles detector system GADAST (GAmma-ray Detectors Around Secondary Target). (v) The OTPC (Optical Time Projection Chamber) for radioactivity studies by the implantation-decay method. (vi) The NeuRad (Neutron Radioactivity) fine-resolution neutron detector.

Moreover, the EXPERTroot framework [5] for simulations of the experiments and processing of the experimental data and the software for simulation of theoretical distributions are indissociable parts of the project.

The neutron detector NeuRad is aimed at providing precise information on angular correlations between the neutrons and the heavy fragment emitted in the decay of the nucleus of interest. The heavy fragment will be identified by Super-FRS beam diagnostic elements situated downstream of the secondary target, whereas trajectories of a heavy fragment and neutrons will be reconstructed by an array of the μSi detectors and the NeuRad detector respectively. This information will be used to determine the decay energy of the investigated nucleus and its lifetime.

The NeuRad will be constructed from a large number of scintillating fibers ($\approx$10000 units). Each fiber will have a square cross section of $3\times3\,\mathrm{mm}^2$ and a length of 1 m. Scintillating fibers will be grouped into bundles fitting the photosensors e.g. multi-anode photomultiplier tubes (MAPMT). The MAPMTs will be mounted to the bundles so that the area of the each pixel will completely overlap the front surface of one fiber. The NeuRad will be placed 28 m downstream of the secondary target in such a way that the fibers will be oriented along the beam direction [6]. It will provide sufficient detection efficiency and high angular resolution for the neutrons with energies 200-800 MeV in the lab frame.

The neutrons penetrating the fibers will undergo different interaction processes. The most important one for the neutron registration is the elastic neutron-proton scattering. In such a case the recoil proton will induce scintillation light inside the fiber. The light emitted within the full reflection angle travels to the MAPMTs located at the both ends of the fiber. With such configuration the total angular acceptance of the detector will be about 12 mrad. Such a value is sufficient for registration of the neutrons with low transversal momenta, as expected if the energy release in the decay is in the range of 0.1-100 keV.

In the case of a single scattering of the neutron inside the NeuRad and sufficiently high energy release in one fiber, the position of the fired fiber gives information about the transversal coordinates of the interaction point. The time stamps of the signals from both the MAPMTs allow to derive the longitudinal position of the hit (spatially restricted energy deposit) providing a correction of the time-of-flight measurement. The time-of-flight information is used for the determination of the neutron kinetic energy.

If several fibers are fired due to the neutron re-scattering in detector volume or sharing of the energy deposited by the recoil proton, the first interaction point may be reconstructed by searching for the earliest signal in the upstream MAPMT. Besides, timing information may be used in more sophisticated analysis of the events with multiple hits.

Feasibility of the detector strongly depends on its timing properties. Each detector pulse is a superposition of several single photoelectron pulses whose distribution in time and amplitude is defined by the spatial distribution of ionization, light propagation inside the fibers, light emission kinetics, transition time jitter in the MAPMT, pulse shape and amplitude spectrum of the single photoelectron.

The pulse shapes produced by the neutrons hitting the exactly known interaction point are difficult to study experimentally. However MC study of the efficiency and precision of the neutron detector is possible provided that the MC is validated in more simple measurements. The first attempt of understanding the pulse formation in the prototype of the NeuRad detector by comparing experimental results and MC simulations is reported in this paper.

2 Test of the NeuRad prototype

The first investigation of the NeuRad detector was performed recently in the Flerov Laboratory of Nuclear Reactions JINR, Dubna. The NeuRad prototype consisted of one bundle of 256 BCF-12 [7] scintillating fibers having a square cross section of 3×3 mm^2, 25 cm – long. Each fiber was covered with a white acrylic paint as the internal layer and a black aerosol paint as an external one for the light insulation and had two MAPMTs H9500 [8] mounted to the both sides of the bundle. Optical grease was used to improve the light transmission.

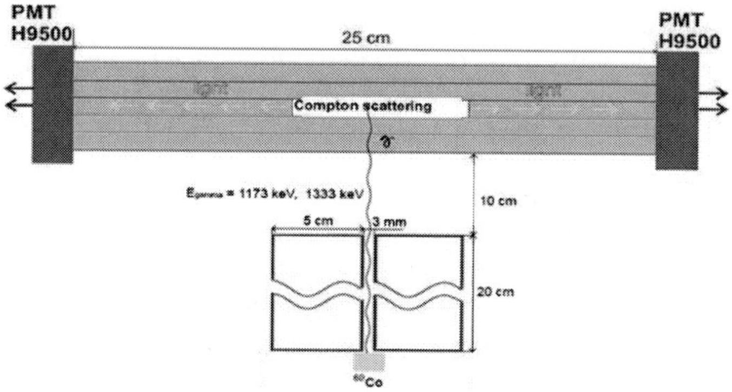

Figure 1: A scheme of the measurements with the NeuRad prototype. The prototype was irradiated by a collimated γ-source ^{60}Co. The gamma rays were focused at the geometrical center of the prototype and signals were collected by a couple of H9500 MAPMTs.

This prototype was irradiated by a collimated ^{60}Co gamma-source with the energies $E_\gamma^{(1)}$=1173 keV and $E_\gamma^{(2)}$=1333 keV as showed in the Fig. 1. Signals obtained from the MAPMTs were collected by the DRS4 evaluation board [9] (12 bit at 5 GS/s) and saved for the each detected event. Two channels of the board were employed for collecting signals from the both sides of one fiber. The channel from one side was used for triggering with the threshold of 20 mV. On the other side of the fiber, two remaining channels of the board were connected to the neighboring fibers.

3 Simulations and data processing

The simulation and data processing were performed within the EXPERTroot framework. A particle transport through the setup was performed using the GEANT4 [12] classes called from the EXPERTroot software. Obtained energy deposits were converted into the MAPMT anode pulses. Each pulse was calculated as a sum of single photoelectron pulses taking into account the following parameters and effects: light output of the scintillating fiber; the efficiency of the light trapping into the fiber due to a total internal reflection; the scintillation decay time; the light attenuation along the fiber; the light collection increase near the ends of the fiber; the light losses at the optical interface; the MAPMT quantum efficiency; the single photoelectron amplitude spectrum, and the pulse shape; the spread of the photoelectron avalanche transition time through the dynode system; the cross-talk as a probability of the single photoelectron avalanche developing in the neighboring pixel. The simulated pulse shapes had the same format as in the experimental data. Implementation details can be found in [10].

Certain effects were not taken into account: any dependence of the pulse shape on the amplitude; the cross-talk as the charge sharing of a single photoelectron avalanche between two or more anodes; the partial collection of the light after a diffuse scattering at the fiber faces; the electronic noises; the pulse shape distortion in the readout line. Most of the parameters for the simulations were taken from the data-sheets of the fibers and MAPMT. The single photoelectron pulse shape was fitted to the experimental data. The single photoelectron amplitude spectrum was borrowed from [11], where a similar MAPMT H12700 has been studied.

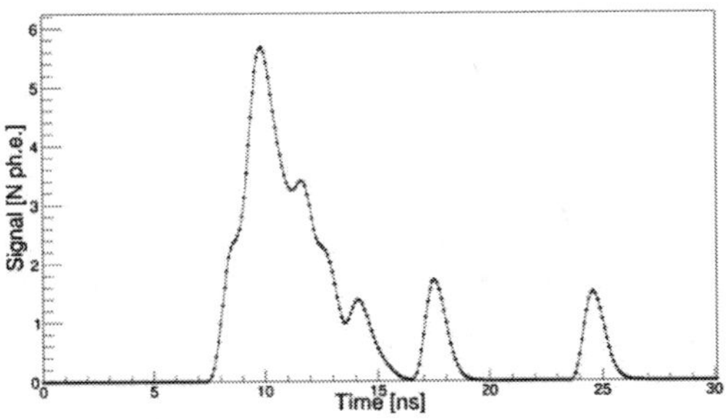

a)

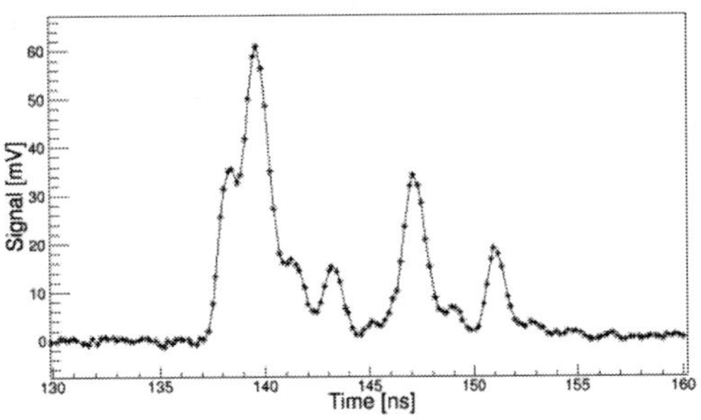

b)

Figure 2: The typical pulse shapes containing about 10 photoelectrons. a) The simulated MAPMT anode output as a result of a composition of several photoelectrons fluctuating according to the assumed distributions b) Typical shape of a measured signal containing about 12 photoelectrons.

Typical pulses obtained in simulations and measurements (shown in the Fig. 2a) and the Fig. 2b) respectively) seem to follow the similar single photoelectron distributions in time and amplitude. However, there is a clear difference between the simulated and the recorded experimental pulses. In the experimental ones one can notice a kind of low-amplitude ringing. It can be due to several reasons, e.g. a stray capacity discharge or influence of the readout line bandwidth, which manifests itself in the damped oscillations caused by the sharp leading edge.

Several standard and rather simple data processing algorithms were implemented into the EXPERTroot framework, such as leading edge discriminator, different event selection filters, alignment and summing up of the pulses, etc. The implemented algorithms allowed us to study the summed pulse and to understand qualitatively its basic features. All algorithms could be applied in the same way to both simulated and experimental data.

Simulated and experimental summed pulses, normalized by their maximum amplitudes, in a logarithmic scale are shown in the Fig. 3. The trigger threshold for the simulated data was set to the 1.7 of the average single photoelectron amplitude. Such a threshold allowed to match the sharp bend at the leading edge (indicated with an arrow in the Fig. 3) in the both pulses. From this picture it can be established that the experimental pulse has the decay profile which includes a fast exponent, faster than the value from the data-sheet, followed by

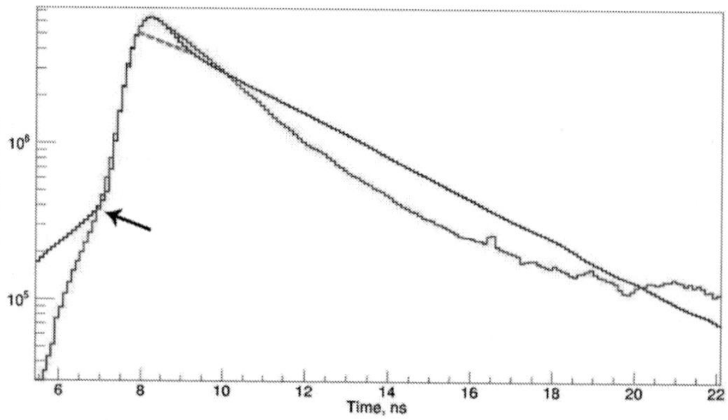

Figure 3: The pulse sums obtained in the measurements (red) and the simulations (blue) in a log scale.

the long non-exponential tail. For simulated pulse there is a bump before the exponential decay. This feature is getting more prominent if the light losses at the optical interface go down or the cross talks go up. In order to get the quantitative fitting of the parameters and validation of the MC model further measurements are needed.

The parameters of the readout chain needs to be precisely measured using a fast laser. The actual single photoelectron pulse shapes and spectrum should be acquired. Bigger number of pixels should be read out at the same time in order to study all the effects more carefully. As the next step, multichannel electronics based on the TOFPET ASIC [13] will be used for the readout of the NeuRad prototype. A transfer function of these electronics will be implemented into the EXPERTroot software in other to compare the simulated and the experimental data.

4 Conclusion

The first small-size prototype of the NeuRad detector was investigated using the γ-source. The pulse shapes have been obtained using the DRS4 waveform digitizer board and have been studied individually and in average. Processing algorithms were implemented into the EXPERTroot software package. In order to understand the pulse formation mechanism, the measurements were simulated using the EXPERTroot.

The influence of such parameters of the detector as light collection efficiency, interfiber cross talk, detection threshold and luminescence kinetics on the signal shape are understood. For a better understanding of the timing properties, quantitative validation of the MC simulations and complete feasibility study it is necessary to conduct new measurements with the NeuRad prototype recording bigger number of channels and using well calibrated electronics. The results of this study will allow to choose the optimal readout electronics for the entire system.

5 Acknowledgement

The authors are grateful to I G Mukha for the stimulating interest and numerous useful advices. This work was partly supported by the Helmholtz Association under grant agreement

IK-RU-002, RSF 17-12-01367 grant and the MEYS Projects (Czech Republic) LTT17003 and LM2015049.

References

[1] L.Grigorenko et al., "Studies of light exotic nuclei in vicinity of neutron and proton drip-lines at FLNR JINR" Phys. Usp. 59 321–366 (2016); DOI: 10.3367/UFNe.0186.201604a.0337

[2] M. Winkler et al., The status of the Super-FRS in-flight facility at FAIR, Nucl. Instr. and Meth. in Phys. Res. B 266 (2008) 4183.

[3] Geissel, H., Kiselev, O., Mukha, I., et al., "Expert (exotic Particle Emission and Radioactivity by Tracking) Studies at the Super-FRS Spectrometer", 2015, Exotic Nuclei: EXON-2014 - Proceedings of International Symposium.

[4] Technical Design Report of the EXPERT setup of the Super-FRS Experiment Collaboration, https;//edms.cern.ch/document/1865700

[5] http://er.jinr.ru/

[6] D. Kostyleva et al., GSI Scientific Report 2016 171 (2017), DOI:10.15120/GR-2017-1

[7] https://www.crystals.saint-gobain.com/

[8] www.hamamatsu.com/

[9] https://www.psi.ch/drs/

[10] http://er.jinr.ru/neurad.html

[11] J. Adamczewski-Musch et al., Inst. and Exp. Tech. (3) 2018

[12] Nuclear Instruments and Methods in Physics Research A 506 (2003) 250-303, and IEEE Transactions on Nuclear Science 53 No. 1 (2006) 270-278.

[13] http://www.petsyselectronics.com/web/

Pairing and $(\frac{9}{2})^n$ configuration in nuclei in the ^{208}Pb region

M. Stepanov[1], *L.* Imasheva[1], *B.* Ishkhanov[1,2], and *T.* Tretyakova[2,⋆]

[1] *Faculty of Physics, Lomonosov Moscow State University, Moscow, 119991 Russia*
[2] *Skobeltsyn Institute of Nuclear Physics, Lomonosov Moscow State University, Moscow, 119991 Russia*

Abstract. Excited states in low-energy spectra in nuclei near ^{208}Pb are considered. The pure $(j = 9/2)^n$ configuration approximation with delta-force is used for ground state multiplet calculations. The multiplet splitting is determined by the pairing energy, which can be defined from the even-odd straggering of the nuclear masses. For the configurations with more than two valence nucleons, the seniority scheme is used. The results of the calculations agree with the experimental data for both stable and exotic nuclei within 0.06-6.16%. Due to simplicity and absence of the fitted parameters, the model can be easily applied for studies of nature of the excited states in a wide range of nuclei.

1 Introduction

Pair interaction of identical nucleons significantly effects on the properties of atomic nuclei [1]. A manifestations of the nucleon pairing is the mass surface splitting and the zero angular momentum of the even-even nuclei ground state. The breaking of two valence identical nucleons pair leads to formation of a characteristic set of the excited states with the positive parity and even spin, the so-called ground state multiplet (GSM). The GSM splitting is determined by the pairing energy (or by the doubled even-odd staggering effect): $\Delta_{nn(pp)} = 2\Delta_{n(p)}$ [2]. The classical example of GSM is the set of low-lying states in the ^{210}Pb and ^{210}Po spectra, which can be considered as the ^{208}Pb core with the pair of the valence neutrons or protons. Since ^{208}Pb is the heaviest double magic stable nucleus, this approximation is quite accurate. Therefore, it is interesting to trace the GSM formation in the systematics of isotopes spectra in the $(N, Z) = (126, 82)$ region with sequential filling of the $2g_{9/2}$ subshell with neutrons and the $1h_{9/2}$ subshell with protons. As an example, figure 1 shows the experimental excitation spectra of the even-even isotones $N = 128$. It can be seen that the neutron pairing energy systematically corresponds to the excited state $J^P = 8^+$, and the proton pairing energy corresponds to the state $J^P = 10^+$. A similar systematic behavior is observed in other chains of isotones and isotopes in the ^{208}Pb region [6].

2 Ground state multiplets and seniority scheme

Since the nucleon force, including the residual pairing interaction, has a short-range attractive character, the δ-potential $V_{12} = -V_0\delta(r_1 - r_2)$ is a good approximation for its description [7]. In this

⋆e-mail: tretyakova@sinp.msu.ru

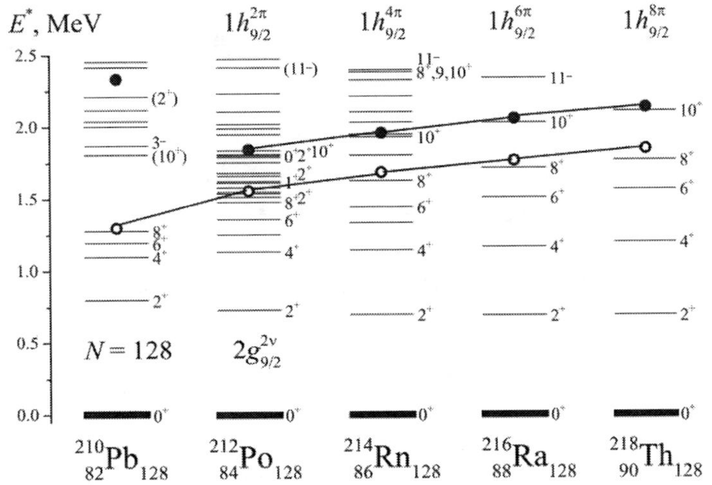

Figure 1: Experimental spectra of low-lying excitations in isotones $N = 128$. The values of the pairing energy of protons $\Delta_{pp}(\bullet)$ and neutrons $\Delta_{nn}(\circ)$ are calculated using nuclear masses [3]. Here and below, the experimental data are from [4, 5]. The figure is from [6].

approximation, the energy of the state with a total angular momentum J depends on V_0 and the radial Slater integral $F_0(nl)$ [7, 8]:

$$\Delta E_{jj:J} = -\frac{1}{2}V_0F_0(nl)(2j+1)^2\begin{pmatrix} j & j & J \\ 1/2 & -1/2 & 0 \end{pmatrix}^2. \tag{1}$$

Since the values of V_0 and $F_0(nl)$ do not depend on J, the structure of the GSM can be determined from the ratio of the energies of levels with different values of J, to the known experimental value of the excitation energy, for example, $E(2^+)$ [9] or $E(J_{max})$, where $J_{max} = 2j - 1$ [10]. However, the GSM structure can be calculated using the ground state energy shift ΔE_0, defined by the even-odd staggering effect: $\Delta E_0 = \Delta_{nn(pp)} = 2\Delta_{n(p)}$ [11]. Thus, the energies of all the excited GSM states are determined by the values of the nuclear masses.

The δ-potential approximation allows one to describe GSM in the cases of several pairs or an odd number of valence nucleons using the seniority scheme. The seniority model makes it possible to express multiplets of states in the case of the nucleon number $N > 2$ on the basis of the known spectrum of states of a nucleon pair through the sets of coefficients of fractional parentage (CFP) $\langle j^{N-1}s_1J_1; j|\}j^N sJ\rangle$ [12]:

$$\Delta E_J = \sum_{J_1}\langle j^{N-1}s_1J_1; j|\}j^N sJ\rangle^2\Delta E_{J_1}, \tag{2}$$

where seniority s is the number of unpaired nucleons, $s_1 = s - 1$, J_1 is the total spin corresponding to s_1, the sum runs over all values of J_1. This approach is actively used for the GSM description in the ^{40}Ca region, because the seniority is a good quantum number for any two-particle interaction for $j \leq 7/2$. Moreover this approach can be used for higher values of j for a wide range of two-particle interactions, including the δ-potential [9].

Table 1: Possible values of the total angular momentum J for the $j = 9/2$ subshell.

Seniority s	J
2	$2, 4, 6, 8$
3	$3/2, 5/2, 7/2, 11/2, 13/2, 15/2, 17/2, 21/2$
4	$0, 2, 3, 4^2, 5, 6^2, 7, 8, 9, 10, 12$

Table 1 shows the possible values of the total angular momentum J for the subshell $j = 9/2$ for different values of the seniority s. With increasing j, the number of possible states essentially increases. Usually in the experimental spectra the whole set of states for $s = 4$ is not observed, but the levels with high values of $J = 10$ and 12 are well identified.

3 Results

The neutron-rich isotopes of lead are ideal for testing of the properties of nucleon-nucleon interaction and the seniority scheme since they have the closed shell core ^{208}Pb. Owing to the development of experimental possibilities, it became possible to determine the position of excited states in the neutron-rich lead isotopes up to ^{216}Pb. In figure 2 the results of our calculations for the seniority $s = 2$ are presented, assuming the filling of the subshell $2g_{9/2}$ by neutrons.

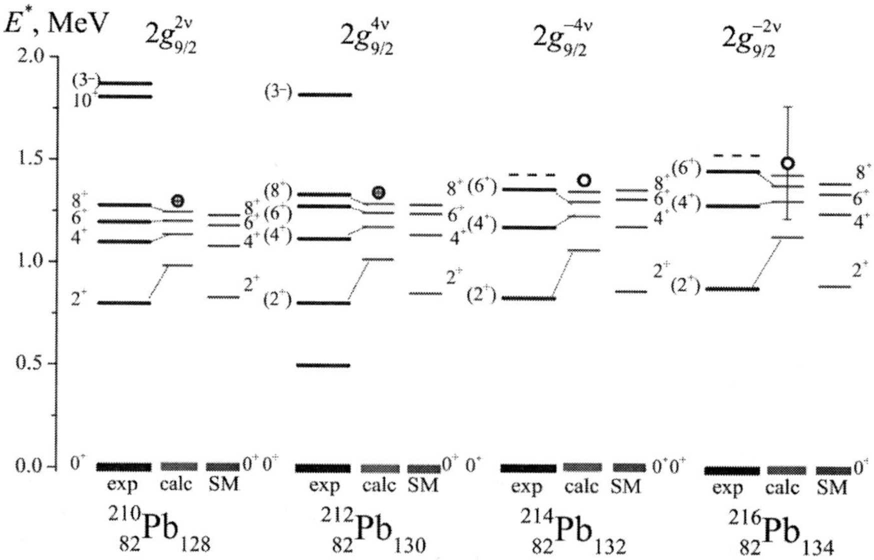

Figure 2: The spectra of the $^{210-216}$Pb isotopes. Black lines are the experimental spectra (exp) from [4, 5], red lines are our calculations (calc) based on Δ_{nn} (○), blue lines are the results from [13] (SM).

Use of Δ_{nn} as the GSM splitting leads to a good agreement for the state $J^P = 6^+$ and allows us to estimate the energy of the state $J^P = 8^+$. The pure-configuration $(2g_{9/2})^n$ approximation with δ-force cannot correctly reproduce positions of the $J^P = 2^+, 4^+$ levels. The regard for the configuration mixing leads to the substantial improvement of the experimental data description. Figure 2 shows the results of the calculations performed in the shell model (SM) in the $(g_{9/2}i_{11/2}d_{3/2}d_{5/2}g_{7/2}s_{1/2}j_{15/2})^n$ configuration space [13]. The wide configuration space in combination with the Kuo-Herlig potential modified for heavy nuclei [14] makes it possible to describe the experimental spectra with an accuracy of about 100 keV [13].

Isotones with $N = 126$ represent an excellent example of the applicability of the senority scheme with $s > 2$ [15]. Calculations in the entire range of proton configurations using the modified Kuo-Herling interaction showed a good agreement with the experimental data and confirmed the seniority conservation for the isotones up to $Z = 91$. The results were obtained in a wide configuration space, and it would be interesting to verify correlation between the experimental spectra and calculations in the pure $(1h_{9/2})^n$ configuration. In figure 3a the experimental spectrum of the even-odd ^{211}At isotope are presented. Our calculations in the configuration approximation $(1h_{9/2})^{3\pi}$ show a good agreement with experiment except for the first state $(7/2)^-$. For comparison, in figure 3a the results of the calculations in the shell model using an effective potential based on the Bonn potential A (SM2) [16] are also present. The states shown in the figure were obtained with the configuration $(1h_{9/2})^{3\pi}$, except for the first state $(7/2)^-$, for which the contribution of the $(1h^2_{9/2}2f_{7/2})^\pi$ configuration dominates.

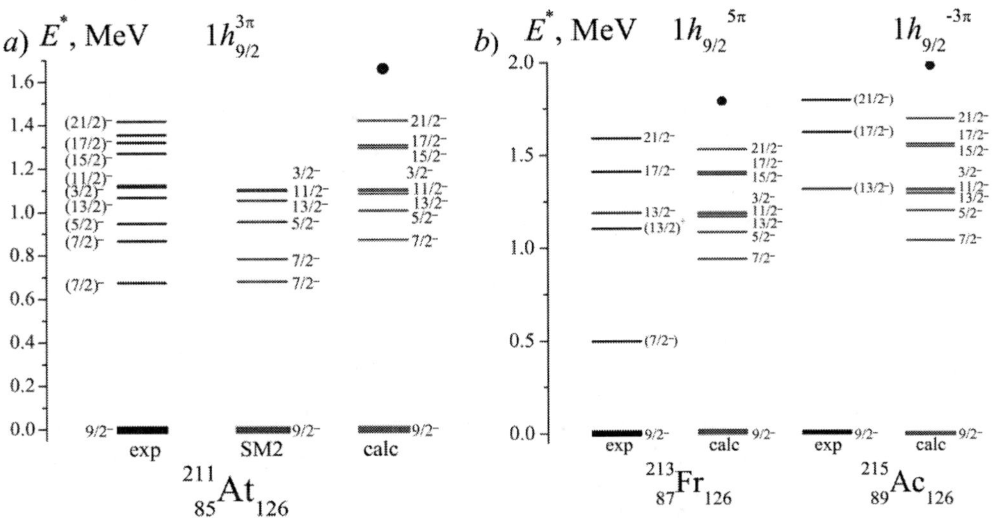

Figure 3: The spectra of the odd isotopes a) ^{211}At, b) ^{213}Fr and ^{215}Ac with Δ_{pp} (•).

In figure 3b, the spectra of the even-odd isotopes ^{213}Fr and ^{215}Ac calculated assuming the pure $(1h_{9/2})^n$ configuration for seniority $s = 1$ and 3 are compared with the experiment. The experimental spectra themselves no longer contain all the levels predicted by the seniority scheme. It should be noted that these isotopes belong to the region of neutron-rich short-lived nuclei and the precise determination of their spectra of excited states is a complex experimental task. Low-lying excitations $J^P = 13/2^-, 17/2^-, 21/2^-$ are defined with good reliability and interpreted as the $(1h_{9/2})^{5\pi}$ configuration in ^{213}Fr [17] and the $(1h_{9/2})^{-3\pi}$ configuration in ^{215}Ac [18]. Our calculations show the authenticity of this interpretation. However, the calculated energies of levels with $J^P = 21/2^-$ lie below the experimental values. Thus, the deviation of the calculated values from the experimental values is 6%. This value corresponds to the accuracy of our model description of the experimental data considered in this paper.

The results of the calculations for $s = 4$ are shown in figure 4 for spectra of the ^{212}Rn and ^{214}Ra nuclei with 4 protons and 4 proton holes in the $1h_{9/2}$ shell. For comparison, the results of calculations [13, 19] and [16] (SM3) are given. One can see that use of pairing energy estimated from the nuclear mass data [3] provide a good description not only for GSM with the senority $s = 2$, but also for experimental states with $J^P = 12^+, 10^+$ with $s = 4$.

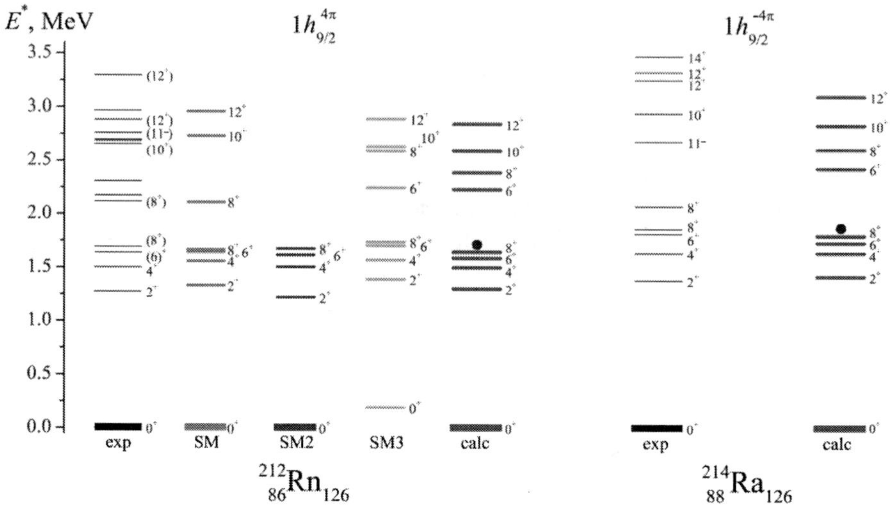

Figure 4: The spectra of even-even isotopes ^{212}Rn and ^{214}Ra. The values of Δ_{pp} are denoted by (•).

4 Conclusions

Low-lying excited states of neutron-rich lead isotopes and the isotopes of the $N = 126$ chain are considered. Basing on the nuclear mass data [3] multiplets for the seniority values $s = 2, 3, 4$ are calculated in the configuration approximation $(2g_{9/2})^n$ for the valence neutrons and $(1h_{9/2})^n$ for the valence protons. The results are compared with the shell model calculations with the effective potentials and the widest possible set of configurations for valence nucleons.

Since the ^{208}Pb core is a closed shell double magic nucleus, the approximation of pure configurations for the valence nucleons is appropriate, and the low-lying part of the spectrum obeys the seniority scheme. The approach of δ-potential considered in this paper is a simplified form of residual interaction and can not lead to an accurate description of the spectrum of excited states. Nevertheless, the application of the pairing energy estimated from the nuclear masses as the splitting of GSM allows us to obtain a agreement with accuracy 6% with the experimental data not only for the seniority $s = 2$, but also for the higher s values without involving fitting parameters.

References

[1] P. Ring, P. Schuck, *The nuclear many-body problem. 3rd ed.* (Springer, Berlin, 2004)
[2] A. Bohr, B. Mottelson, *Nuclear Structure, vol. 2* (Benjamin, NY, 1975)
[3] G. Audi et al. Chin. Phys. C **41**, 3, 030001 (2017)

[4] Database «Centre for Photonuclear Experiments Data SINP MSU», http://cdfe.sinp.msu.ru/
[5] National Nuclear Data Center, Brookhaven, Evaluated Nuclear Structure Data File, http://ie.lbl.gov/ensdf/.
[6] B.S. Ishkhanov, M.E. Stepanov, T.Yu. Tretyakova, Bull. of the RAS: Physics **78**, 5, 405 (2014)
[7] M.G. Mayer, Phys. Rev. **75**, 1969 (1949)
[8] A. De-Shalit, Phys. Rev. **91**, 6 (1953)
[9] I. Talmi, *Simple Models of Complex Nuclei* (Harwood Ac. Publ. Chur., 1993)
[10] A. Heusler, P. von Brentano, Eur. Phys. J. A. **38**, 9 (2008)
[11] B.S. Ishkhanov, M.E. Stepanov, T.Yu. Tretyakova, Moscow Univ. Phys. Bull. **69**, 1 (2014)
[12] B.F. Bayman, A. Lande, Nucl. Phys. **77**,1 (1966)
[13] A. Gottardo et al, Phys. Rev. Lett. **109**, 162502 (2012)
[14] E. K. Warburton, B. A. Brown, Phys. Rev. C **43**, 602 (1991)
[15] E. Caurier, M. Rejmund, and H. Grawe, Phys. Rev. C **67**, 054310 (2003)
[16] G.D. Dracoulis et al, Phys. Rev. C **80**, 054320 (2009)
[17] M. S. Basunia, Nucl. Data Sheets **108**, 633 (2007)
[18] B. Singh et al, Nucl. Data Sheets **114**, 2023 (2013)
[19] A. Covello et al, Acta Physica Polonica B **30**, 3 (1999)

New test of the dynamic theory of neutron diffraction by a moving grating.

Maxim Zakharov[1, 2,*], *Alexander* Frank[1], *German* Kulin[1], and *Semyon*Goryunov[1]

[1]FLNP, JINR, Joliot-Curie 6, Dubna, Moscow region, Russian Federation, 141980
[2]MIPT, Institutskiy per. 9, Dolgoprudny, Moscow Region, Russian Federation, 141701

Abstract. Recently, multiwave dynamical theory of neutron diffraction by a moving grating was developed. The theory predicts that at a certain height of the grating profile a significant suppression of the zero-order diffraction may occur. The experiment to confirm predictions of this theory was performed. The resulting diffracted UCNs spectra were measured using time-of-flight Fourier diffractometer. The experimental data were compared with the results of numerical simulation and were found in a good agreement with theoretical predictions.

1. Introduction

Diffraction of neutron waves, like most other neutron-optical phenomena, has been considered for a long time from the point of the demonstration of the wave properties of a neutron and for the test of the predictions of stationary quantum mechanics. At the same time, it is known that neutron optics acquires new qualities in the cases when any parameters describing the interaction of a neutron beam with objects begin to depend on time. Nonstationary actions on a neutron wave allow variate significantly its characteristics such as the energy spectrum, phase, spin, intensity, direction of propagation, etc.

Neutron diffraction under certain conditions can also be considered as a nonstationary phenomenon. It was shown [1-3] that in the case of an absorbing or a phase grating moving across a monochromatic beam of ultracold neutrons (UCN), the grating can serve as a quantum modulator of a transmitted wave and the spectrum of neutrons passed through it is discrete.

The phenomenon of neutron diffraction on a moving grating has found its application in experiments on the test of the weak equivalence principle for a neutron [4]. The continuation of this investigation [5-7] and recent theoretical results [8] required a more rigorous investigation of the UCN spectra after diffraction by a moving grating.

2. Theory of neutron diffraction from a moving grating

The problem of diffraction on a moving grating can be resolved in the coordinate system where the grating is at rest (fig 1b). In the laboratory coordinate system (fig 1c) the neutron energy is quantized, energy spectrum consists of discrete lines, and the neighboring lines of the spectrum differ by a value $\hbar\Omega$. The frequency Ω is determined by the velocity V and the period d of the lattice $\Omega = 2\pi d / V$.

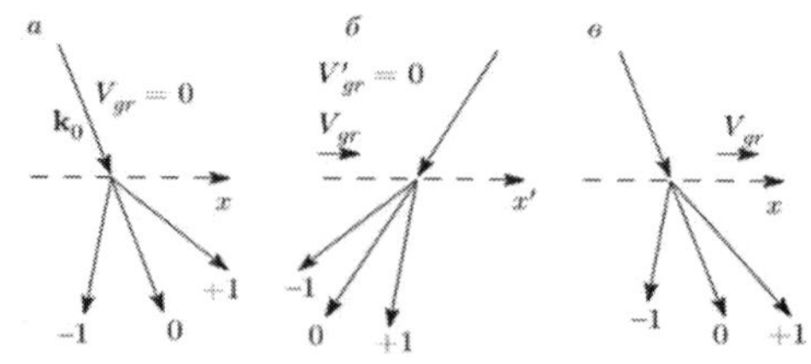

Fig.1. Schemes of neutron diffraction on a grating: a – the grating at rest, $\omega_n = \omega_0$; b – the coordinate system moves together with the grating, $\omega_n = \omega'$; c – moving grating in the laboratory coordinate system, $\omega_n = \omega_0 + n\Omega$.

[*]Corresponding author: zakharovmax@jinr.ru

In Refs. [1-3], neutron diffraction from the moving grating was considered within the framework of the kinematic theory. An essential disadvantage of this theory is the fact that it is valid only for thin gratings where there is no any mutual effect of waves of different diffraction orders. To describe diffraction on real gratings, a more rigorous dynamic theory should be used.

The dynamic theory of diffraction developed in [8] is based on the method of slowly varying coupled amplitudes of wave functions. It takes into account the mutual effect of waves of different orders in the process of penetration of neutrons into the three-dimensional phase grating and is based on the solution of the Schrödinger equation in the coordinate system where the grating at rest:

$$\Delta\psi(\mathrm{r})+\left[k^2-\chi(\mathrm{r})\right]\psi(\mathrm{r})=0 \tag{1}$$

where Δ – the Laplace operator, k – neutron wave number in vacuum, $\chi(\mathbf{r}) = 4\pi N(\mathbf{r})b(\mathbf{r})$, $N(\mathbf{r})$ – nuclear density, $b(\mathbf{r})$ the coherent scattering length of neutrons in the medium. Periodic in the range $0 \le z \le h$ function $\chi(x)$ is represented in the form of the series:

$$\chi(x) = \sum_{n=-\infty}^{\infty} \chi_n \exp(ig_n x)\cdot \tag{2}$$

where $g_n = ng_0$, $g_0 = 2\pi/d$, $n = 0, \pm1, \pm2, \ldots$ - integer numbers, and

$$\chi_n = \frac{1}{d}\int_0^d \chi(x)\exp(-ig_n x)dx\cdot \tag{3}$$

The zero Fourier amplitude χ_0 determines the value of the average refractive index of the grating$n_e = (1 - \chi_0/k^2)^{1/2}$ in the layer of thickness h.

The solution of Schrödinger equation in the range $0 \le z \le h$, in view of (2) and (3), is represented in the form of the sum of Bloch functions with the amplitudes $\Psi_m(z)$ depending on the vertical coordinate z:

$$\Psi'(x',z) = \sum_{m=-\infty}^{\infty} \Psi_m(z)\exp[i(q_{mx}x' + q_{0z}z)], \tag{4}$$

where the projections of the wave vectors

$$\mathrm{q_{mx}} = \mathrm{k_{0x}} - \mathrm{k_V} + \mathrm{g_m},\ \ \mathrm{q_{0z}} = (\mathrm{k_{0z}^2} - \chi_0)^{1/2}. \tag{5}$$

Substituting (3) and (4) into (1) and equating terms with the same exponents, the following infinite system of coupled second-order differential equations was obtained:

$$\frac{d^2\Psi_m}{dz^2} + 2iq_{0z}\frac{d\Psi_m}{dz} - \alpha_m\Psi_m - \sum_{n\neq0}\chi_n\Psi_{m-n} = 0\cdot \tag{6}$$

where

$$\alpha_m = g_m[g_m - 2(k_V - k_{0x})]\cdot \tag{7}$$

The solution of system (6) is a rather laborious task. However, it is substantially simplified if the second derivatives in this system are neglected. As shown in [4], under conditions corresponding to real experiments, such approximation does not lead to significant errors. As a result, the following system of truncated differential equations of the first order was obtained:

$$\frac{d\Psi_m}{dz} = -i\gamma_m\Psi_m - i\sum_{n\neq0}\beta_n\Psi_{m-n}, \tag{8}$$

where $\gamma_m = \alpha_m/2q_{0z}$, $\beta_n = \chi_n/2q_{0z}$. The system (8) should be supplemented by the following boundary conditions: $\Psi_0(z = 0) = 1$, $\Psi_{m\neq0}(z = 0) = 0$.

The coefficients γ_m in (8), and, thereby, the functions $\Psi_m(z)$, depend on the velocity of the grating V_{gr}, the horizontal projection of the neutron velocity V_{0x}, the grating period d and the diffraction order m. They are zero for both at m = 0 and under the Bragg condition $2(k_V - k_{0x}) = g_m$. The system (8) is solved numerically.

3. Experiment

One of the predictions of the dynamic theory of diffraction [8] is that at a certain height of the grating profile a significant suppression of the zero-order diffraction may occur with a corresponding increase in the intensity of the lines of other orders (Fig.2). The aim of the experiment was to confirm this prediction. The experiment was performed at the PF2 UCN source of the Laue-Langevin Institute (Grenoble, France). As in the previous experiments [3- 7], a rotating grating was used instead of linearly moving grating (Fig. 3). The grating was prepared on the surface of a silicon disk. Radial grooves were made in the peripheral region of the disk. Two gratings were used. New grating (Grating 1) with 84000 grooves and the depth of 0.22 μm was specially manufactured for this experiment. Grating 2, with 94500 grooves and the depth of 0.14 μm was the same as in experiment [7].

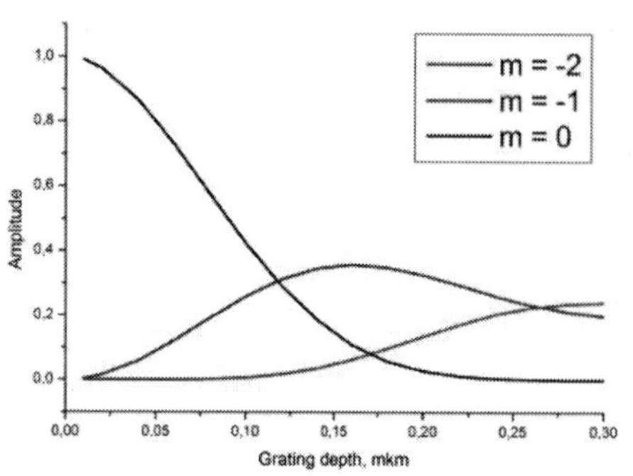

Fig.2. The dependence of the diffraction orders intensities on the depth of grooves at the grating rotation frequency f = 4800 rpm.

Fig.3. Diffraction grating with radial groves.

Ultracold neutrons were fed to the entrance chamber through the UCN neutron guide and, after several reflections, fell down the annular channel with the lower section closed by a monochromator, which is a five-layer Ni-Ti neutron interference filter (NIF) [9]. To suppress the background of neutrons with energies higher than the effective potential of nickel, the filter-monochromator was combined with a multilayer "superwindow" filter [10]. Two NIFs with maximum of resonant transmission at 107 (NIF1) and 115 neV (NIF2) and FWHF 4 neV were used. The neutron spectrum formed by the filter-monochromator and transformed by the moving grating was measured using time-of-flight Fourier spectrometry technique [11].

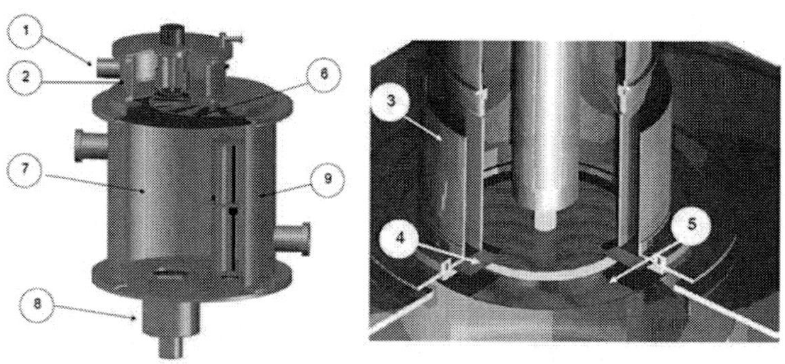

Fig. 4. Time-of-flight Fourier spectrometer: general view (left) and its upper part (right). 1 - feeding neutron guide, 2 - entrance chamber, 3 - annular corridor, 4 - filter-monochromator, 5 - grating, 6 - rotor of the Fourier modulator, 7 - vertical glass guide, 8 - detector, 9 - vacuum vessel.

The measurements were done for different frequencies of the grating rotation from 2400 to 6000 rpm. In Fig.5 the results of the time of flight measurements of the diffraction spectra for the grating 1 and NIF1 are displayed. It is clearly seen in this picture that intensity of the zeroth diffraction order decreasing when the rotation speed is increasing from 2400 to 4800 rpm but increasing again for the rotation frequency 6000 rpm. To compare spectra formed by grating 1 and grating 2, the measurements with the grating 2 and NIF 1 have been also made with the rotation frequency 4800 (See Fig.6).

The measurement with grating 1 were repeated with NIF2 after some modification of the experimental installation (fig. 7). The results of this measurement were compared with the results of detailed calculation. To do that a computational-mathematical model of the experiment was created. The model includes a numerical calculation of the spectrum of UCN transmitted through a rotating radial grating using formalism of dynamic theory of neutron diffraction on moving grating [8], as well as a Monte Carlo simulation of the process of measuring of the neutron spectrum obtained after diffraction.

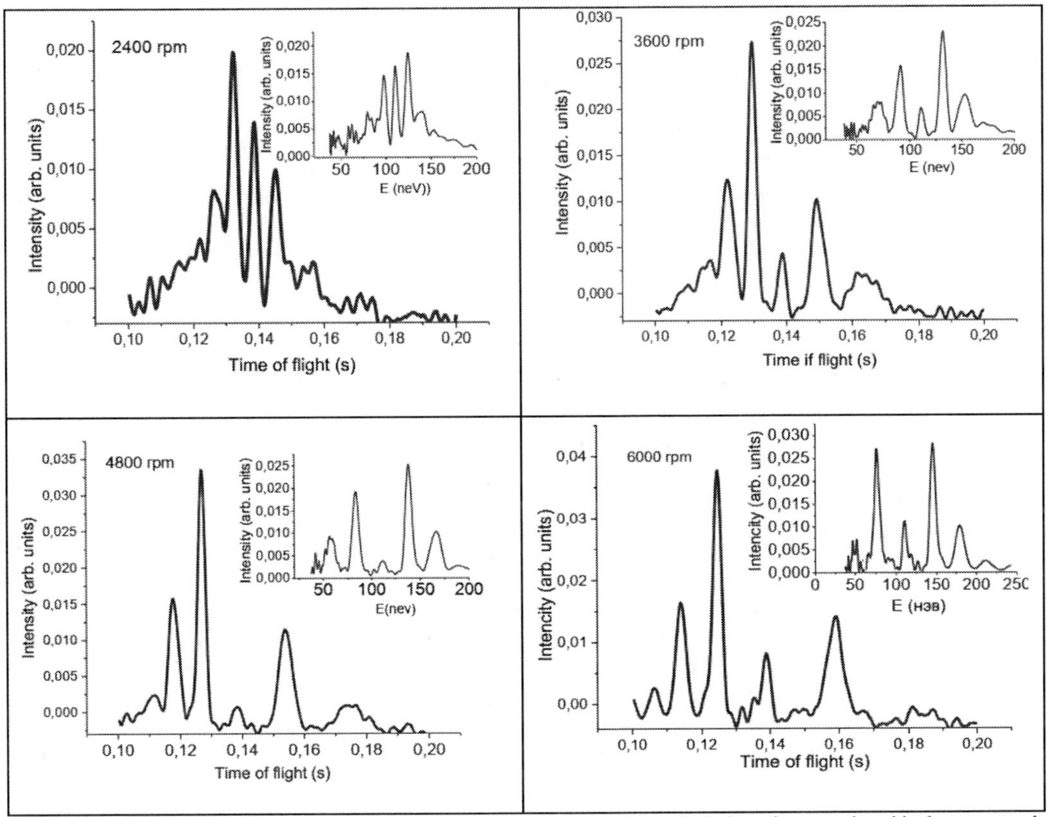

Fig. 5. Time of flight spectra appeared at neutron diffraction by a moving grating 1. In the inset the same plotted in the energy scale

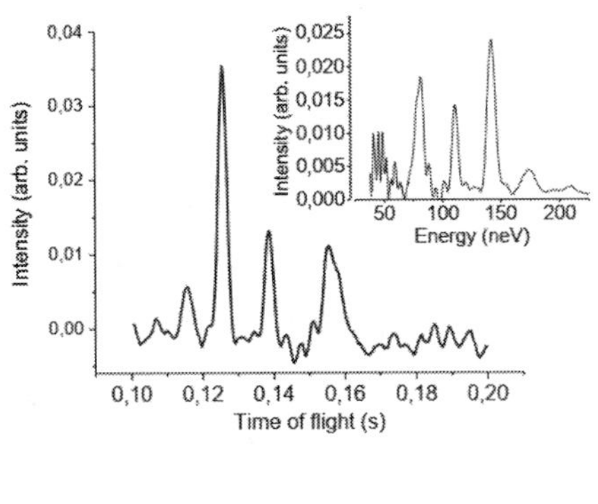

Fig. 6. Time of flight spectrum appeared at neutron diffraction by a moving grating 2. In the inset the same plotted in the energy scale

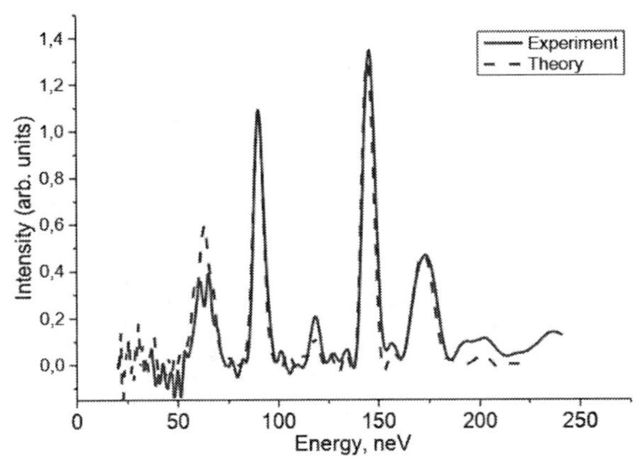

Fig.7 Comparison of the measured spectrum with the theoretical model

Whenexperimental spectrum of fig 7 is compared with theoretical calculation shown in fig.2 it is necessary to consider that in this last the total intensity of the diffraction order has to be correlated with the integral under the correspondent line in fig.7. Notice that the width of the line in the spectrum depends on the initial horizontal velocities and additional spreading due to this effect is proportional to the number of the diffraction order.

Conclusion

The experiment to test the dynamic theory of neutron diffraction by a moving grating was performed. The resulting diffracted UCNs spectra were measured using time-of-flight Fourier diffractometer. The experimental data were compared with the results of numerical simulation and were found in a good agreement with theoretical predictions.

References

1. A. I. Frank and V. G. Nosov, Phys. Lett. A **188**, 120 (1994).
2. A. I. Frank, S. N. Balashov, I. V. Bondarenko et al., Phys. Lett. A **311**, 6 (2003).
3. A. I. Frank, P. Geltenbort, G. V. Kulin et al., JETP Lett. **81**, 427 (2005).
4. A. I. Frank, P. Geltenbort, M. Jentschel et al., JETP Lett. **86**, 225 (2007).
5. A. I. Frank, P. Geltenbort, M. Jentschel et al., NIM A. A **611**, 314 (2009).
6. G.V. Kulin, A.I. Frank, S.V. Goryunov,et al. NIM A **792** 38 (2015)
7. G.V. Kulin, A.I. Frank, S.V. Goryunov, et al .Phys. Rev. A 93, 033606(2016)
8. Bushuev V. A., Frank A.I., Kulin G.V., JETP, **122** (1) (2016) 32
9. I. V. Bondarenko, V. I. Bodnarchuk, and S. N. Balashov. Phys.At. Nucl. **62**, 721 (1999).
10. A. I. Frank, S. V. Balashov, V. I. Bodnarchuk *et al.*, Proc. SPIE**3767**, 360 (1999).
11. G.V.Kulin, A.I. Frank, S.V. Goryunov, D.V. Kustov et al. Nucl. Instr. Meth. Phys. Res. A**819** 67(2016).

Electronic catalogue of muonic X-rays

Daniya Zinatulina[1,*], *Chantal* Briançon[2], *Victor* Brudanin[1], *Viacheslav* Egorov[1], *Lev* Perevoshchikov[1], *Mark* Shirchenko[1], *Igor* Yutlandov[1], and *Claude* Petitjean[3]

[1]Joint Institute for Nuclear Research, DLNP, 141980 Dubna, Russia
[2]Centre de Spectrometrie Nucleaire et de Spectrometrie de Masse, 91405 Orsay, France
[3]Paul Sherrer Institute, CH 5232 Villingen, Switzerland

Abstract. μX-ray spectra for Z=9-90 were measured with HPGe detectors and muonic beams of PSI (Villigen, Switzerland) [1]. The results are presented as electronic atlas composed of graphic plots. The atlas is available at JINR site [2].

1 Experiment and purposes

Stopping of negative muons in mater is accompanied by emission of so-called "muonic X-rays". This radiation, similar to conventional electronic characteristic X-rays, is unique for each chemical element, but has much higher energy (up to ~10 MeV). Contrary to electronic X-rays, there was lack of systematic information about μX-rays. We have tried to fill this gap.

Idea of the experiment is based on the precise measurement of time- and energy-distribution of γ-rays following μ-capture. Low-momentum negative muons from the μE4 or μE1 beam-line of the PSI "meson factory" were stopped in thin targets (about 0.5 g/cm^2). In case of gas target, we used a special construction, described in our previous works [3, 4].

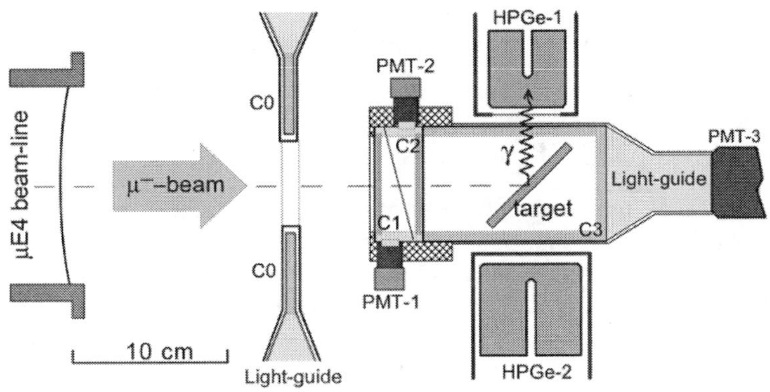

Fig. 1. Measurement set-up.

* Corresponding author: d.zinatulina@gmail.com

Muonic X-rays accompanying muon stops were registered with several HPGe detectors (description of the acquisition system was given in [3]). The experimental set-up is shown in Figure 1.

There are ~75 elements presented in our catalogue. We always tried to use the simple chemical compositions for the investigation. In case when the element was too chemically active, we used it as a passive compound. The second "companion" was chosen to have Z which differs a lot from the main one, in order to avoid the overlapping of corresponding muonic spectra.

Below there is a Figure 2 with the periodic table (from the web-page with our catalogue), where the different colors of the elements stand for the different type of the target: "green" – target is comprised from the pure element, "red" – target is used as an oxide, "orange" – halogen compound, "violet" – nitrate, "grey" – not investigated. To see the spectra of the interesting element, one need to click on it in Periodic table.

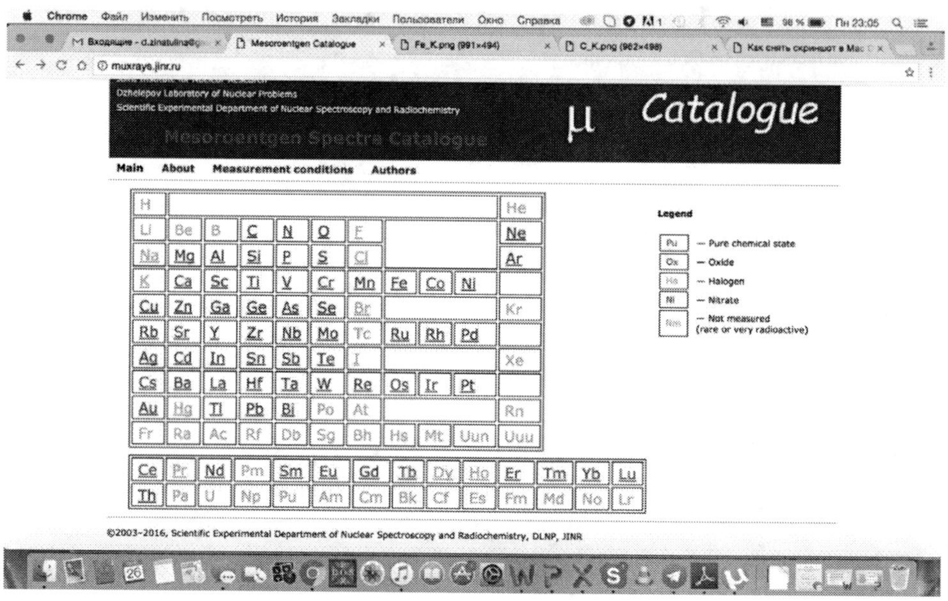

Fig.2 Periodic table from the web-page of the http://muxrays.jinr.ru

2 Results

As a result of the measurements, we have what we call "prompt energy spectra" ($T_{\mu\text{-}\gamma}$< 20-50 ns) which contain mainly muonic X-rays, fast part of delayed gammas and scattered beam electrons. Normally, muon can be captured by nucleus from the upper orbits only in very heavy muonic atoms, so for the light and intermediate Z it could be stated that at-least 99% of μ-stops are accompanied by any of *np-1s* muonic transition. Our interactive catalogue contains the spectra of *np-1s* muonic transition for all of the measured elements. For many of the elements we also present L-, M-series and sometimes even more (N, O etc). As example, such spectra are presented in the Figures 3, 4. Figure 3 is a total μX-ray spectrum of Cd. The structure of the spectrum sometimes quite complicated. There are the Compton bump, pair production, and also SE (single escape) and DE (double escape) peaks due to high energy of the resulting gamma. Figure 4 – is a zoom of the K-series from the total μX-ray spectrum in Figure 3 (that could be done as well for other L-, M- etc. series also by clicking them in Total spectrum).

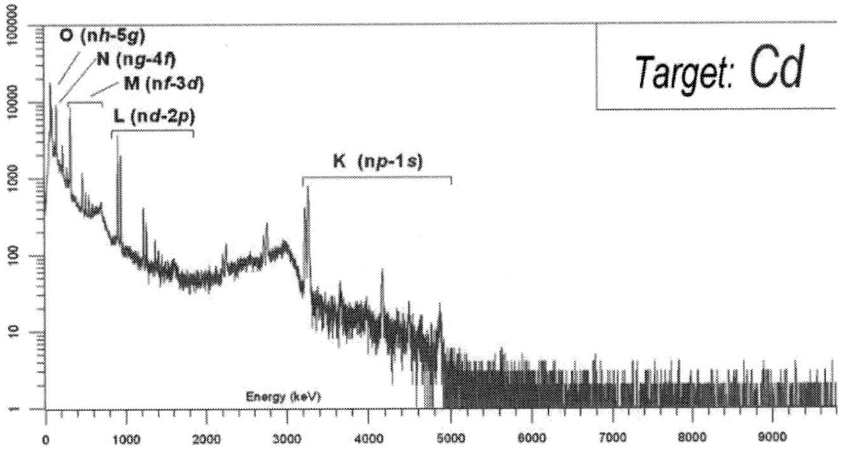

Fig. 3. Total μX-ray spectrum for Cd.

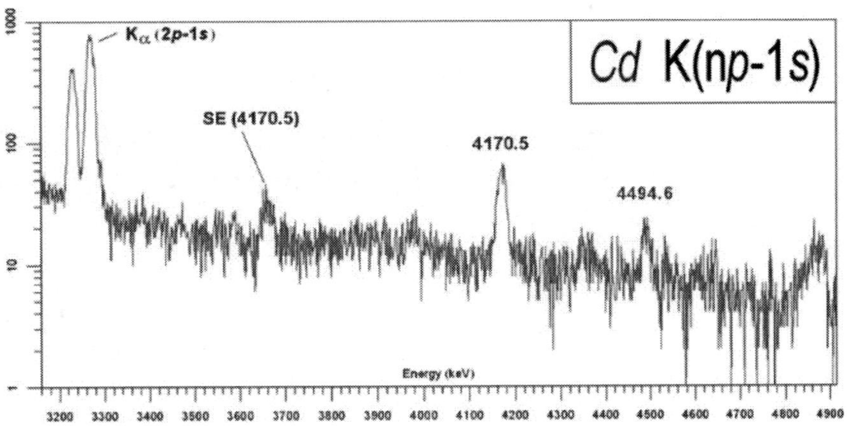

Fig. 4. K-series of Cd.

Whereas for muons the radius of K-shell comparable of the nuclei radius, then this K-shell very sensitive to the nucleus parameters. That is why μX-transition to K-shell depend on the properties of the nucleus. Thus, for K-series of the Cd, for example, we have a complex structure of the Kα, because there is an isotopic shift. The region of such Kα is presented in Figure 5.

3 Conclusions

Finally, we combined all the information into the catalogue and put it on the web-page [2]. In the Figure 6 the main web-page of the Electronic catalogue is shown. There is the information about the conditions of the measurements, contacts of the authors, users guide etc. The information from the μX-ray spectra catalogue is important for the precise identification of the structure of the background and for the correct selection of the targets and construction materials for different experiments with muons. Total intensity of the μX(Z)-ray K-series also give the number of muons stopped exactly in the target with specific Z.

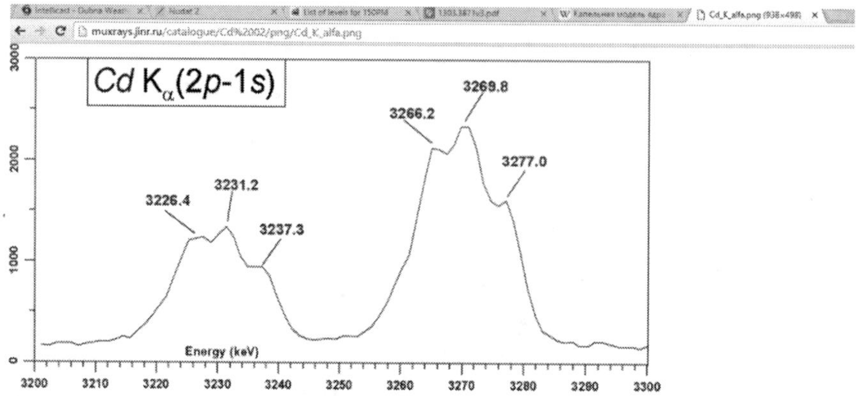

Fig. 5. Kα of Cd.

The enrichment of the measured isotopes could be found knowing the intensities of the Kα from the isotopic shifts. The information from that catalogue could be helpful for people who connected with mesoroentgen measurements.

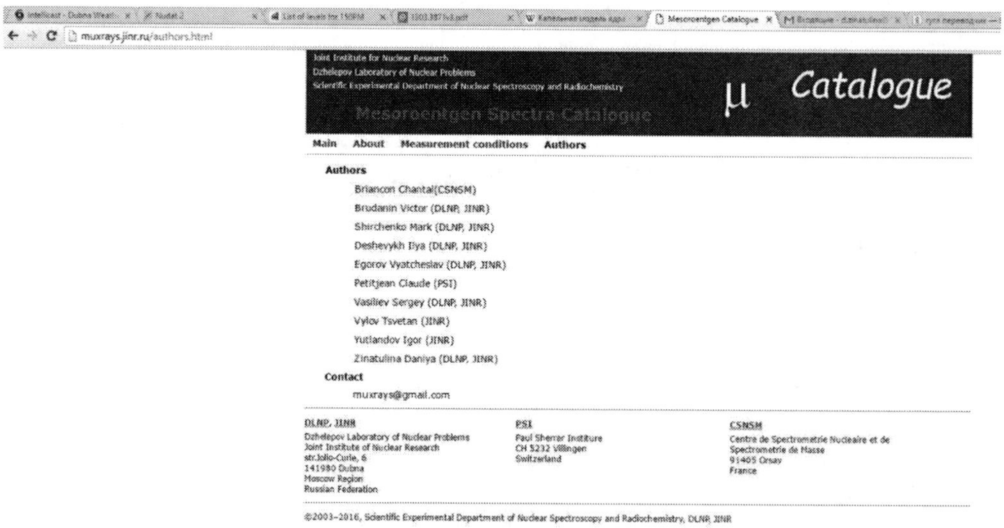

Fig. 6. Main web-page of the http://muxrays.jinr.ru

References

1. PSI experiment R-02-02
2. http://muxrays.jinr.ru
3. Yu. Shitov et al., Nucl. Phys., A**699**(2002) 917
4. D. Zinatulina et al., Bulletin of the Russian Acadeny of Science, V**76**(2008)6, pp.737-743

$\overline{\text{P}}$ANDA **Muon System Prototype**

Victor Abazov[1], *Gennady* Alexeev[1], *Maxim* Alexeev[2], *Vladimir* Frolov[1], *Georgy* Golovanov[1], *Sergey* Kutuzov[1], *Alexei* Piskun[1], *Alexander* Samartsev[1], *Valeri* Tokmenin[1], *Alexander* Verkheev[1,⋆], *Leonid* Vertogradov[1], *Nikolai* Zhuravlev[1]

[1] *Joint Institute for Nuclear Research, Dubna, Russia*
[2] *INFN Sezione di Torino, Torino, Italy*

Abstract. The $\overline{\text{P}}$ANDA Experiment will be one of the key experiments at the Facility for Antiproton and Ion Research (FAIR) which is under construction now in the territory of the GSI Helmholtz Centre for Heavy Ion Research in Darmstadt, Germany. $\overline{\text{P}}$ANDA is aimed to study hadron spectroscopy and various topics of the weak and strong forces. Muon System is chosen as the most suitable technology for detecting the muons. The Prototype of the $\overline{\text{P}}$ANDA Muon System is installed on the test beam line T9 at the Proton Synchrotron (PS) at CERN. Status of the $\overline{\text{P}}$ANDA Muon System prototype is presented with few preliminary results.

1 Introduction

$\overline{\text{P}}$ANDA (antiProton ANnihilation at Darmstadt) experiment is one of the four major scientific pillars of FAIR and is dedicated to perform fundamental research with antiproton beam with the momentum between 1.5 GeV/c and 15 GeV/c interacting with a fixed target inside the $\overline{\text{P}}$ANDA detector. The purpose of this experiment is hadron spectroscopy with the search for exotic particles and nucleon structure measurements for studying in-media effects of hadronic particles or hypernuclei. The envisaged physics program of the $\overline{\text{P}}$ANDA experiment requires large acceptance, high resolution for tracking as well as excellent particle identification and calorimetry. To obtain good momentum resolution the $\overline{\text{P}}$ANDA setup is divided into two parts: the Target Spectrometer (TS) based on a superconducting solenoid magnet surrounding the interaction point and the Forward Spectrometer (FS) based on a conventional dipole magnet for detecting particles emitted at forward angles.

Muons are an important probe for J/ψ identification, semi-leptonic D-meson decays and the Drell-Yan process. To measure muons within a wide momentum range, from hundreds MeV up to 10 GeV inside $\overline{\text{P}}$ANDA the Muon System [1] will be installed. The $\overline{\text{P}}$ANDA Muon System uses a Range System (RS) technique to register muons in a laminated iron absorber. Mini Drift Tubes (MDT) [2, 3] with two-coordinate readout of track information (the wires and the orthogonal strips) are used for charged particle detection in the RS. The resolution of Muon System is enough to use it as a coarse sampling hadron calorimeter. The Muon System consists of four parts: a Barrel (13 sensitive layers), an Endcap (5 layers), a Muon Filter (4 layers) and a Forward RS (16 layers). In the barrel part the granularity of the iron absorber is 3 cm (6 cm for the first and last layers) while in other parts of the

⋆e-mail: alver@jinr.ru

Muon System – 6 cm (for better detection/absorption of muons with higher momenta). The aim of the Muon System is to identify the primary muons with maximal possible separation from background contamination originating mostly from primary low-momenta pions and secondary decay muons.

2 Range System Prototype

To test and optimise different aspects of the $\overline{\mathrm{P}}$ANDA Range System performance based on MDTs with two-coordinate readout, a Range System prototype of the Muon System has been constructed at JINR (Dubna) and installed at CERN. The prototype was placed at the end of the secondary particles test beam line T9 at the Proton Synchrotron. It consists of steel absorber plates, detecting layers of MDTs, strip boards, corresponding front-end electronics for wire and strip readout, trigger and data acquisition system (DAQ), high-voltage and low-voltage supplies, and a gas system. The prototype is designed as '4 in 1' device: the structure of absorber plates reproduces the different parts of the Muon System mentioned above. The physical dimensions of the prototype are 220 x 150 x 170 cm^3. The weight of the prototype is about 10 tons. The total number of readout channels is about 4000, i.e. 2000 for wires and 2000 for strips. The prototype has equal spatial granularity in both coordinates – 1 cm given by the wire pitch of MDT and the strip width. The wires and strips are running perpendicular to each other.

The information is read out by the trigger and the data acquisition system of the COMPASS (CERN) experiment [4] adapted for the $\overline{\mathrm{P}}$ANDA Muon System prototype.

The main tasks of the beam tests are: 1) study the system response to all types of particles in the full energy range of $\overline{\mathrm{P}}$ANDA; 2) test the algorithms for muon-to-pion separation; 3) tune the Monte Carlo (MC) model parameters.

3 Modeling of Muon System Prototype

It is important to have the model of the Muon System prototype in PandaRoot package [5] for full MC simulation of $\overline{\mathrm{P}}$ANDA setup. Transfer of the detector geometry from Computer-Aided Design (CAD) systems to particle transports MC codes like GEANT4 [6] and ROOT [7] is always an issue due to possible geometry description incompatibility. To describe the geometry of the existing prototype of $\overline{\mathrm{P}}$ANDA Muon System, its model was developed using CATIA-GDML geometry builder package [8]. The package allows exchanging of the CAD-geometry to GEANT4 or ROOT compatible code using Geometry Description Markup Language (GDML) [9]. The geometrical model of the prototype consists of four parts as well as the real setup (see Sec. 2). Each part comprises detector layers consisting of 8-12 MDTs, which are separated by the absorbers of varying thickness. Each MDT consists of 8 sensitive elements, aluminium profiles and plastic envelope. In total, it equals to 24 layers. The geometrical model of the prototype was transferred and integrated into PandaRoot framework. Different colors represent different parts of the $\overline{\mathrm{P}}$ANDA Muon System, see Fig. 1. Also new program classes (MdtPrototype, MdtPoint etc.) have been developed and can be added into PandaRoot framework.

4 Experimental Data and Results

The main goal of the tests is calibration of the prototype as a particle identification (PID) device - for further use of this information for the Muon System. At the energies above 2 GeV the identification of muons versus hadrons is rather simple: the muon is represented by a 'straight' track, the hadron gives obvious sample of a shower. It is demonstrated in Figs. 2(a) and 2(b) for the full depth of the

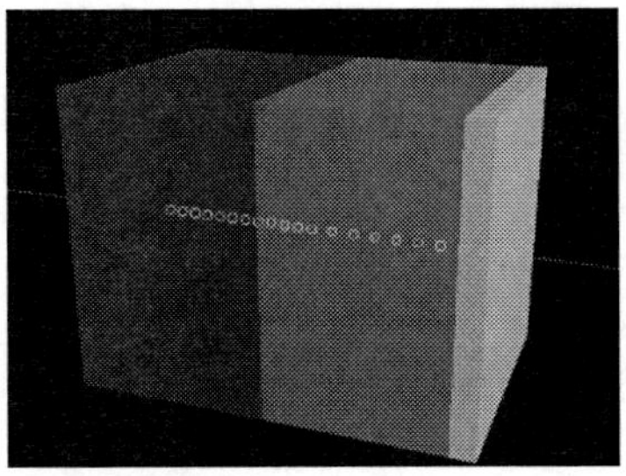

Figure 1. The geometrical model of the prototype in PandaRoot framework with hits (wire cells) fired by 5 GeV/c muon. Different colors represent different parts of the Muon System.

prototype. The patterns (muon and hadron) were readout from the early version of prototype when it was equipped with only one-dimensional (wire) readout.

It is important to mention that since 2017 we have equipped the prototype with the second coordinate (strip) readout system. Fig. 2(c) shows the muon track sample (6 GeV/c) for two-coordinate readout (wires and orthogonal stripes) obtained for the barrel structure of prototype.

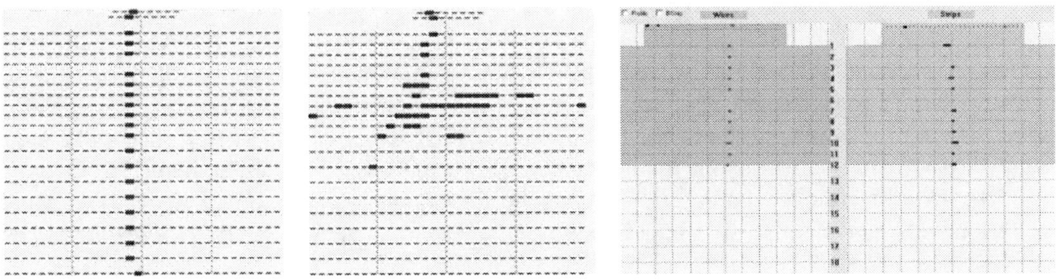

Figure 2. (a) Muonic sample, (b) hadronic sample: 5 GeV/c with one-coordinate readout (wires only), one bar represents one fired MDT cell (1 cm); (c) two-coordinates readout (wires - left, strips - right) for 6 GeV/c muon momentum; the particles on the figures enter the prototype from top.

The goal of 2017 test beam runs was aimed to identify low energy particles (by measuring their time-of-flight (TOF) in a beam) and study the corresponding response in the prototype structure.

Two test beam runs have been conducted in 2017. The prototype was pre-tested on cosmic rays and then it was installed at the end (focus) of the T9/PS test beam line and experimental data were collected.

Two scintillation counters of the TOF system have been mounted into the beam line (at the distance $\sim$ 26 m of each other) which allowed observing three timing peaks in a beam (electrons, muons and pions). Fig. 3 shows results of particles separation. The beam momentum was set up to 0.5 GeV/c. Setting the selection criteria for the time-to-digital converter (used by DAQ system) channel numbers allow separating events to two different data samples (with μ and π-mesons) interesting for further analysis.

The obtained data sets were processed, converted to a root format and analysed. At the same time MC simulation of events with muons and pions with the same momentum was performed using prototype geometrical model in the PandaRoot framework. FairBoxGenerator generator is used [5]. To study the prototype response to different particles and find the way to separate the signal muons from background pions at low energies the set of variables has been investigated. As an illustration, Figs. 4(a) and 4(b) show the normalized distributions of events containing muons (green line) and

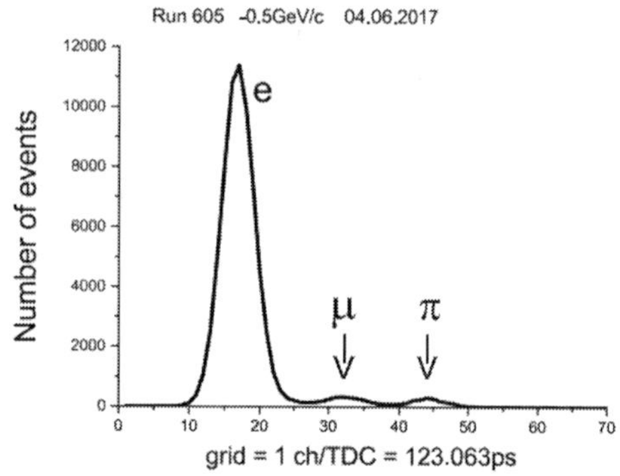

Figure 3. Results of particle TOF separation (identification): three peaks correspond to electrons, muons and pions.

pions (black triangles) as a function of the last number of triggered (fired) layer in the prototype in data and MC, respectively. Distributions in data and MC have a similar shape, although they are different in numerical terms. The experimental data permit us to make an initial / preliminary rough estimate of the pion-to-muon separation at low (~ 0.5 GeV) momentum. So, the overlap of the pionic and muonic spectra after layer # 7 (to keep high efficiency for muon registration, 93%) gives us ~ 22% of pion contamination. Whereas for the MC model we have contamination of pions 27% at 99% of muon efficiency. We assume now that the observed difference is caused by the secondary muons present in the beam due to pion-to-muon decay, and we plan to discriminate them in further runs by direct measurement of incoming momenta at the final focus of the beam. The depth of particle penetration will be used for particle identification in the Muon System at low (≈ 1 GeV/c) momenta.

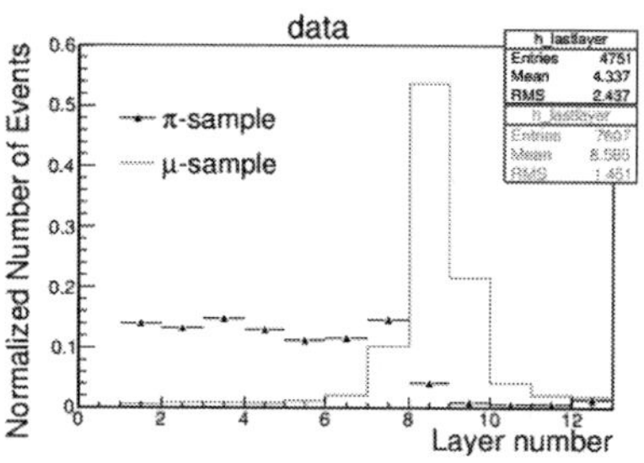

(a)

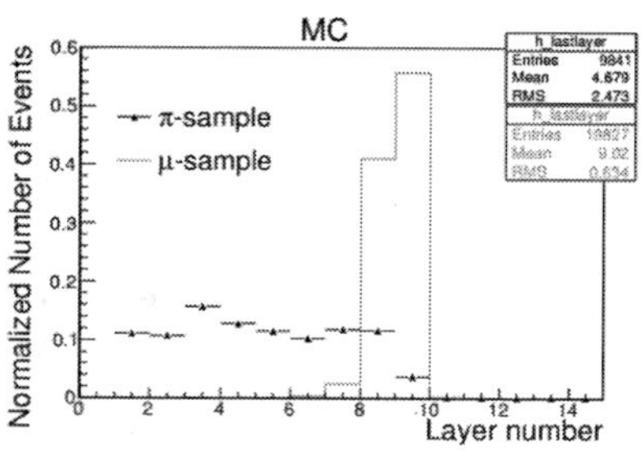

(b)

Figure 4. Normalized distributions of events with muons (green line) and pions (black triangles) as a function of the last hitted layer in the prototype: (a) in data; (b) in MC.

Some methodical tests have been also performed in 2017 using muon samples with the momenta of 2.5 GeV/c. For instance, the efficiency of particle's registration by MDT as a function of applied high voltage and the input current threshold was obtained, see Fig. 5(a). The detection efficiency of muon by single MDT at the operating voltage and current threshold (2250V and 1μA) is 92%, which corresponds to its geometrical efficiency. Fig. 5(b) shows the counting rate of MDT (normalized to trigger rate) as a function of the voltage and the current threshold. These two plots indicate rather good high voltage operational plateau of MDTs with the small threshold (~ 1μA) of about 150 – 200 Volts.

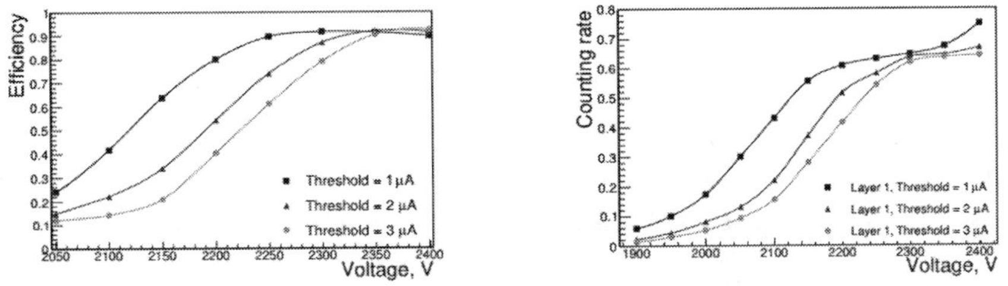

(a) (b)

Figure 5. (a) MDT efficiency and (b) MDT counting rate as a function of voltage and current thresholds.

5 Summary

The geometrical model of the $\overline{\text{P}}$ANDA Muon System Prototype is done and ready to be implemented into the PandaRoot package. It is used to perform MC simulations of events with muons and hadrons. Also in 2017, two test beam runs were conducted at CERN for the $\overline{\text{P}}$ANDA Muon System prototype. It allowed making experimental tests and estimate the muon-to-pion separation in data and MC at low energy. The $\overline{\text{P}}$ANDA Muon System prototype will be modified for 2018 test beam run.

Authors would like to thank Dr. S. Belogurov and E. Ovcharenko - for help with modeling of the $\overline{\text{P}}$ANDA Muon System prototype, I. Konorov - for help and advising on DAQ system and V. Malyshev - for the participating in the beginning of the work. Authors are particularly grateful to L. Gatignon for the multiple and fruitful discussions of the beam features. The work is partially carried out with the financial support of FAIR-Russia Research Center. One of the authors (A. Verkheev) also would like to acknowledge support by JINR, grant for JINR Young Scientists and Specialists no. 17-201-01.

References

[1] $\overline{\text{P}}$ANDA Collaboration, TDR for the: $\overline{\text{P}}$ANDA Muon System http://www.fair-center.eu/fileadmin/fair/publications_exp/Muon_TDR.pdf.
[2] V. Abazov *et al.*, Instrum. Exp. Tech. **53**, 356-361 (2010).
[3] V. Abazov *et al.*, Instrum. Exp. Tech. **53**, 648-652 (2010).
[4] H. Fischer *et al.*, IEEE Trans. Nucl. Sci. **49**, 443-447 (2002).
[5] https://fairroot.gsi.de/.
[6] S. Agostinelli *et al.*, Nucl. Inst. Meth. Phys. Res. A **506**, 250-303 (2003).
[7] R. Brun and F. Rademakers, Nucl. Inst. Meth. Phys. Res. A **389** 81-86 (1997).
[8] S. Belogurov *et al.*, J. Phys.: Conf. Ser. **513**, 022003 (2014).
[9] R. Chytracek *et al.*, IEEE Trans. Nucl. Sci. **54**, 2892-2896 (2006).

EPJ Web of Conferences **177**, 04002 (2018) https://doi.org/10.1051/epjconf/201817704002
AYSS-2017

Slow magnetic monopoles search in NOvA

Alexander Antoshkin,[1]* Martin Frank[2]
[1] Dzhelepov Laboratory of Nuclear Problems, JINR
[2] University of South Alabama, USA

ABSTRACT

The NOvA far detector is well suited for finding exotic particles due to its technical features (see [1]). One type of those exotic particles is a "slow" magnetic monopole. It is assumed that the energy deposition of such monopoles should be enough to be registered (see [2]). Measurement of the expected signals was performed on the NOvA test bench at JINR (see [3]). Result of this measurement allows us to perform slow monopole's research using NOvA software and hardware with high efficiency. As a whole, the research can lead to a discovery, or it can limit the existence of monopoles in a wide range of parameters, previously unreachable in other experiments (MACRO, SLIM, RICE, IceCube). Several special software tools have been developed. Slow Monopole Trigger has been created and implemented in the NOvA Data-Driven-Trigger system. Also, an offline reconstruction algorithm has been developed and tested on 5% of the data. A technical description of these tools and current results of the analysis are presented in this work.

INTRODUCTION

NOvA is a long-baseline neutrino experiment that is studying various parameters of neutrinos (see [4]). The detailed description and specific reports could be found on the project site [5].

The NOvA experiment uses two detectors: a 330 metric-ton near detector at Fermilab and a much larger 14 metric-kiloton far detector in Minnesota just south of the U.S.-Canada border. The detectors are made up of 344,000 cells of extruded, highly reflective plastic PVC filled with liquid scintillator. Light from the scintillator is captured by the fibers and is transmitted to an Avalanche Photodiode (APD). Fibers from 32 cells are grouped on one APD board. Each cell in the far detector measures 3.9 cm wide, 6.0 cm deep and 15.5 meters long. The far detector itself measures $15.6\times15.6\times60$ meters.

Such a large and sensitive detector has never been constructed so close to the Earth's surface. It is a good opportunity to detect lighter monopoles which do not penetrate far into the earth, even with the problem of cosmic background due to the large size of our detector and sufficient accessible angular region (see [6]).

WHY ARE WE INTERESTED IN MONOPOLES?

Quantum mechanical formulation of the magnetic monopoles was made by Paul Dirac in 1931. Searches for these particles are very important for several reasons:

* Their existence would explain the quantization of electric charge.

* It is possible to restore symmetry between electricity and magnetism by means of their introduction into the theory of electromagnetism.

* Magnetic monopoles naturally appears in Grand Unification Theories (GUT).

*antoshkinalig@gmail.com, Work supported by RFBR project No. 14-22-03090

SEARCHING STRATEGIES IN NOVA

The Dirac's electric charge quantization relation says $e \times g = n \times \frac{\hbar c}{2}$, where e is a basic electric charge and n is an integer. It means that magnetic monopoles could have a magnetic charge (g) 68.5 times greater than the charge of the electron. As the result they are expected to be very highly ionizing (See Fig. 1 and [2]). For example, for $\beta = 10^{-3}$ one can expect ionisation 10 times larger then for the cosmic muon. We call this muon MIP (Minimum ionizing particle) due to it's mean energy loss rate through matter.

"Slow" monopoles with $\beta < 10^{-2}$ and less can be identified due to their linear tracks with long transit times through the detector (See Fig. 2). In this figure one can see a software visualization of the Far detector and the monopole in it's center. The monopole's energy in this case is almost at the ADC saturation limit (12 bits – 4096 counts). In terms of energy it is more then 650 MeV.

Monopoles with $\beta = 10^{-3}$ take 5 μs to cross the whole detector (one can see the temporal distribution on the bottom of Fig. 2) in comparison with cosmic muons, which take only 50 ns.

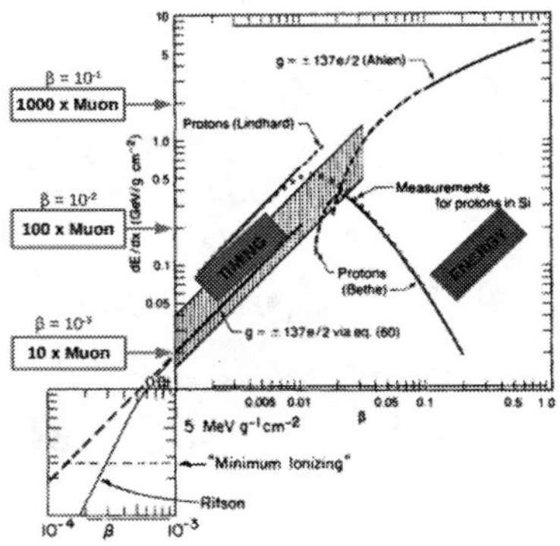

Figure 1. Monopole ionization [2]

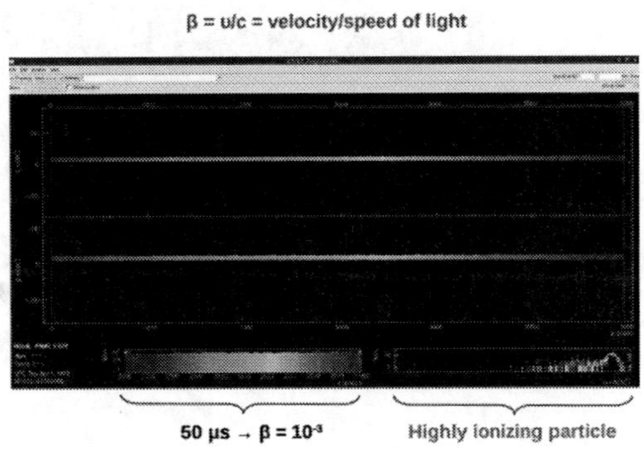

Figure 2. Event display with simulated monopole

In order to proceed to actual measurements using NOvA setup one needs to determine whether it is possible to achieve sufficient sensitivity first. Ordinary signals in NOvA detectors have time width less than 100 ns and energy less than 300 *ADC counts*. Actual tests could be performed utilizing the NOvA test bench at JINR.

TEST BENCH AT JINR AND IT'S ROLE

NOvA test bench @JINR consists of a few components. First is the native NOvA electronics: Avalanche PhotoDiode and Front-End Board. Second is the special hardware: download cable, DCM-emulator, LED, Low and High Voltage Sources, Pulse generator and cooling system. Third is a PC with necessary software.

All devices were put into a black metal box. The black box allows one to perform all the measurements with photodetectors like APDs and PMTs, and screens all external electromagnetic noises.

A pulse generator waits for the FEB trigger and sends the electric pulse to the LED. Then the light pulse from the LED comes through the fibers to the APD, and the FEB is reading at this moment (See Fig. 3).

Measurements performed on this bench demonstrated that it is possible to use NOvA electronics for the search of monopoles. It means that one can see long signals ($5\mu s$) with high energy distributions and reconstruct them with high efficiency. For more details see [7].

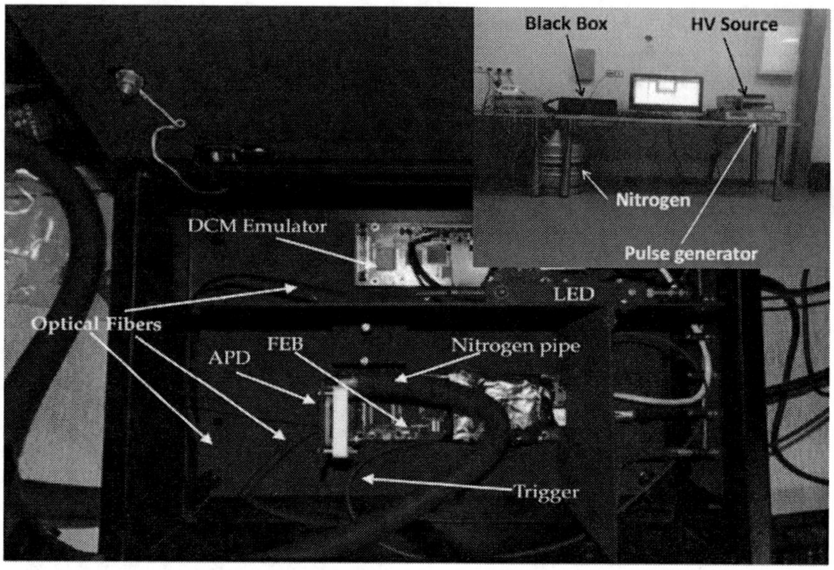

Figure 3. NOvA test bench at JINR

SLOW MONOPOLE TRIGGER

The slow monopole trigger was implemented in June 2015. It allows for the identification of slow ($\beta < 10^{-2}$ and less) monopole tracks by checking the number of plane gaps between the entry and exit hits. We look at all of the hit planes in the contained area and look for gaps. This trigger helps us to choose primary candidates for our further analysis.

OFFLINE RECONSTRUCTION

We have a special procedure for "catching" (See Fig. 4) a monopole track in the real data. Basically the algorithm proceeds by first removing all of the relativistic cosmic rays to form a single monopole cluster. This cluster is then divided into several monopole slices. Finally, we look for track-like objects in the monopole slices and form monopole tracks.

The tracks are then sorted from slowest to fastest track. In other words, the first track is the slowest track. The current reconstruction implementation retains the information of the three slowest tracks for each event. Generally, we expect only one track per event, but this algorithm may also identify several tracks in one event.

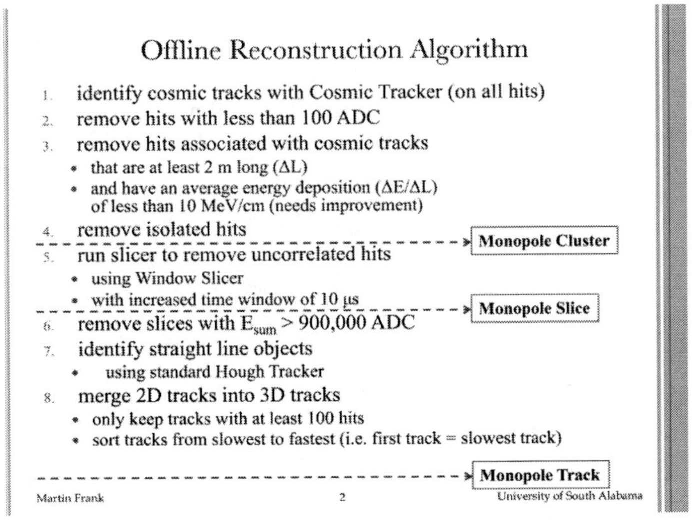

Figure 4. **"Catching" algorithm**

Data Samples

We have several datasets for testing of our algorithm:

Monte Carlo (Simulated monopoles + 5 ms long non-bias data produced by the daily SNEWS trigger – true monopole and nominal detector activity). Four velocities β_{sim}: $5 \times 10^{-4}, 1 \times 10^{-3}, 5 \times 10^{-3}, 1 \times 10^{-2}$.

Slow Monopole Triggered Events (Slow Monopole Trigger – first run is 19728, last one is 20752 for Low Gain (**100**) and 20753 like the first one for the new Data Set with High Gain (**150**).) These numbers (100 and 150) refer to the APD gain.

FURTHER ALGORITHM AND PRELIMINARY RESULTS

Currently we have tested 10% of the Low Gain data. Our main tools for post selection are: Linear Regression coefficient and Time Gap Fraction.

Linear Regression coefficient

We fit a line to the collection of hits associated with the monopole track. Using standard linear regression techniques (see [8]), we can calculate the squared regression coefficient (r^2) for hits in the xt and yt spaces separately. A value of r^2 equal to unity means that the hits lie on a straight line while a value of zero means that the hits are uncorrelated. For a high-quality track, we require the linear regression coefficient to be close to unity. This calculation is performed separately for each view, yielding ${r^2}_{xt}$ and ${r^2}_{yt}$ for each track. The minimum of these two quantities is called ${r^2}_{min}$. Therefore, if we require ${r^2}_{min}$ to have a large value close to unity, both ${r^2}_{xt}$ and ${r^2}_{yt}$ will have to exceed this requirement.

Time Gap Fraction

First, we sort all of the hits associated with the monopole track by time. We then look for the largest gap in time between adjacent hits (δt_{max}). The time gap fraction is then defined by: $f = \frac{\delta t_{max}}{\delta t_{track}}$, where δt_{track} is the difference in time between the first and last hits of the monopole track sorted by time. A high-quality track will have a value of f close to zero indicating that there are no gaps in time along the track. A track that consists of two independent clusters of hits will have a value of f close to unity. We therefore require f to be close to zero. This calculation is performed separately for each view, yielding f_{xt} and f_{yt} for each track. The maximum of these two quantities is called f_{max}. Therefore, if we require f_{max} to have a small value close to zero, both f_{xt} and f_{yt} will have to be below this requirement.

Track Requirements

For a track to be linear, it has to meet the following requirements: ${r^2}_{min} \geq 0.95$ and $f_{max} \leq 0.2$. We consider an event linear if its primary monopole track is linear.

Right now we observed **no** slow events in the 10% data sample.

CONCLUSION

Our research shows that it is possible to measure lighter monopoles on NOvA far detector. The detector itself has the unique potential to "touch" a new region of phase space due to it's location on the surface and large surface area. These factors give one very high chance to "catch" the magnetic monopoles, or make the limit on the monopole flux (see [6]). Preliminary calculation gives one upper flux limit $\Phi = 4.1 \times 10^{-15}\ cm^{-2}s^{-1}sr^{-1}$.

The Slow Monopole Trigger is proved to provide sufficient precision. A special "cutter" and "selector" were developed. We tested only 10% of the Low Gain data. A technote about the general search strategy and description of tools is ready (see [9]) and right now we have asked to allow us apply the "cutter" for the remaining data. The collaboration gave us useful comments and we solved the majority of the issues and shortcomings. We still do not see any good candidates in real data, but our cutter and selector work well on Monte Carlo events.

ACKNOWLEDGMENTS

We would like to thank the NOvA Collaboration for providing us with unpublished and public information.

REFERENCES

[1] Martin Frank, 2017 SESAPS NOvA Slow Monopole Talk, http://nova-docdb.fnal.gov/cgi-bin/ShowDocument?docid=24903

[2] Donald E. GROOM, In search of the supermassive magnetic monopole, PHYSICS REPORTS (Review Section of Physics Letters) 140. No. 6 (1986) 323373. North-Holland, Amsterdam

[3] Alexander Antoshkin et all., NOVA test bench at JINR http://nova-docdb.fnal.gov/cgi-bin/ShowDocument?docid=14987

[4] Site of the NOvA experiment. URL: https://www-nova.fnal.gov/

[5] NOvA Public Document Database, http://nova-docdb.fnal.gov/cgi-bin/DocumentDatabase

[6] Enhao Song, Monopole trigger efficiency and upper flux limit, NOvA DocDB – 15938-v1 (unpublished)

[7] Alexander Antoshkin, NOvA DocDB – 14261-v1 (unpublished).

[8] D. A. McQuarrie, Mathematical Methods for Scientists and Engineers, University Science Books (2003).

[9] Martin Frank et al., NOvA DocDB – 22610-v2 (unpublished).

EPJ Web of Conferences **177**, 04003 (2018) https://doi.org/10.1051/epjconf/201817704003
AYSS-2017

Determination of the phase ϕ_s at LHCb

Varvara Batozskaya[1,⋆,⋆⋆]
on behalf of the LHCb Collaboration
[1]*National Centre for Nuclear Research (NCBJ), Warsaw, Poland*

Abstract. The determination of the mixing-induced *CP*-violating phase ϕ_s in the $B_s^0 - \bar{B}_s^0$ system is one of the key goals of the LHCb experiment. Using several B_s^0 decay modes it has been measured by the LHCb collaboration exploiting the Run I data set. The first observation of the $B_s^0 \to \eta_c\phi$ and $B_s^0 \to \eta_c\pi^+\pi^-$ decay modes which can be used to measure ϕ_s with Run II data is presented.

1 Introduction

The *CP*-violating phase ϕ_s originates from the interference between the mixing and direct decay of the B_s^0 mesons to *CP* eigenstates. Ignoring subleading penguin contributions, the phase ϕ_s within the Standard Model (SM) is predicted to be $-2\beta_s$ where $\beta_s = \arg(-V_{ts}V_{tb}^*/V_{cs}V_{cb}^*)$ [1]. An indirect determination of $2\beta_s = 0.0376^{+0.0008}_{-0.0007}$ rad is obtained using a global fit to experimental data [2]. Any deviation from this prediction would be a clear sign, so-called New Physics effects, strongly motivating the need for precise experimental measurements of this quantity [3]. The measurement of *CP*-violating phase ϕ_s has been independently performed using $B_s^0 \to J/\psi K^+K^-$, $B_s^0 \to J/\psi\pi^+\pi^-$ and $B_s^0 \to \psi(2S)\phi$ decay channels. All measurements shown in the proceedings use 3 fb^{-1} of data collected by the LHCb experiment [4] in pp collisions during 2011 and 2012.

2 Status of ϕ_s measurement

2.1 ϕ_s from $B_s^0 \to J/\psi\phi$ and $B_s^0 \to J/\psi\pi^+\pi^-$

The $B_s^0 \to J/\psi\phi$ candidates are reconstructed as the decay $J/\psi \to \mu^+\mu^-$ and $\phi \to K^+K^-$ [5]. After applying a full offline and trigger selection, 95690 ± 350 signal candidates of the $B_s^0 \to J/\psi\phi$ are obtained. The decay time and angular acceptances, decay time resolution as well as flavour tagging efficiency are taken into account in the fitting procedure. Using a prescaled unbiased trigger sample and a tag and probe technique the decay time acceptance is determined from data. The angular acceptance is determined using simulated events. The decay time resolution is estimated using a sample of prompt $J/\psi K^+K^-$ combinations produced directly in the pp interactions and is found to be 46 fs. The combined effective tagging power is $(3.73 \pm 0.15)\%$ [5].

A weighted unbinned maximum likelihood fit is performed using a signal-only Probability Density Function (PDF), as described in [6]. Fig. 1 shows the projections of the decay time and angular

⋆e-mail: varvara.batozskaya@cern.ch
⋆⋆On leave of absence from JINR, Dubna

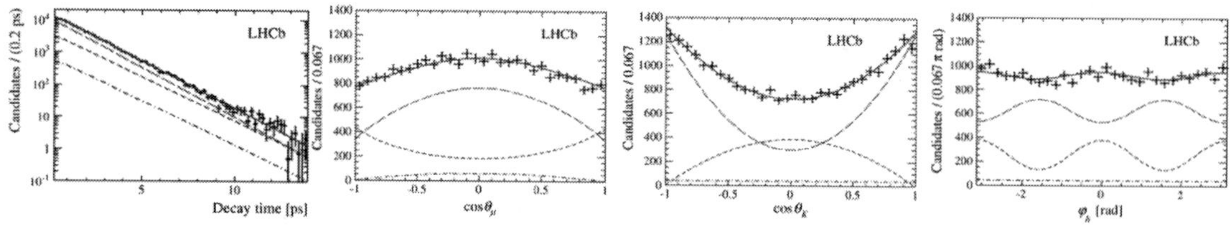

Figure 1. Decay time and angle distributions for $B_s^0 \to J/\psi\phi$ decays (black markers) with the one-dimensional projections of the PDF. The solid blue line shows the total signal contribution, which is composed of *CP*-even (long-dashed red), *CP*-odd (short-dashed green) and S-wave (dotted-dashed purple) contributions.

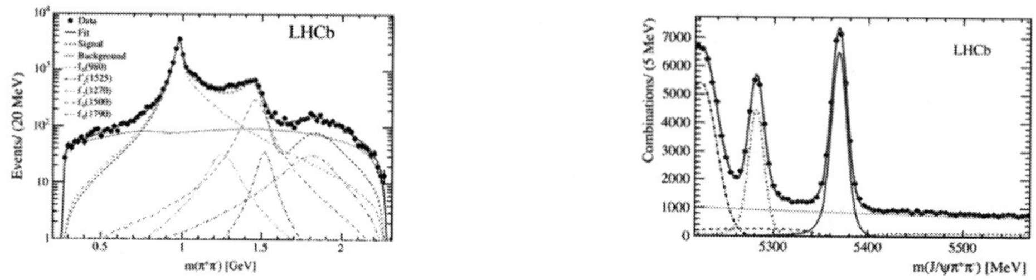

Figure 2. (left) Distribution of $m(\pi^+\pi^-)$ invariant mass with contributing components. (right) Invariant mass of $J/\psi\pi^+\pi^-$ combinations where the (red) solid curve shows the B_s^0 signal, the (brown) dotted line shows the combinatorial background, other colour lines indicate different reconstructed background contributions.

distributions. The final results are $\phi_s = -0.058 \pm 0.049 \pm 0.006$ rad, $\Gamma_s = 0.6603 \pm 0.0027 \pm 0.0015$ ps^{-1} and $\Delta\Gamma_s = 0.0805 \pm 0.0091 \pm 0.0032$ ps^{-1} [5]. The dominant contribution to the systematic uncertainty is given by the decay time and angular efficiency and background subtraction.

The $B_s^0 \to J/\psi\pi^+\pi^-$ decays has also been used by the LHCb collaboration [7] to measure ϕ_s. The decay is similar to the $B_s^0 \to J/\psi\phi$ one with a noticeable simplification: the final state being *CP*-odd, there is no need for the angular analysis. Five interfering $\pi^+\pi^-$ states dominated by $f_0(980)$ component are shown in Fig. 2. After trigger and selection chain 27100 ± 200 signal $B_s^0 \to J/\psi\pi^+\pi^-$ candidates are reconstructed (Fig. 2). With the time-dependent amplitude analysis, the measured value of the phase ϕ_s is $0.070 \pm 0.068 \pm 0.08$ rad. The dominant systematic uncertainty is the one coming from knowledge about $\pi^+\pi^-$ resonance model. The combination of the $B_s^0 \to J/\psi\phi$ and $B_s^0 \to J/\psi\pi^+\pi^-$ fit results gives $\phi_s = -0.010 \pm 0.039$ rad [5].

2.2 ϕ_s from $B_s^0 \to \psi(2S)\phi$

Another B_s^0 decay mode with $\bar{b} \to \bar{c}c\bar{s}$ transition that has been exploited by the LHCb collaboration to measure the phase ϕ_s is $B_s^0 \to \psi(2S)(\to \mu^+\mu^-)\phi(\to K^+K^-)$ [8]. The formalism used for this analysis is very close to that of $B_s^0 \to J/\psi\phi$ decay [5] where the J/ψ meson is replaced with $\psi(2S)$. The number of signal candidates selected from a fit to the data sample is ~ 4700 (Fig. 3). The decay time acceptance is determined using a control $B^0 \to \psi(2S)K^{*0}(\to K^+\pi^-)$ decay mode as shown in Fig. 3. The first measurement of the *CP*-violating parameters in a final state containing the $\psi(2S)$ resonance is $\phi_s = -0.23^{+0.29}_{-0.28} \pm 0.02$ rad, $\Gamma_s = 0.668 \pm 0.011 \pm 0.006$ ps^{-1} and $\Delta\Gamma_s = 0.066^{+0.041}_{-0.044} \pm 0.007$ ps^{-1}. The fit result is consistent with $B_s^0 \to J/\psi\phi$ measurement and the SM predictions. The systematic uncertainty is less than 20% of the statistical uncertainty, except for Γ_s where it is close to 60%.

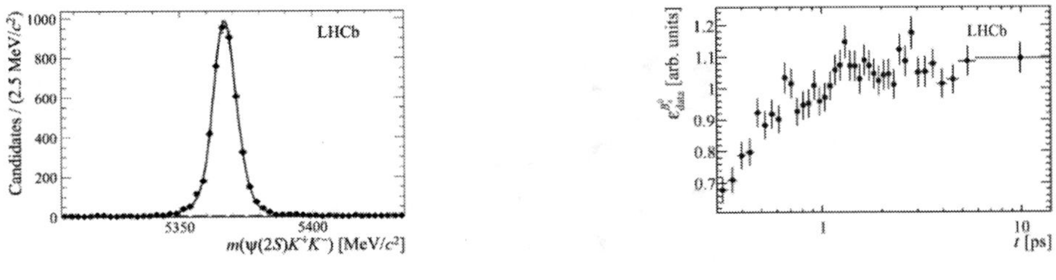

Figure 3. Distribution of the $m(\psi(2S)K^+K^-)$ invariant mass for the selected $B_s^0 \to \psi(2S)\phi$ candidates and decay time acceptance in arbitrary units.

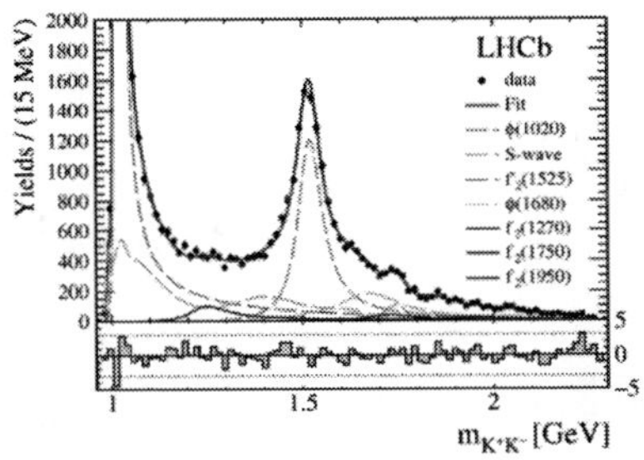

Figure 4. Distribution of the $m(J/\psi K^+K^-)$ invariant mass with contributing components.

2.3 ϕ_s from $B_s^0 \to J/\psi K^+K^-$ in high $m(K^+K^-)$ range

The first measurement of the phase ϕ_s has been performed in the $B_s^0 \to J/\psi K^+K^-$ decay with K^+K^- invariant mass higher than 1050 MeV/c^2 [9] that is above the $\phi(1020)$ resonance region. The important difference between both decay analyses is that modelling of the $m(K^+K^-)$ distribution is included to distinguish different resonant and nonresonant contributions. The decay time acceptance is determined with the same method as described in [8] by using a control channel $B^0 \to J/\psi K^{*0}$. The K^+K^- mass spectrum is fitted by considering the different contributions found in the time-dependent amplitude analysis as shown in Fig. 4. The final fit has been performed allowing eight independent sets of *CP*-violating parameters: three corresponding to $\phi(1020)$ transversity states, K^+K^- S-wave, $f_2(1270)$, $f_2'(1525)$, $\phi(1680)$ and the combination of the two high-mass $f_2(1750)$ and $f_2(1950)$ states. The *CP*-violating parameters measurement of $B_s^0 \to J/\psi K^+K^-$ in high $m(K^+K^-)$ region is $\phi_s = 0.119 \pm 0.107 \pm 0.034$ rad, $\Gamma_s = 0.650 \pm 0.006 \pm 0.004$ ps^{-1} and $\Delta\Gamma_s = 0.066 \pm 0.018 \pm 0.006$ ps^{-1}. The largest contribution to systematic uncertainty results from the resonance fit model. The combination with the B_s^0 decay fit results in the $\phi(1020)$ region gives $\phi_s = -0.025 \pm 0.045 \pm 0.008$ rad that improves a precision of the ϕ_s measurement by more than 9%.

2.4 Global combination

The *CP*-violating phase and lifetime parameters have been measured by several experiments, namely four analysis using the $B_s^0 \to J/\psi\phi$ final state from CDF [10], D0 [11], ATLAS [12] and CMS [13] collaborations and five analysis using different final states performed by the LHCb collaboration, four of which discussed here. The world average result of ϕ_s and $\Delta\Gamma_s$ measurements from the Heavy Flavour Averaging Group [14] is shown in Fig. 5. They found $\phi_s = -0.021 \pm 0.031$ rad and $\Delta\Gamma_s =$

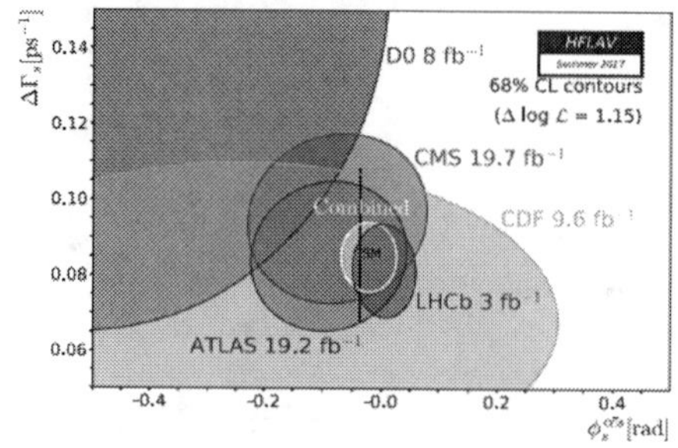

Figure 5. 68% confidence level regions in $\Delta\Gamma_s$ and ϕ_s plane obtained from individual contours of CDF, D0, CMS, ATLAS and LHCb measurements and the combined contour (solid line and shaded area) [14]. The expectation within the SM [2] is shown as a black thin rectangle.

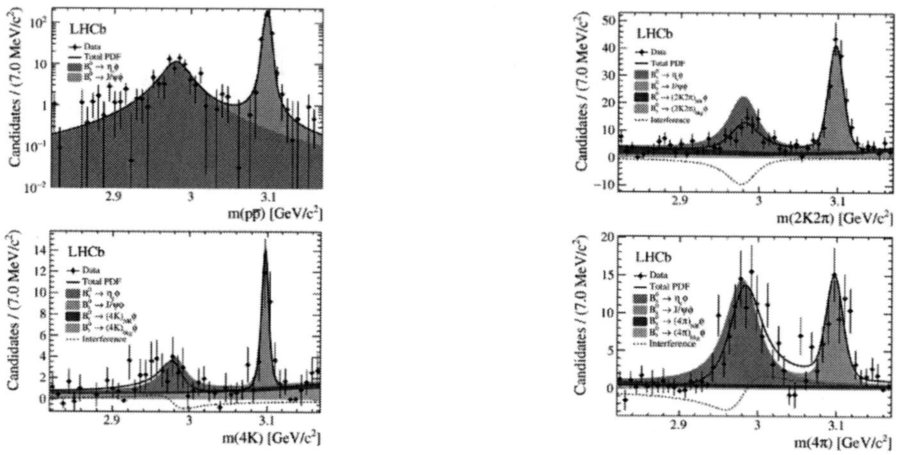

Figure 6. Invariant mass distributions for selected $p\bar{p}$, $K^+K^-\pi^+\pi^-$, $K^+K^-K^+K^-$ and $\pi^+\pi^-\pi^+\pi^-$ combinations.

0.085 ± 0.006 ps^{-1} that is dominated by the measurements from LHCb collaboration and is consistent with the SM predictions.

3 Future contributions for measuring ϕ_s

3.1 ϕ_s from $B_s^0 \to \eta_c\phi$

For the first time the LHCb collaboration has observed the $B_s^0 \to \eta_c\phi$ decay mode, with $\eta_c \to K^+K^-\pi^+\pi^-$, $K^+K^-K^+K^-$, $\pi^+\pi^-\pi^+\pi^-$ and $p\bar{p}$ [19]. This decay is an another $b \to c\bar{c}s$ transition that could be used to measure ϕ_s. The interference between the η_c and purely nonresonant contributions is taken into account using an amplitude model for simultaneously fit the four hadrons and $p\bar{p}$ mass distributions (Fig. 6). The branching fraction is normalized to the J/ψ mode and found to be $\mathcal{B}(B_s^0 \to \eta_c\phi) = [5.01 \pm 0.53(\text{stat}) \pm 0.27(\text{syst}) \pm 0.63(\mathcal{B})] \times 10^{-4}$. First evidence for the $B_s^0 \to \eta_c\pi^+\pi^-$ decay mode has also been reported, with a branching fraction of $\mathcal{B}(B_s^0 \to \eta_c\pi^+\pi^-) = [1.76 \pm 0.59(\text{stat}) \pm 0.12(\text{syst}) \pm 0.29(\mathcal{B})] \times 10^{-4}$.

4 Summary

The most precise measurement of *CP*-violating phase ϕ_s and lifetime parameters in the B_s^0 system has been performed using Run I data collected by LHCb experiment. So far all results are compatible with the SM prediction. In order to reach an uncertainty of the measurement comparable or even better than the theoretical uncertainty of the SM prediction aside from improvements in available luminosity for the $B_s^0 \to J/\psi\phi$ channel, inclusion of new decay modes have been investigated. For example, the $B_s^0 \to J/\psi(\to e^+e^-)\phi$ channel not only could bring about 10% of the $\mu^+\mu^-$ mode statistics, but it will be also an important verification of the $B_s^0 \to J/\psi(\to \mu^+\mu^-)\phi$ as kinematics for both channels are expected to be identical. The statistical sensitivity to ϕ_s measurement after the LHCb upgrade, with an integrated luminosity of 46 fb^{-1}, is expected ~0.01 rad that will be close to the present theoretical uncertainty [20]. As the measurement precision improves, the penguin polluion contributions to the B_s^0 meson decays have to been kept under control [21, 22].

Acknowledgments

The work has been supported by the Polish National Science Centre (NCN) with Preludium grant UMO-2015/17/N/ST2/04056. I would like to thanks the organizers of the AYSS'2017 conference for the invitation to present this work and my LHCb colleagues who helped in the preparation of this talk.

References

[1] M. Kobayashi and T. Maskawa, Prog. Theor. Phys. **49**, 652 (1973).; N. Cabibbo, Phys. Rev. Lett. **10**, 531 (1963).
[2] J. Charles *et al.*, Phys.Rev. **D91** (2015) no.7, 073007.
[3] Z. Ligeti, M. Papucci and G. Perez, Phys. Rev. Lett. **97**, 101801 (2006); P. Ball and R. Fleischer, Eur. Phys. J. C **48**, 413 (2006); A. Lenz, Phys. Rev. D **76**, 065006 (2007); R. Fleischer, eConf C **0610161**, 020 (2006); U. Nierste, Int. J. Mod. Phys. A **22**, 5986 (2007).
[4] A. A. Alves, Jr. *et al.* [LHCb Collaboration], JINST **3**, S08005 (2008).
[5] R. Aaij *et al.* [LHCb Collaboration], Phys. Rev. Lett. **114**, no. 4, 041801 (2015).
[6] Y. Xie, arXiv:0905.0724.
[7] R. Aaij *et al.* [LHCb Collaboration], Phys. Lett. B **736**, 186 (2014).
[8] R. Aaij *et al.* [LHCb Collaboration], Phys.Lett. **B762** (2016) 253.
[9] R. Aaij *et al.* [LHCb Collaboration], JHEP **1708** (2017) 037.
[10] T. Aaltonen *et al.* [CDF Collaboration], Phys. Rev. Lett. **109**, 171802 (2012).
[11] V. M. Abazov *et al.* [D0 Collaboration], Phys. Rev. D **85**, 032006 (2012).
[12] G. Aad *et al.* [ATLAS Collaboration], JHEP **1608** (2016) 147.
[13] V. Khachatryan *et al.* [CMS Collaboration], Phys.Lett. **B757** (2016) 97.
[14] Y. Amhis *et al.*, arXiv:1612.07233 [hep-ex].
[15] R. Aaij *et al.* [LHCb Collaboration], Phys.Rev. **D90** (2014) no.5, 052011.
[16] M. Beneke, J. Rohrer and D. Yang, Nucl.Phys. **B774** (2007) 64.
[17] M. Bartsch, G. Buchalla and C. Kraus, arXiv:0810.0249 [hep-ph].
[18] H. Y. Cheng and C. K. Chua, Phys.Rev. **D80** (2009) 114026.
[19] R. Aaij *et al.* [LHCb Collaboration], JHEP **1707** (2017) 021.
[20] R. Aaij *et al.* [LHCb Collaboration], Eur. Phys. J. C **73**, no. 4, 2373 (2013).
[21] R. Aaij *et al.* [LHCb Collaboration], JHEP **1511** (2015) 082.
[22] R. Aaij *et al.* [LHCb Collaboration], Phys.Lett. **B742** (2015) 38.

Study of the GEM detector performance in BM@N experiment

Sergei Bazylev[1], *Mikhail* Kapishin[1], *Kacper* Kapusniak[2], *Vladimir* Karjavine[1], *Sergei* Khabarov[1], *Alexander* Kolesnikov[1], *Elena* Kulish[1], *Vasilisa* Lenivenko[1], *Alexander* Makankin[1], *Anna* Maksymchuk[1,*], *Bertrand* Mehl[2], *Rui* De Oliveira[2], *Vladimir* Palchik[1], *Gleb* Pokatashkin[1], *A* Rodriguez[2], *Igor* Rufanov[1], *Alexander* Shutov[1], *Ilya* Slepnev[1], *Vyacheslav* Slepnev[1], *Sergei* Vasiliev[1], and *Alexander* Zinchenko[1]

[1]JINR, Joliot-Curie 6, 141980 Dubna, Moscow region, Russia
[2]CERN, Route de Meyrin 385, 1217 Meyrin, Switzerland

Abstract. BM@N is the fixed target experiment at the accelerator complex NICA-Nuclotron aimed to study nuclear matter in the relativistic heavy ion collisions. Triple-GEM detectors were identified as appropriate for the BM@N tracking system located inside the analyzing magnet. Seven GEM chambers are integrated into the BM@N experimental setup and data acquisition system. GEM construction, main characteristics and first obtained results of the GEM tracking system performance in the technical run with the deuteron beam are shortly reviewed.

1 Introduction

The BM@N (Baryonic Matter at the Nuclotron) is the first experiment at the accelerator complex of NICA-Nuclotron [1]. Its main purpose is to investigate properties of nuclear matter under extreme density and temperature. The Nuclotron beam energy range corresponds to $\sqrt{s_{NN}} = 2.3 - 3.5$ GeV, which is of great interest for scientific studies, but still poorly investigated.

A scheme of the BM@N experimental setup is shown in Fig. 1. The experiment combines high precision track measurements with time-of-flight information for particle identification and calorimetry for the analysis of the collisions centrality. The magnetic field of the analyzing magnet can be varied up to 1 T to get the optimal detector acceptance and momentum resolution for different reactions and beam energies. The charged particle momentum and multiplicity are measured with the set of Gas Electron Multipliers (GEM) located inside the analyzing magnet and by the Drift Chambers (DCH) situated outside the magnetic field. Double-sided Silicon Micro-strip Sensors (SiMS) are installed between the GEM tracker and the target to improve the track and vertex reconstruction in heavy ion collisions.

* Corresponding author: anna_maksymchuk@mail.ru

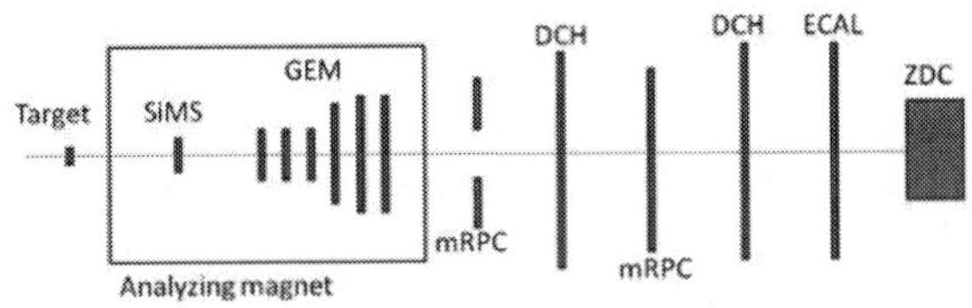

Fig. 1. Schematic view of the BM@N experimental setup: SiMS – Silicon Micro-strip Sensors, GEM – Gas Electron Multipliers, mRPC – Multi-gap Resistive Plate Chambers, DCH – Drift Chambers, ECAL - Electromagnetic Calorimeter, ZDC – Zero Degree Calorimeter.

2 BM@N tracking system

Physics measurements will be performed in conditions of high beam intensities up to 10^6 ions per second in collisions with large multiplicity of charged particles. This requires the use of detectors with the capacity to resolve multiple tracks produced at very high rate. Detectors based on the GEM technology posses all the mentioned characteristics combined with the capability of stable operation in a strong magnetic field. For this reason, two coordinate Triple-GEM detectors were chosen for the central part of the tracking system, which is located inside the analyzing magnet downstream of the target. At present, the GEM tracking system of the BM@N experiment consists of seven detectors: five GEM detectors of the size of 66 x 41 cm^2 (small GEM detectors) and two GEM detectors of the size of 163 x 45 cm^2 (big GEM detectors). These detectors were installed into the BM@N set-up for technical runs with the deuteron beam in December 2016 and with the carbon beam in spring 2017.

2.1 GEM detector technology and construction

One GEM foil is made of a 50 μm Kapton foil clad on both sides with 5 μm copper electrodes. The GEM foil is optically semi-transparent due to chemical perforation by a large number of holes of about 70 μm diameter, separated by a distance of 140 μm (manufactured by CERN PH Detector Technologies (DT) and Micro-Pattern Technologies (MPT) workshop) [2].

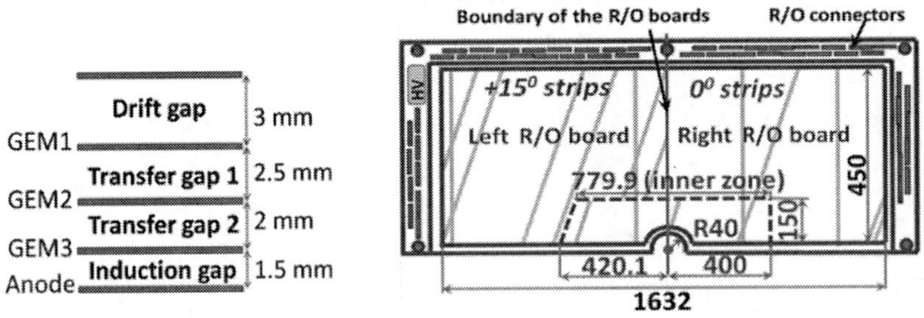

Fig. 2. Left: Schematic cross section of BM@N triple GEM detector. Right: Schematic view of the big GEM detector.

When the GEM foil is placed between a drift cathode and a readout anode plate and a potential applied between the two sides of the foil, it behaves like a charge amplifier. It is possible to use several GEM foils to achieve gains appropriate for efficient detection of minimum ionizing particles. The BM@N GEM detectors consist of three GEM multipliers, with the following gaps between the electrodes: the drift gap of 3 mm, the first transfer gap of 2.5 mm, the second transfer gap of 2 mm and the induction gap of 1.5 mm, as it is shown in Fig. 2 (Left).

A two-dimensional projective readout of the electron charge signal is performed on a multilayer readout board with two types of straight parallel metal strips. Their inclination angles to the vertical axis are 0 (X coordinate) and 15 (X′ coordinate) degrees (Fig.2 Right). There is a hole of 80 mm in diameter in the central region of the big GEM detector in order to install a vacuum beam pipe for heavy ion beam particles. Because of a large multiplicity near the beam line, the readout is divided into the outer and the inner ("hot") zones. Separate readout of X and X′ strips of each zone is performed by multilayer printed circuit readout board. The rigidity and support of big GEMs is achieved by using two honeycomb plates. Assembly of the BM@N GEM detectors is performed at the CERN PH DT and MPT workshop.

2.2 FEE and DAQ

Front-end electronics is based on the charge sensitive pre-amplifier chip VA162 (IDEAS) [3]. Chips are joined in groups of 4 in one front-end board. The multiplexed data from each board are transmitted through 13 m of twisted pair flat cable to the 12-bit analog-to-digital converter (ADC) readout by the BM@N data acquisition system [4].

2.3 GEM gas gain measurements

To minimize the discharge probability, $Ar(90)/C_4H_{10}(10)$ was selected as a gas mixture for the GEM routine operation within the BM@N experiment, due to its lower, compared to $Ar(70)/CO_2(30)$, values of the electric field in the GEM gas gaps. Measurements of the gas gain as a function of the current in the high voltage divider, designed for the BM@N GEM detectors, were performed using the small GEM detector equipped with the calibrated VA162 FEE irradiated with Fe^{55} radioactive source (Fig. 3). The measurements show, that operational gain of $3 \cdot 10^4$ for the $Ar(90)/C_4H_{10}(10)$ gas mixture is achieved at a significantly lower value of the divider current compared to the $Ar(70)/CO_2(30)$ gas mixture.

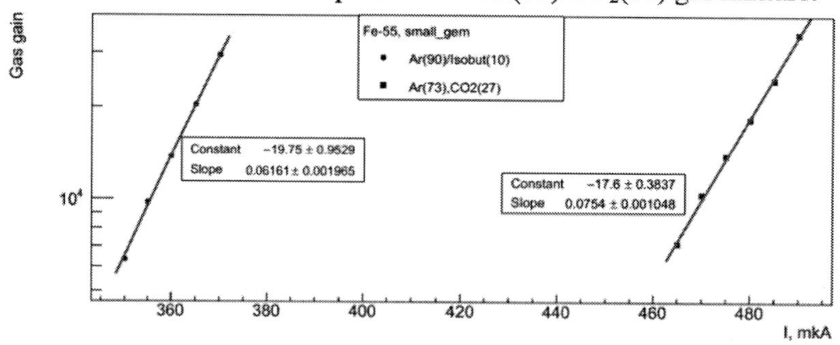

Fig. 3. The GEM gas gain for the $Ar(70)/CO_2(30)$ and $Ar(90)/C_4H_{10}(10)$ gas mixtures as a function of the current in the high voltage divider

2.4 GEM response uniformity

Studies with 163 × 45 cm^2 GEM chamber irradiated by ^{106}Ru radioactive source were performed to measure the response uniformity across the detectors (Fig.4). Gas gain was measured for 62 points, which were chosen uniformly over the surface of the chamber

(Fig.4 Left). The position of the points is shown on XY plane of the distribution. Gap in 3D plot corresponds to the hole for vacuum beam pipe.

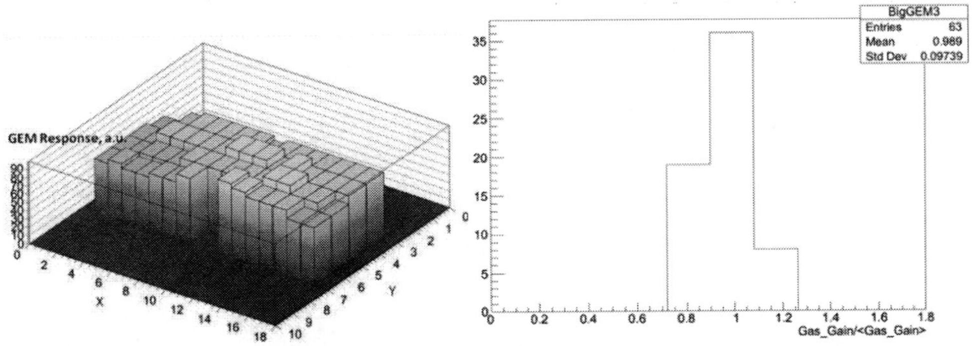

Fig. 4. Gas gain uniformity measurements for 163 × 45 cm^2 GEM chamber. Left: Response uniformity 3D plot. Right: Gas gain distribution normalized on average gas gain.

3 Beam results

First beam tests of 7 GEM chambers were performed in December of 2016 with the deuteron beam at the Nuclotron accelerator in JINR. The GEM detectors were arranged in 6 planes as it is shown in Fig.1. The left and right parts of the Big GEM detectors were treated as independent detectors. The main goal of the tests was to study the performance of the GEM detectors and the FEE and readout electronics as a part of the BM@N experimental setup. The GEM detectors were filled with the flowed Ar(90)/C_4H_{10}(10) gas mixture and equipped with the VA162 based electronics. The gas gain was around $3 \cdot 10^4$ for all the chambers. The measurements were made within the magnetic field varied from 0 up to 0.9 T. The recorded ADC data undergo the pedestal subtraction procedure. The clusters are determined as the groups of adjacent strips with amplitudes of signals above the threshold. For reconstructed clusters, the center of gravity, the width and the total charge are calculated. After applying the alignment procedure for the GEM detector positions, the cluster data were converted into the spatial coordinates of hits. Finally, the tracks of charged particles were reconstructed and track detection efficiency of the GEM chambers was calculated. For the efficiency determination four GEM planes were used as reference detectors and two other were tested. The magnetic field was switched off. The efficiency of the chambers under study was defined from the presence of a cluster with a centre of gravity within ±2 mm of the predicted position.

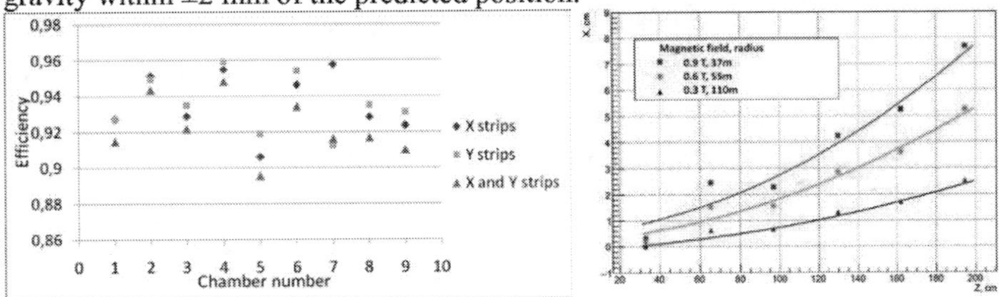

Fig. 5. Left: GEM chamber efficiencies measured in the deuteron beam run in December 2016. Right: The average trajectories of the deuteron beam and the average Lorentz shifts of an electron avalanche in 6 GEM planes measured for three values of the magnetic field.

The efficiencies are presented in Fig.5 (Left) for the vertical X strips, the inclined X′(“Y”) strips and for the coincidence of the X and X′(“Y”) strips. The difference in efficiencies can be explained by the dispersion of FEE amplification factor and by the spatial response non-uniformity of some chambers.

The average shifts of the electron avalanche in the direction perpendicular to the magnetic field (Lorentz shift) were measured from the deviation of hits from the averaged trajectory of the beam as it is illustrated in Fig.5 (Right). Subsequent GEM planes had opposite directions of the electric field in the detectors. The average positions of hits in 6 GEM planes relative to the average trajectory of the beam are shown in Fig. 6 for different values of the magnetic field. For first two chambers the Lorentz shift value exceeds the average shift by factors 2-3. The reason is in the insufficient rigidity of the support construction of these chambers. That leads to an enlargement of the drift gap due to gas flow, which in turn leads to an increase of the Lorentz shift of the electron avalanche. This construction defect was fixed by installation of an additional support structures on the GEM frames.

Coordinate and momentum resolutions were calculated using data, collected without target and with magnetic field of 0.79 T. Gaussian fit of hit residuals distribution (Fig.6 Left) gives the standard deviation of 0.067, which conform to the Monte-Carlo simulation. Momentum resolution for deuteron beam (9.7 GeV/c) is ~9% (Fig.6 Right). The resolution for protons from deuteron defragmentation is ~6%.

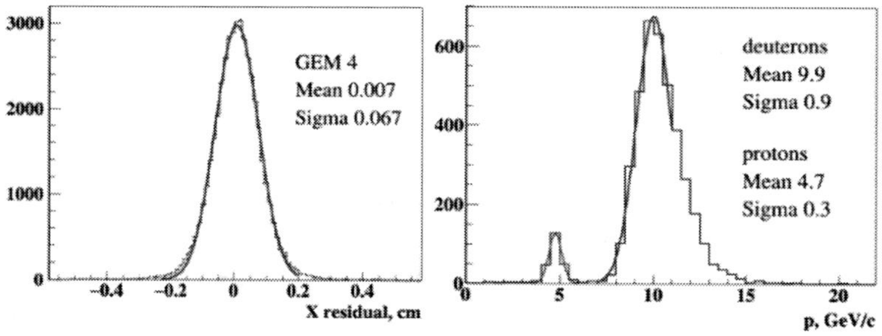

Fig. 6. Left: GEM hit residuals for experimental data. Magnetic field 0.79 T. Right: Momentum resolution for deuteron beam ~9%. Momentum resolution for proton spectators ~6%.

4 Conclusions

Seven triple GEM detectors of the BM@N tracker system have been assembled and studied in the deuteron beam of the Nuclotron accelerator. They comprise one third of the central tracking system of the BM@N experiment. Detectors, readout electronics, cables and readout services were successfully integrated into the BM@N experimental setup and tested during the technical run in December 2016. The measured parameters of the GEM detectors are consistent with the design specifications.

This work would not have been possible without the help of V. Spaskov, N. Zamyatin, O. Tarasov, E. Zubarev, A. Smirnov, I. Filippov, N. Azorskiy, V. Sidorenko, S. Piyadin, A. Terletsky, L. Ropelevsky, E. Oliveri.

References

1. M. Kapishin, Eur. Phys. J. **A52** (2016) no.8, 213
2. F. Sauli, Nucl. Instrum. Methods **A 805** (2016) 2–24
3. http://ideas.no/products/ide1162/
4. https://afi.jinr.ru/

EPJ Web of Conferences **177**, 04005 (2018) https://doi.org/10.1051/epjconf/201817704005
AYSS-2017

Search for new neutral gauge bosons with the CMS Experiment at the LHC

Alexander Lanyov[1,*], *Sergei* Shmatov[1,**], and *Ilia* Zhizhin[1,2,***]

[1] *Joint Institute for Nuclear Research, Joliot-Curie 6, 141980 Dubna, Russia*
[2] *Dubna State University, Universitetskaya 19, 141982 Dubna, Russia*

Abstract. A search for narrow resonances in dimuon invariant mass spectra has been performed using 13 fb^{-1} data obtained in 2016 from proton-proton collisions at $\sqrt{s}$ = 13 TeV with the CMS experiment at the LHC. No evidence for physics beyond standard model is found. Limits on the production cross section and the masses of hypothetical particles that could appear in the scenarios of new physics have been set.

1 Introduction

The CMS physics research programme on the detection of new physics beyond the standard model (SM) is aimed at detecting deviations from SM predictions in various channels [1]. The CMS searches signals from new physical objects and phenomena predicted by supersymmetric models [2], models of the extended gauge sector [3], models of technicolor [4], scenarios of extra-dimensions theories [5], and to solve the hierarchy problems [6, 7] (fine tuning of SM) and many others. The variety of new particles and phenomena predicted by these models requires the use of a wide range of experimental studies in various channels:

- Heavy resonance states and nonresonant signals in the spectrum of pairs of leptons, jets, photons, top quarks, gauge bosons to search for new heavy gauge bosons Z' [3], Kaluza-Klein excited states of the graviton [8], technicolor particles [4], etc.
- The single production of particles (jets, photons, leptons), accompanied by a large fraction of the missing transverse energy. A deviation in the behavior of such signals from SM can indicate an existence of a new heavy charged gauge boson W', extra spatial dimensions [9, 10], weakly interacting particles of dark matter [11].

Many physical scenarios outside the framework of the SM predict the existence of new resonance states in the spectra of the SM particles [12]. For example, heavy (with a mass of several TeV) resonances in the spectrum of pairs of leptons and jets appear in four-dimensional models with an extended gauge sector (extra gauge boson Z' beyond the SM with spin $S = 1$). In its characteristics, Z' is similar to the usual SM neutral gauge boson Z, the main differences are the larger expected mass and different coupling constants. The extra-dimension gravity scenario Randall-Sundrum (RS1) predicts other massive (also with m ~ TeV) particles — the Kaluza-Klein excitations (KK modes) of

*e-mail: alexander.lanyov@cern.ch
**e-mail: sergei.shmatov@cern.ch
***e-mail: ilia.zhizhin@cern.ch

the graviton G_{KK} (RS graviton) [9, 10]. Due to a different spin value ($S = 2$), RS gravitons can decay not only into leptons and jets originated from quarks or gluons, but also into two SM gauge bosons ($\gamma\gamma$, WW, ZZ). One of the experimental signatures of these processes can be a pair of leptons, photons or jets with large transverse momenta from the same vertex [1].

In this paper, we will give results of the search for new heavy resonances outside the framework of the SM, using the data obtained by the CMS experiment at $\sqrt{s} = 13$ TeV on statistics corresponding to the integral luminosity of 13 fb^{-1} in 2016 [13].

2 CMS detector

The search for new particles is carried out using the CMS detector (Compact Muon Solenoid), one of the four main LHC detectors [14, 15, 17]. It has an overall length of 22 m, a diameter of 15 m, and weighs 14 000 tonnes. The central feature of the CMS apparatus is a superconducting solenoid of 6 m internal diameter, providing a magnetic field of 3.8 T. Within the solenoid volume are a silicon pixel and strip tracker, a lead tungstate crystal electromagnetic calorimeter, and a brass and scintillator hadron calorimeter, each composed of a barrel and two endcap sections. Muons are detected in gas-ionization chambers embedded in the steel flux-return yoke outside the solenoid, in the pseudorapidity range $|\eta| < 2.4$, with detection planes made using three technologies: drift tubes, cathode strip chambers, and resistive plate chambers. Matching muons to tracks measured in the silicon tracker results in a relative transverse momentum resolution for muons with $20 < p_T < 100$ GeV of 1.3–2.0% in the barrel ($|\eta| < 1.2$) and better than 6% in the endcaps ($0.9 < |\eta| < 2.4$), the p_T resolution in the barrel is better than 10% for muons with p_T up to 1 TeV [15]. Events of interest are selected using a two-tiered trigger system [16]. The first level (L1), composed of custom hardware processors, uses information from the calorimeters and muon detectors to select events at a rate of around 100 kHz within a time interval of less than 4 μs. The second level, known as the high-level trigger (HLT), consists of a farm of processors running a version of the full event reconstruction software optimized for fast processing, and reduces the event rate to around 1 kHz before data storage. A more detailed description of the CMS detector can be found in Ref. [14].

3 Dimuon resonances

In the CMS analysis the typical models [3] predicting extra Z-like bosons decaying to muon pairs have been considered. These models extend the gauge group of the SM by additional $U'(1)$ gauge groups. The $U'(1)$ gauge groups, or a linear combination of them, are maybe broken near the TeV scale giving rise to massive Z' gauge bosons. These models are generalized using an angle that continuously parametrizes the mixing of the respective $U'(1)$ generators. Specific choices of the mixing angle produce different models. One of them, the so-called Sequential Standard Model (SSM) in which heavy bosons (Z'_{SSM} and W'_{SSM}) are assumed to have the same coupling constants as for Z and W bosons in the SM. Another one, ψ-model, is a special case of an extension of the SM to one of the gauge groups of the grand unification theory [3].

4 Event selection

The reconstruction algorithms and event selection employed are refined versions of those used for previous high-mass dilepton searches [18]. To reconstruct a muon candidate, hits from the inner tracker and the muon system are first fitted separately to have inner tracker and standalone muon tracks and then combined in a global muon track hypothesis.

Candidate muon pair events are required to have at least one muon with track segments reconstructed in the muon detectors and with transverse momentum p_T above 16 GeV at L1 trigger, and at least one muon with $p_T > 50$ GeV and $|\eta| < 2.4$ at the HLT. In order to extract the normalization factor in the control region ($60 < m_{ll} < 120$ GeV) around the Z peak, a prescaled HLT trigger with a p_T threshold of 27 GeV is used without changing other L1 requirements. The muon candidates are required to have $p_T > 53$ GeV and be within the sensitive region of $|\eta| < 2.4$.

The muon candidates are also required to pass a dedicated high-momentum muon identification selection [19]. Finally, isolated muon candidates are selected by requiring the p_T sum of tracks within a cone of radius $\Delta R = \sqrt{(\Delta\eta)^2 + (\Delta\phi)^2} < 0.3$ around the candidate direction to be less than 10% of the p_T of the candidate. The sums exclude the lepton candidate under consideration.

The selected opposite-charge muons are combined to form dimuon candidates. To ensure that the two muons originate from the same vertex a fit is performed to a common vertex. This vertex fit is required to have $\chi^2/\text{dof} < 20$. To suppress background from cosmic ray muons that pass near the interaction point, the three-dimensional angle between the two track momentum vectors is required to be less than $\pi - 0.02$. In case several dimuon pairs are found in an event, only the one with the two highest-p_T muons is retained.

Description of the dielectron selection can be found in [13, 20].

5 Analysis strategy

The search for new resonances was carried out by a combined analysis of the shape of muon pair invaraiant mass distributions. The analysis in the muon channel is performed in three different categories to isolate detector alignment and calibration effects in the endcap region, while have a category with the better resolution and calibration that provides higher sensitivity to new physics. Three categories are defined on this purpose considering events with: two muons in the barrel region ($|\eta| < 1.2$); at least one muon in the negative endcap ($\eta < -1.2$); one muon in the barrel region and one muon in the positive endcap ($\eta > 1.2$). This separation allows the assignment of separate uncertainties to each category.

Using a Bayesian approach with an unbinned extended likelihood function, limits are derived for the production of a narrow spin-1 resonance [13, 21]. The likelihood function is based on probability density functions (pdf) describing the invariant mass spectra as a sum of signal and background pdfs.

The signal pdf is modeled as the convolution of a non relativistic Breit-Wigner (BW) with a Gaussian that accounts for the detector resolution effects. This analysis is designed for narrow resonances having a small BW intrinsic width (Γ) compared to the detector resolution. For such narrow widths, variations in Γ have typically little effect on the derived limits. The intrinsic widths of the Z'_{SSM} and Z'_ψ resonances are 3% and 0.6%, respectively. The resolutions are predicted by Monte Carlo (MC) as a function of the generated dilepton mass and are adjusted using Z boson and cosmic ray data events. The resolution for dimuon (dielectron) pairs with a mass of 2 TeV is 5.5% (1%) for barrel-barrel pairs, 8.5% (1.5%) for barrel-endcap pairs. [13, 21]

The functional form of the background pdf is defined by considering the complete background. The dominant and irreducible SM background to a Z' decaying to a muon or electron pair arises from Drell–Yan $Z/\gamma^* \rightarrow l^+l^-$ process. The irreducible $Z/\gamma^* \rightarrow l^+l^-$ background and real prompt leptons backgrounds such as $t\bar{t}$, tW, pair production of gauge bosons (ZZ, ZW, WW) are estimated using MC simulated events.

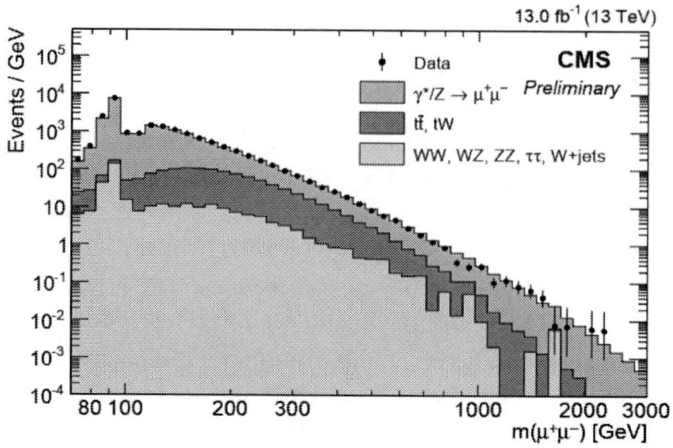

Figure 1. Invariant mass spectrum of dimuon events at $\sqrt{s} = 13$ TeV.

6 Analysis and results

Search for narrow heavy dimuon resonances has been performed with the data recorded in 2016, corresponding to an integrated luminosity of 13 fb^{-1} at $\sqrt{s} = 13$ TeV [13]. The dimuon invariant mass spectrum is shown in Fig. 1. To calculate the mass limits, normalization to $\sigma(Z)$ is performed:

$$R_\sigma = \frac{\sigma(Z' \to \mu^+\mu^-)}{\sigma(Z \to \mu^+\mu^-)} = \frac{N(Z')}{N(Z)} \times \frac{A(Z)}{A(Z')} \times \frac{\varepsilon(Z)}{\varepsilon(Z')}$$

In this way the luminosity uncertainty is removed and other systematic effects reduced. The search for resonances is based on a shape analysis of dilepton mass spectra, in order to be robust against uncertainties in the absolute background level. Existence of signal is established by performing unbinned maximum likelihood fits to the observed spectrum. No significant deviations from the SM expectation have been observed, therefore lower bounds are set on the masses of hypothetical particles that arise in new-physics scenarios. The procedure of setting the confidence level by maximization of the Bayes likelihood function using the Metropolis-Hastings method [22] was similar to the analysis at 8 TeV [19]. Using the dimuon analysis, we could exclude at 95% C.L. the following Z' masses: the Z_{SSM} can be excluded below 3.75 TeV, the Z_ψ below 3.20 TeV. Combining the measurements from both dimuon and electron channels (Fig. 2), the limits for production cross section at $\sqrt{s} = 13$ TeV for Z' were improved: they exclude a Z'_{SSM} with a mass less than 4.0 TeV and Z'_ψ with a mass less than 3.5 TeV [13].

7 Conclusions

The CMS Collaboration has searched for narrow resonances in the invariant mass spectrum of dimuon and dielectron final states in the event samples corresponding to an integrated luminosity of 13 fb^{-1} at $\sqrt{s} = 13$ TeV. This search channel benefits from high signal selection efficiencies and relatively small, well-understood backgrounds. The spectra are consistent with expectations from the SM and the upper limits have been set on the cross section times branching fraction for Z' into lepton pairs relative to SM Z boson production.

We would like to thank the management of the CMS theme at JINR for support.

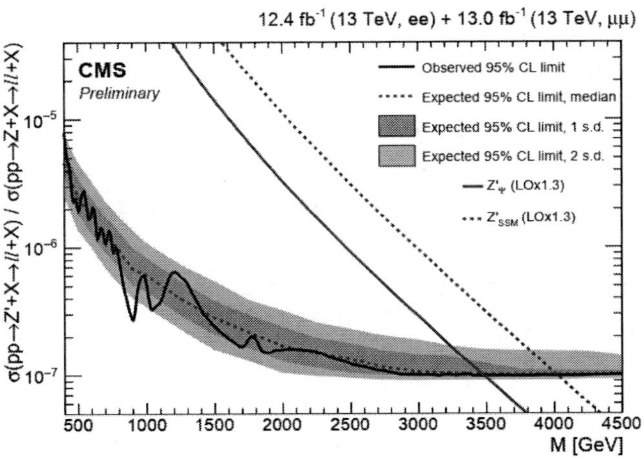

Figure 2. The cross section limits as a function of dilepton mass. Black curve is data, bands around the curve show the expected limits. The solid curve shows the theoretical prediction of cross section for Z'_{ψ} model as a function of dilepton mass and dotted curve corresponds to the prediction for Z'_{SSM} model.

References

[1] CMS Collaboration, CMS Physics TDR, Vol. II: Physics Performance, J. Phys. G **34**, **995** (2007).
[2] H. Baer, M. Brhlik, C. H. Chen, and X. Tata, Phys. Rev. D **55**, 4463 (1997).
[3] A. Leike, Phys. Rept. **317**, 143 (1999).
[4] L. Susskind, Phys. Rev. **D20**, 2619 (1979).
[5] V. A. Rubakov, Phys. Usp. **44**, 871 (2001).
[6] N. Arkani-Hamed, S. Dimopoulos, and G. Dvali, Phys. Lett. B **429**, 263 (1998).
[7] N. Arkani-Hamed, S. Dimopoulos, and G. Dvali, Phys. Rev. D **59**, 086004 (1999).
[8] J. Hewett, Phys. Rev. Lett. **82**, 4765 (1999).
[9] L. Randall and R. Sundrum, Phys. Rev. Lett. **83**, 4690 (1999).
[10] L. Randall and R. Sundrum, Phys. Rev. Lett. **83**, 3370 (1999).
[11] V. A. Ryabov, V. A. Tsarev, and A. M. Tskhovrebov, Phys. Usp. **51**, 1091 (2008)
[12] A. V. Lanyov, Phys. Usp. **57**, 923 (2014).
[13] CMS Collaboration, CMS-PAS-EXO-16-031 (2016).
[14] CMS Collaboration, JINST **3**, S08004 (2008).
[15] CMS Collaboration, JINST **8**, P11002 (2013).
[16] CMS Collaboration, JINST **12**, P01020 (2017).
[17] CMS Collaboration, CMS Physics TDR, Vol. I: Detector Performance and Software, CERN/LHCC 2006-001 (2006).
[18] CMS Collaboration, Phys. Lett. **B768**, 57 (2017).
[19] CMS Collaboration, JHEP **04**, 025 (2015).
[20] CMS Collaboration, JINST **10**, P06005 (2015).
[21] E. Accomando, A. Belyaev, L. Fedeli et al., Phys. Rev. **D83**, 075012 (2011).
[22] W. Hastings, Biometrika **57**, 97 (1970).

EPJ Web of Conferences **177**, 04006 (2018) https://doi.org/10.1051/epjconf/201817704006
AYSS-2017

EMC effect in the Drell-Yan process at COMPASS

Evgenii Mitrofanov (On behalf of COMPASS Collaboration)[1,⋆]

[1] *Joint Institute for Nuclear Research, Dubna*

Abstract.
The EMC effect or a modification of parton distributions in bound nucleons as compared to free ones, has been extensively studied during the last 30 years but its full understanding is still lacking. The COMPASS experiment at CERN will provide new results on the EMC effect, originating from the Drell-Yan process and studied in the 190 GeV/c π^- beam scattering on the ammonia and tungsten targets. The present understanding of the EMC effect and experimental possibilities of COMPASS in this context are discussed.

1 Introduction

The momentum distribution functions of the partons within the hadron ($q(x, Q^2)$, $\bar{q}(x, Q^2)$, $g(x, Q^2)$), called Parton Distribution Functions (PDFs) describe the internal structure of hadrons in the quark-parton model. They represent the probability densities to find a parton carrying a proton momentum fraction x at a negative four-momentum transfer squared Q^2. At the leading order in QCD the PDFs are related to the structure function $F_2(x, Q^2)$ which is directly accessible in deep-inelastic charged lepton-nucleon scattering experiments:

$$F_2(x, Q^2) = \sum_{q=u,d,s...} x e_q^2 \left[q(x, Q^2) + \bar{q}(x, Q^2) \right], \qquad (1)$$

where e_q is the quark charge (in units of elementary charge $|e|$).

2 EMC effect

The EMC effect is a modification of quark and gluon distributions in bound nucleons by the nuclear environment. It was discovered by the European Muon Collaboration in 1983 [1] in the deep inelastic scattering of muons. The ratio of structure function F_2 per nucleon for iron and deuterium targets was measured as a function of x. It was found to be different from unity and at least for $x \lesssim 0.7$ not following expectations from Fermi-motion calculations. This result demonstrated for the first time that the structure function F_2 is modified, when nucleons are embedded in a nucleus. Since then the EMC effect has been observed in the numerous experiments. Figure 1 shows a typical behavior of the ratio σ^A/σ^D as a function of x.

Dependences of nuclear modifications of F_2 on kinematics and various nuclear properties like mass, density or radius of a nucleus are rather well-known, but, nevertheless, the origin of the effect

⋆e-mail: evgenii.mitrofanov@cern.ch

is still not fully understood. Numerous explanations of the EMC effect have been proposed: nuclear binding, pion excess in nuclei, multi-quark clusters, dynamical rescaling, medium modification, short-range correlations, etc. However, there is no generally accepted model for the effect over all A and x. Comprehensive reviews of the EMC effect can be found in Refs. [5, 12–15].

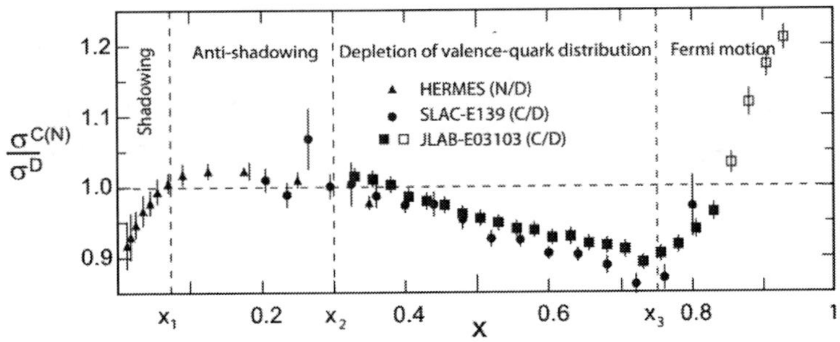

Figure 1. Ratio $\sigma^{C(N)}/\sigma^{D}$ as a function of x from HERMES [2], SLAC-E139 [3], and JLAB-E03103 [4]. Figure comes from [5].

3 EMC effect in the Drell-Yan process

Apart of numerous deep inelastic experiments, the EMC effect has also been experimentally verified in the time-like region using both the pion- and the proton-induced Drell-Yan reactions in the E772 [6], E866 [7], NA3 [8], and NA10 [9] experiments. In the Drell-Yan process a quark (anti-quark) with the momentum fraction x_1 from the beam hadron and an anti-quark (quark) of the target nucleon with the momentum fraction x_2 annihilate via a virtual photon (contribution of the Z-boson exchange at the discussed energies is extremely small) into a charged-lepton pair: $q(x_1)\bar{q}(x_2) \to \gamma^* \to l^+l^-$. The Drell-Yan cross section is

$$\frac{d^2\sigma_{DY}}{dx_1 dx_2} = K\frac{4\pi\alpha^2}{9x_1x_2}\frac{1}{s}\sum_{q=u,d,s...} e_q^2\left[q(x_1)\bar{q}(x_2) + \bar{q}(x_1)q(x_2)\right], \tag{2}$$

where $K \approx 2$ is a factor representing the deviation from the simple parton model due to QCD corrections, s is the squared total energy of the $q(x_1)\bar{q}(x_2)$ system, α - the fine-structure constant.

The pion-induced Drell-Yan process is complementary to the DIS process and can provide another experimental tool that is sensitive to flavour-dependent effects in the nuclear quark distributions [10]. Keeping only the dominant terms in the cross section, one obtains

$$\frac{\sigma_{DY}^{\pi^- A}}{\sigma_{DY}^{\pi^- D}} \approx \frac{u_A(x_2)}{4u_D(x_2)}, \tag{3}$$

where $u_A(x_2)$ and $u_D(x_2)$ are PDFs for u-quark in a nucleus with a mass A and deuterium, respectively. The previous measurements of a nuclear dependence of the pion-induced Drell-Yan cross section have been performed in the NA3 [8] and NA10 [9] experiments. The ratio (3) measured by NA10 for tungsten and deuterium targets using a negative pion beam of 140 GeV/c and 286 GeV/c, is shown in the left and right panels of Fig 2 respectively. However the accuracy is too low for detailed tests of various theoretical models.

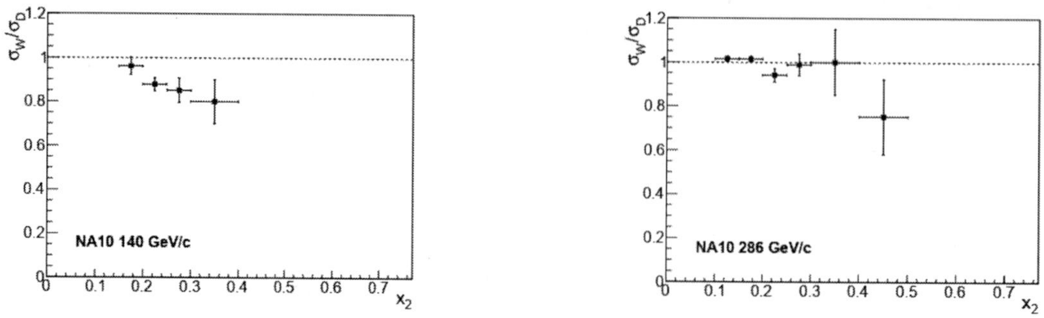

Figure 2. Ratio of the Drell-Yan cross sections for W and D measured by the NA10 Collaboration using a π^- beam of 140 GeV/c (left) and 286 GeV/c (right) [9].

4 COMPASS experiment

COMPASS is a fixed-target experiment at a secondary beam of the Super Proton Synchrotron at CERN. The purpose of this experiment is a study of hadron structure and hadron spectroscopy with muon and hadron beams of high intensity. The experiment has an intensive physics programme which includes such topics as a study of nucleon spin structure in the semi-inclusive deep inelastic scattering and Drell-Yan process, search for new exotic hadron states in diffractive, central, Primakoff and lepto-production, tests of the chiral theory predictions, measurement of generalized parton distributions, etc.

4.1 COMPASS setup

The COMPASS setup (Fig. 3) can be divided into four parts along the beam axis. It starts with the beam line telescope and the detectors that identify the incoming beam particles. It is followed by the target region, which is specific for each of the COMPASS physics programmes. The third part called the Large Angle Spectrometer, built around the first dipole magnet SM1, detects scattered particles with polar angles of up to 180 mrad. The fourth part is the Small Angle Spectrometer, which occupies the downstream part of the setup and detects particles at small angles (less than 30 mrad) and momenta larger than 5 GeV/c.

4.2 EMC effect at COMPASS

The ratio (3) of the cross sections calculated for the COMPASS kinematic coverage using the PDFs for deuterium and gold from the model of Ref. [10] is presented in Fig. 4 as a function of x_2 for the fixed mass $M_{l^+l^-}$=5 GeV/c^2 (left) and as a function of $M_{l^+l^-}$ for the fixed value $x_1 = 0.5$.

Data taking for the Drell-Yan process study with a negative pion beam of 190 GeV/c was performed at COMPASS in 2014 and 2015. The main ammonia target, a tungsten beam plug of the hadron absorber and an additional aluminium target present in the setup can be used to study the EMC effect. The observed dimuon mass spectrum and the kinematic range of x_1 and x_2 for $M_{\mu\mu} > 4$ GeV/c^2 covered by COMPASS for the ammonia target are shown in the left and right panels of Fig. 5, respectively. Dimuon mass resolution for J/ψ peak is about 0.2 MeV/c^2 for the ammonia, while for the tungsten plug it is a bit worse. Thus only the mass range above 5 GeV/c^2 can be used for investigation of the EMC effect, which defines a lower limit of x_2 range accessible by COMPASS to be 0.07.

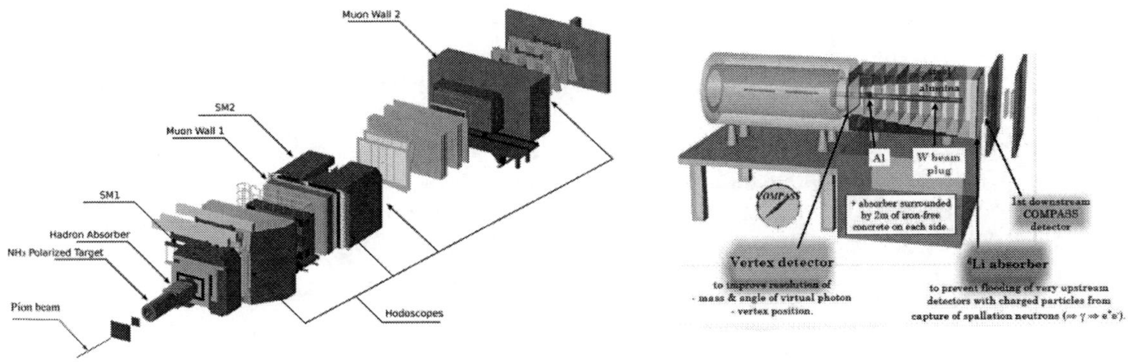

Figure 3. Left: sketch of the COMPASS setup for data taking with pion beam. Right: the ammonia target region.

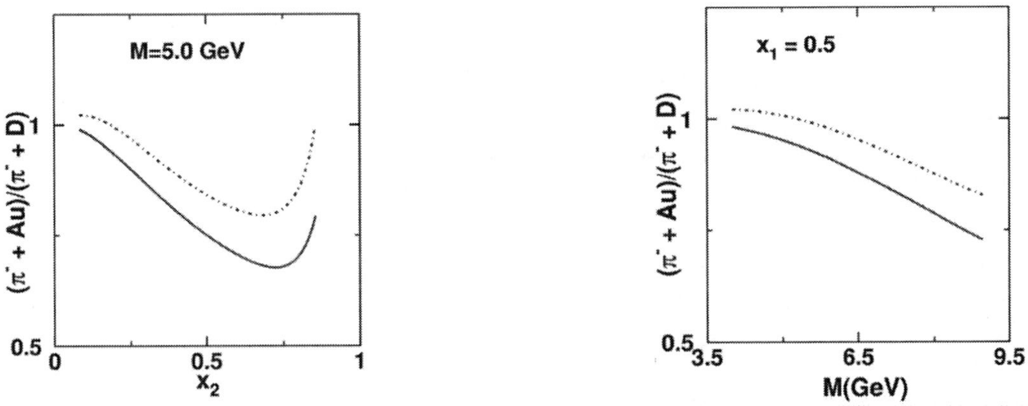

Figure 4. Ratio of the Drell-Yan cross sections for Au and D calculated for the COMPASS kinematic coverage using the PDFs from the model of Ref. [10] as a function of x_2 for the fixed mass $M_{l^+l^-}$=5 GeV/c^2 (left) and as a function of $M_{l^+l^-}$ for the fixed value x_1 = 0.5.

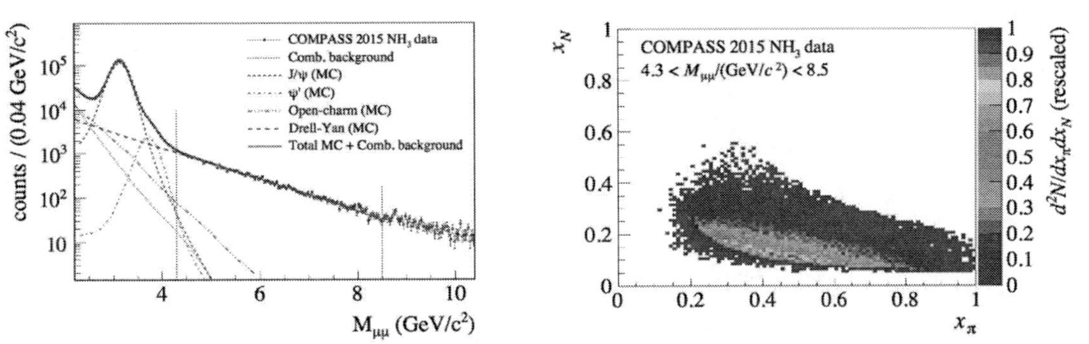

Figure 5. Dimuon mass spectrum (left) and kinematic range of x_1 and x_2 for $M_{\mu\mu}$ > 4 GeV/c^2 covered by COMPASS (right) for the ammonia target [11].

5 Summary

After 30 years the origin of the EMC effect is still not fully understood and there is no no model which describes the whole range of x. An experimental tool that is sensitive to flavour-dependent effects in the nuclear quark distributions is the pion-induced Drell-Yan process which allows to measure EMC effect separetly for u and d – quarks.

The COMPASS experiment at CERN will provide new results on the EMC effect originating from the Drell-Yan process $\pi^- + A \rightarrow \mu^+ + \mu^- + X$ (where A is proton, tungsten and aluminium) covering values of x down to 0.07, i.e. lower than in other experiments

References

[1] J. J. Aubert et al. (EMC Collaboration), Phys. Lett. B **123**, 275 (1983).
[2] A. Airapetian et al. (HERMES Collaboration), Phys. Lett. B **567**, 339 (2003).
[3] J. Gomez et al. (SLAC-E139 Collaboration), Phys. Rev. D **49**, 4348 (1994).
[4] J. Seely et al. (JLAB-E03103 Collaboration), Phys. Rev. Lett. **103**, 202301 (2009).
[5] K. Rith, Subnucl. Ser. **51**, 431 (2015).
[6] D. M. Alde et al. (FNAL-E772 Collaboration), Phys. Rev. Lett. **64**, 2479 (1990).
[7] M. A. Vasiliev et al. (FNAL-E866 Collaboration), Phys. Rev. Lett. **83**, 2304 (1999).
[8] J. Badier et al. (NA3 Collaboration), Phys. Lett. B. **104**, 335 (1981).
[9] P. Bordalo et al. (NA10 Collaboration), Phys. Lett. B. **193**, 368 (1987).
[10] D. Dutta et al., Phys. Rev. C. **83**, 042201 (2011).
[11] M. Aghasyan et al. (COMPASS Collaboration), Phys. Rev. Lett. **119**, 112002 (2017).
[12] M. Arneodo, Phys. Rep. **240**, 301 (1994).
[13] D. Geesaman, K. Saito, A. W. Thomas, Ann. Rev. Nucl. Part. Sci. **45**, 337 (1995).
[14] P.R. Norton, Rep. Prog. Phys. **66**, 1253 (2006).
[15] S. Malace et al., Int. J. Mod. Phys. E. **23**, 1430013 (2014).

Measurement of multiplicities of charged hadrons, pions and kaons in DIS at COMPASS

Nikolai Mitrofanov (On Behalf of COMPASS Collaboration)[1,⋆]

[1] *Joint Institute for Nuclear Research, Dubna*

Abstract. Precise measurements of multiplicities of charged hadrons, pions and kaons in deep inelastic scattering were performed. The data were obtained by the COMPASS Collaboration by scattering 160 GeV muons off an isoscalar ^{6}LiD target. The results were obtained in three-dimensional bins of the Bjorken scaling variable x, the relative virtual-photon energy y, and the fraction z of the virtual-photon energy carried by the produced hadron. A leading-order pQCD analysis was performed using the pion multiplicity results to extract quark fragmentation functions into pions. The results for the sum of the z-integrated multiplicities for pions and for kaons, differ from earlier results from the HERMES experiment. The results from the sum of the z-integrated K^+ and K^- multiplicities at high x point to a value of the non-strange quark fragmentation function larger than obtained by the earlier DSS fit.

1 Introduction

Hadron production in semi-inclusive measurements of deep-inelastic lepton–nucleon scattering (SIDIS) is one of the most powerful tools to investigate the structure and formation of hadrons. Within the standard framework of leading-twist perturbative QCD (pQCD), factorisation theorems [3] allow one to write the SIDIS cross section as a convolution of hard scattering cross sections, which are calculable in pQCD, with non-perturbative Parton Distribution Functions (PDFs) and Fragmentation Functions (FFs). The PDFs account for the partonic structure of hadrons in the initial state. The FFs encode the details about the hadronisation mechanism that describes the transition from final-state partons into colour-neutral hadrons. Both types of functions are believed to be process independent, and can be interpreted in leading-order (LO) as number densities. While PDFs have been studied in detail for several decades and are hence known with good precision, new accurate measurements are necessary to constrain the FFs.

The DIS process $lN \to l'hX$ is described by the negative square of the four-momentum transfer $Q^2 = -q^2$, the Bjorken variable $x = -q^2/(2P \cdot q)$ and the fraction of the virtual-photon energy that is carried by the final-state hadron, $z = (P \cdot p_h)/(P \cdot q)$. Here, l and l' are beam and scattered leptons respectively, $q = k - k'$, P and p_h denote the four-momenta of the virtual photon, the nucleon N and the observed hadron h respectively, with k (k') the four momentum of the incident (scattered) lepton. Additional variables used are the lepton energy fraction carried by the virtual photon, $y = (P{\cdot}q)/(P{\cdot}k)$, and the invariant mass of the final hadronic system, $W = \sqrt{(P + q)^2}$. hadronisation mechanism in

⋆e-mail: nikolai.mitrofanov@cern.ch

SIDIS, the relevant observable is the differential multiplicity for hadrons of a specific type h, which is defined as the differential cross section for hadron production normalised to the differential inclusive DIS cross section:

$$\frac{dM^h(x,z,Q^2}{dz} = \frac{d^3\sigma^h(x,z,Q^2)/dxdQ^2dz}{d^2\sigma^{DIS}(x,Q^2)/dxdQ^2}. \tag{1}$$

Interpreted in pQCD, the cross sections on the right-hand side are expressed in terms of PDFs and FFs and read at leading-order (LO)

$$\frac{d^2\sigma^{DIS}}{dxQ^2} = C(x,Q^2)\sum_q e_q^2 q(x,Q^2),$$
$$\frac{d^2\sigma^h}{dxQ^2dz} = C(x,Q^2)\sum_q e_q^2 q(x,Q^2)D_q^h(z,Q^2). \tag{2}$$

Here, $C(x,Q^2) = 2\pi\alpha^2(1+(1-y)^2/Q^4$ with α the fine structure constant, $q(x,Q^2)$ is the quark PDF for the flavour q and $D_q^h(z,Q^2)$ the quark-to-hadron FF. In LO, D_q^h denotes the number density of hadrons h produced in the hadronisation of partons of species q.

2 COMPASS experiment

COMPASS [4] is a fixed-target experiment at a secondary beam of the Super Proton Synchrotron at CERN. The purpose of this experiment is the study of hadron structure and hadron spectroscopy with muon and hadron beams of high intensity. The experiment has an intensive physics programme which includes topics such as a study of nucleon spin structure in the semi-inclusive deep inelastic scattering and Drell-Yan process, search for new exotic hadron states in diffractive, central, Primakoff and lepto-production, tests of the chiral theory predictions, measurement of generalized parton distributions (GPDs), etc.

The measurement was performed in 2006 with the naturally polarised muon beam of the CERN SPS using positive muons of 160 GeV/c. The beam momentum had a spread of 5%. The intensity was 4×10^7 s^{-1} with spills of 4.8 s and a cycle time of 16.8 s. The momentum of each incoming muon was measured before the COMPASS experiment with a precision of 0.3%. The target consisted of three cells, which were located along the beam one after the other and filled with ^{6}LiD immersed in a liquid ^{3}He/^{4}He mixture. The admixtures of H, ^{3}He and ^{7}Li in the isoscalar target lead to an effective excess of neutrons of about 0.2%. The direction of the polarisation in the 60 cm long middle cell was opposite to that in the two 30 cm long outer cells and was reversed once per day. In the analysis, the data are averaged over the target polarisation for the determination of multiplicities.

3 Data Analysis

Differential multiplicities are determined in 3-dimensional (x, y, z) bins from the acceptance-corrected hadron yields N^h normalised by the number of DIS events,

$$\frac{dM^h(x,y,z)}{dz} = \frac{1}{N^{DIS}(x,y)}\frac{dN^h(x,y,z)}{dz}\frac{1}{A(x,y,z)}. \tag{3}$$

The acceptance correction factor A takes into account the limited geometric and kinematic acceptance of the spectrometer and the efficiency of event reconstruction. The choice of the z and x variables is natural because multiplicities depend mostly on these variables. Because of the strong correlation between x and Q^2 in the COMPASS fixed-target kinematics, it appears more appropriate to use y instead of Q^2 as the third variable.

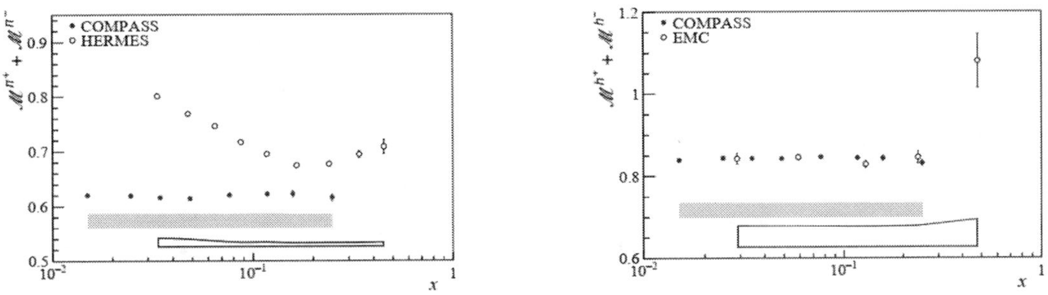

Figure 1. Left: Sum of $\mathcal{M}^{\pi^+} + \mathcal{M}^{\pi^-}$ versus x. The COMPASS data (160 GeV, closed circles) are compared to HERMES results (27.5 GeV, open circles); Right: Sum of $\mathcal{M}^{h^+} + \mathcal{M}^{h^-}$ versus x. The COMPASS data (160 GeV, closed circles) are compared to EMC results (100 GeV to 280 GeV, open circles). The systematic uncertainties are shown as bands at the bottom.

4 Results for multiplicities

4.1 Charged pion and charged hadron multiplicities

On the Fig. 1 (left) is shown the result [1] for the sum $\mathcal{M}^{\pi^+} + \mathcal{M}^{\pi^-}$ of π^+ and π^- multiplicities, integrated over z from 0.2 to 0.85 and averaged over y between 0.1 and 0.7, as a function of x. The expected weak x dependence is indeed observed in the data. In the same figure, the results of HERMES [5] integrated over z from 0.2 to 0.8 are shown using the so-called x representation. The HERMES multiplicities are larger and show a different dependence on x. Note however that the HERMES data were measured at a lower energy and correspond to different kinematics, in particular accepting lower W values. In order to compare the COMPASS results also with the EMC ones [6], the sum of charged hadron multiplicities is shown in Fig. 1 (right). The results from COMPASS and EMC, which correspond to comparable kinematics, are found in excellent agreement.

Another quantity of interest is the x dependence of the ratio $\mathcal{M}^{\pi^+}/\mathcal{M}^{\pi^-}$.The results are shown in Fig. 2 (left) as a function of x. They are in reasonable agreement with the HERMES values in the measured range. The values obtained from the JLab E00-108 [7] experiment. for $z > 0.3$ at higher x and lower W values are also shown for completeness. In Fig. 2 (right), the ratio $\mathcal{M}^{h^+}/\mathcal{M}^{h^-}$ calculated for charged hadron multiplicities is shown for COMPASS and EMC data. These results are in excellent agreement.

Direct extraction of FFs from pion multiplicities was done [1]. The results from this direct extraction of FFs are in good agreement with the results obtained from the LO fit. This is illustrated in Fig. 3 for one (x, y) bin.

4.2 Charged kaon multiplicities

The result [2] for $\mathcal{M}^{K^+} + \mathcal{M}^{K^-}$ is presented in Fig.4 (left) as a function of x at the measured values of Q^2. The data are integrated over z in the range 0.20 to 0.85 and averaged over y in the range 0.1 to 0.7. Only those eight x bins which have a sufficient z coverage are shown. Fig. 4 (left) also shows the HERMES results [8] that were taken at 27.5 GeV beam energy and correspond to different kinematics in particular accepting lower W values. They lie well below the COMPASS points and exhibit a different x behaviour. The x dependence of the HERMES results for pion and kaon multiplicities gave rise to some dispute [9–11]. Note that COMPASS multiplicities are computed in bins of $(x, y,$

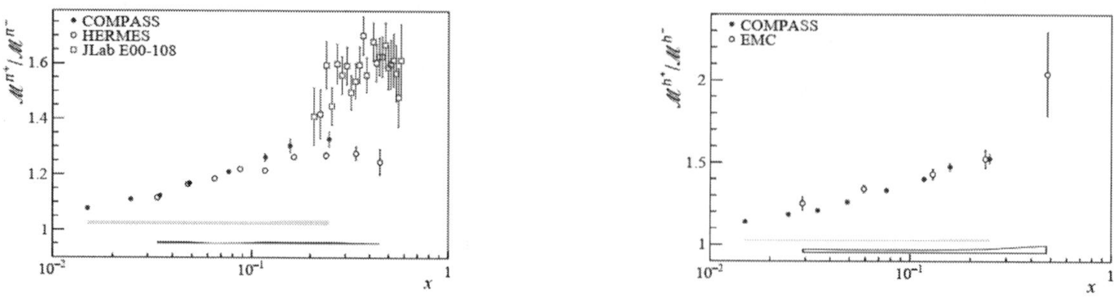

Figure 2. Left: Ratio $\mathcal{M}^{\pi^+}/\mathcal{M}^{\pi^-}$ versus x from COMPASS (160 GeV, closed points), HERMES (27.5 GeV, open circles) and JLab (5.5 GeV, open squares). Right: Ratio $\mathcal{M}^{h^+}/\mathcal{M}^{h^-}$ versus x for COMPASS (160 GeV, closed circles) and EMC (100 GeV to 280 GeV, open circles) results. The systematic uncertainties are shown as bands at the bottom.

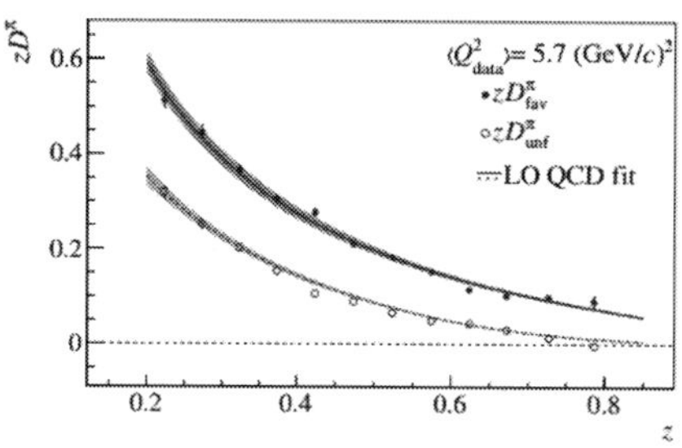

Figure 3. The z dependence of the favoured (closed symbols) and unfavoured (open symbols) quark-to-pion fragmentation functions, zD^{π}_{fav} and zD^{π}_{unf}, extracted directly from pion multiplicities at the measured Q^2 for one (x, y) bin $(0.04 < x < 0.06, 0.3 < y < 0.5)$. For comparison, the result from the present LO QCD fit to the pion multiplicities is shown at $Q^2 = 6$ (GeV/c)^2 . The bands represent the total uncertainties of the QCD fit.

z) before integration over z and averaging over y while in the case of HERMES, the hadron and DIS yields are obtained and integrated over separately, and finally combined in a ratio depending on x only [11].

The results for the ratio are shown in Fig. 4 (right) as a function of x. In the region of overlap, COMPASS results are found to be systematically lower than those of HERMES.

5 Summary and conclusions

Precise results for the multiplicities of charged-kaons, charged-pions and hadrons are obtained from SIDIS measurements using a muon beam scattering off an isoscalar target. The data are given in a three-dimensional x, y, and z binning and cover a wide kinematic range: 1 (GeV/c)^2 $< Q^2 <$ 60 (GeV/c)^2, $10^{-3} < x < 0.4$, $0.1 < y < 0.7$, and $0.20 < z < 0.85$ with W > 5 GeV/c^2 . These high precision multi-dimensional data provide an important input for future NLO QCD fits of fragmentation functions.

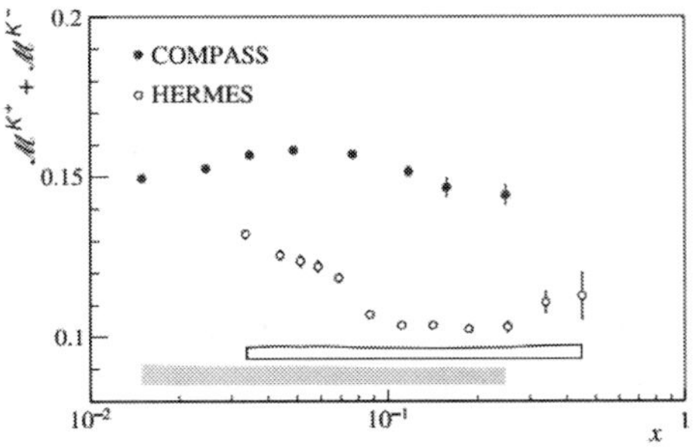

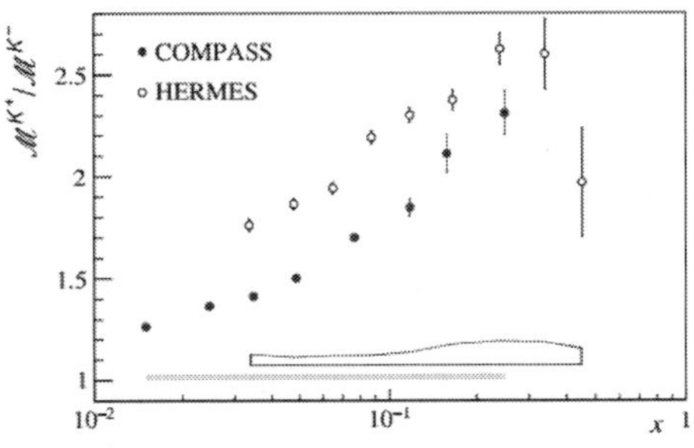

Figure 4. Left: sum of z-integrated multiplicities, $\mathcal{M}^{K^+} + \mathcal{M}^{K^-}$. COMPASS data (160 GeV, full points) are compared to HERMES data [8] (27.5 GeV, open points). Right: Ratio of z-integrated multiplicities, $\mathcal{M}^{K^+} / \mathcal{M}^{K^-}$. COMPASS data (160 GeV, full points) are compared to HERMES data [12] (27.5 GeV, open points). The systematic uncertainties are shown as bands at the bottom.

The sum of the z-integrated positively and negatively charged hadron and pion multiplicities shows a flat x behaviour, as expected in LO pQCD. For charged hadrons this sum is in agreement with EMC results obtained at comparable kinematics, whereas some inconsistency is observed when comparing to HERMES pion data that were taken at different kinematics. The ratio of the z-integrated positive and negative hadron and pion multiplicities as a function of x nicely confirms the previous measurements from HERMES and EMC.

The measured charged pion multiplicities were used for a LO extraction of the favoured and unfavoured pion FFs. While both FFs are significantly different from those obtained in the HKNS fit to only the electron–positron annihilation data, they are in good agreement with those obtained in recent NLO fits that also include a preliminary release of the present data.

The sum of K^+ and K^- multiplicities integrated over z shows a flat distribution in x, with values significantly higher than those measured by HERMES at lower energy. By covering lower x values, the multiplicity sum data will significantly improve the constraint on the product of the strange quark PDF and FF.

References

[1] C. Adolph et al., (COMPASS Collaboration) Physics Letters B **764**, 1-10 (2017).
[2] C. Adolph et al., (COMPASS Collaboration) Physics Letters B **767**, 133–141 (2017).
[3] J.C. Collins, D.E. Soper, G.F. Sterman, Nucl. Phys. B **261**, 104 (1985).
[4] P. Abbon, et al., (COMPASS Collaboration) Nucl. Instrum. Methods A **577**, 455 (2007).
[5] A. Airapetian, et al., (HERMES Collaboration) Phys. Rev. D **87**, 074029 (2013).
[6] EMC, J. Ashman, et al., Z. Phys. C **52**, 361 (1991).
[7] R. Asaturyan, et al., (JLab E00-108) Phys. Rev. C **85**, 015202 (2012).
[8] A. Airapetian, et al., (HERMES Collaboration) Phys. Rev. D **89**, 097101 (2014).
[9] E. Leader, A.V. Sidorov, D.B. Stamenov, Phys. Rev. D **93**, 074026 (2016).
[10] M. Stolarski, Phys. Rev. D **92**, 098101 (2015).
[11] E. Aschenauer, et al., Phys. Rev. D **92**, 098102 (2015).
[12] A. Airapetian, et al., (HERMES Collaboration) Phys. Rev. D **87**, 074029 (2013).

EPJ Web of Conferences **177**, 04008 (2018) https://doi.org/10.1051/epjconf/201817704008
AYSS-2017

Studying the strong interaction with baryon-(anti)baryon femtoscopy in Pb-Pb collisions measured by ALICE

Jeremi Niedziela[1,2,*] for the ALICE Collaboration

[1] *Warsaw University of Technology*
[2] *CERN*

Abstract.
The shape of the baryon-(anti)baryon femtoscopic correlation function is influenced by the size of the emission source, strong and Coulomb interactions and Quantum Statistics. Another factor introducing additional correlation structures is related to the residual correlations, which are related to the fact that baryons may come from decays of heavier particles. The correlation function of a given pair of baryons (for example $p\bar{p}$) is closely connected with correlation functions of other particles (such as $p\bar{\Lambda}$).
Analysing correlation functions of multiple baryon pairs simultaneously can further constrain measured strong scattering parameters. A newly developed fitting procedure allowing for a combined fit of several correlation functions is presented. The procedure was applied to 2.76 TeV and 5.02 TeV Pb-Pb data measured by ALICE. The analysis was performed for all combinations of (anti)protons and (anti)lambdas for different collision centralities.
Measured strong interaction parameters for $p\bar{\Lambda}$, $\Lambda\bar{\Lambda}$ and heavier baryon-antibaryon ($b\bar{b}$) pairs are shown and possible underlying physical processes are discussed.

1 Introduction

The strong interaction for (anti)baryons (particles composed of three quarks) is not well know. There have been only a few cross section measurements performed for baryon-baryon pairs: pp, pn, pd, $p\Lambda$ and $p\Sigma^-$ [1]. For baryon-antibaryon pairs there are even less data - measurements exist only for $p\bar{p}$, $\bar{p}n$ and $d\bar{p}$ [1]. This is mainly caused by the technical difficulty to form, accelerate and collide beams of heavier, short lived (anti)particles.

Heavy-ion collisions at relativistic energies allow for a study of interactions between (anti)baryons by looking at the collisions products, rather than at the particles scattered from the incoming beams. Experiments such as ALICE at the LHC detect nearly the same number of baryons and antibaryons at midrapidity, hence they are well suited for the study of the strong interaction.

What motivate the research presented in this work are several unanswered questions concerning the interaction between baryons and antibaryons, such as "is the strong potential attractive or repulsive for $b\bar{b}$ pairs?", "do pairs of non-identical baryons annihilate?" or "do bound states of baryon-antibaryon exist?". It will be shown that the recent measurements of ALICE and analysis method based on the femtoscopic correlations may help to find answers to these questions.

*e-mail: jeremi.niedziela@cern.ch

1.1 Two-particle correlations at low relative momenta

The analysis performed in this study is based on femtoscopy - a technique of measuring correlation functions with respect to the relative momenta of particle pairs. Such distributions depend on the shape and size of the emission source, as well as on the interactions between the particles. The two-particle femtoscopic correlation function can be formally written as [2]:

$$C(k^*) = \int S(\bar{r}^*) \left|\Psi(\bar{k}^*, \bar{r}^*)\right|^2 d^4\bar{r}^* \quad (1)$$

where $S(\bar{r}^*)$ is a source emission function, defining probability of producing two particles with given space-time separation $\bar{r}^*$ and $\Psi(\bar{k}^*, \bar{r}^*)$ is a pair wave function. The $\bar{k}^*$ vector is a momentum of one of the particles in the Pair Rest Frame (PRF).

In case of small relative momenta and spherically-symmetric potential, the pair wave function can be defined as the sum of the incoming plane wave and the outgoing spherical wave [3]:

$$\Psi(\bar{k}^*, \bar{r}^*) = e^{-i\bar{k}^*\bar{r}^*} + f(k^*)\frac{e^{ik^*r^*}}{r^*} \quad (2)$$

where $f(k^*)$ is the scattering amplitude that for small momenta and small potential length scales (the so-called effective range approximation) can be expressed in the following way [4]:

$$f(k^*) = \left[\frac{1}{f_0} + \frac{d_0 k^{*2}}{2} - ik^*\right]^{-1} \quad (3)$$

The scattering length f_0 and the effective range d_0 are parameters describing the strong potential. Measurement of these parameters is crucial to understand whether the strong interaction for the studied pairs is attractive or repulsive, whether annihilation channels for non-identical particles exist or not, and whether the creation of bound states is possible.

1.2 Measuring correlation functions in ALICE

ALICE [5] is one of the four major experiments at the LHC, designed to study heavy-ion collisions at relativistic energies. With its six-layer silicon tracker, Time Projection Chamber, Time Of Flight and Transition Radiation Detector, ALICE excels in the precise particle identification at low-momenta. As an example, a visualisation of a Pb-Pb collision registered by ALICE is shown in Fig.1.

Thanks to the similar amount of matter and antimatter being produced in heavy-ion collisions at the LHC and the great capability of the detector to measure particle momenta and to perform precise identification, ALICE is a perfect experiment for the study of interactions between baryons and antibaryons.

The experimental two-particle correlation function can be measured for particles produced in heavy-ion collisions using:

$$C(p_1, p_2) = \frac{P(p_1, p_2)}{P(p_1)P(p_2)} \quad (4)$$

where $P(p_1, p_2)$ is a probability of observing particles with momenta p_1 and p_2 in the same event, while $P(p_1)$ and $P(p_2)$ are probabilities of observing particles with such momenta independently. With this definition of the correlation function, some detector effects (e.g. connected to the limited detector's acceptance) cancel out in the ratio.

Correlation functions were measured for all combinations of (anti)protons and (anti)lambda baryons, at two collision energies in six centrality bins. This gives in total 72 correlation functions, including 36 baryon-antibaryon correlation functions. An example of the measured functions is presented in Fig.2.

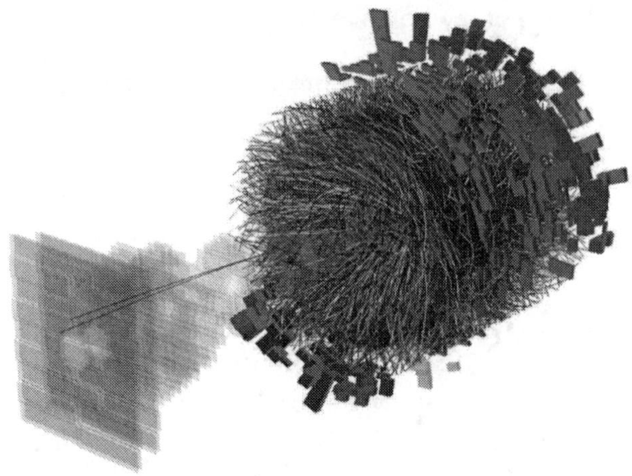

Figure 1. Visualisation of one of the first lead-lead collisions at 5.02 TeV per nucleon pair as seen by ALICE.

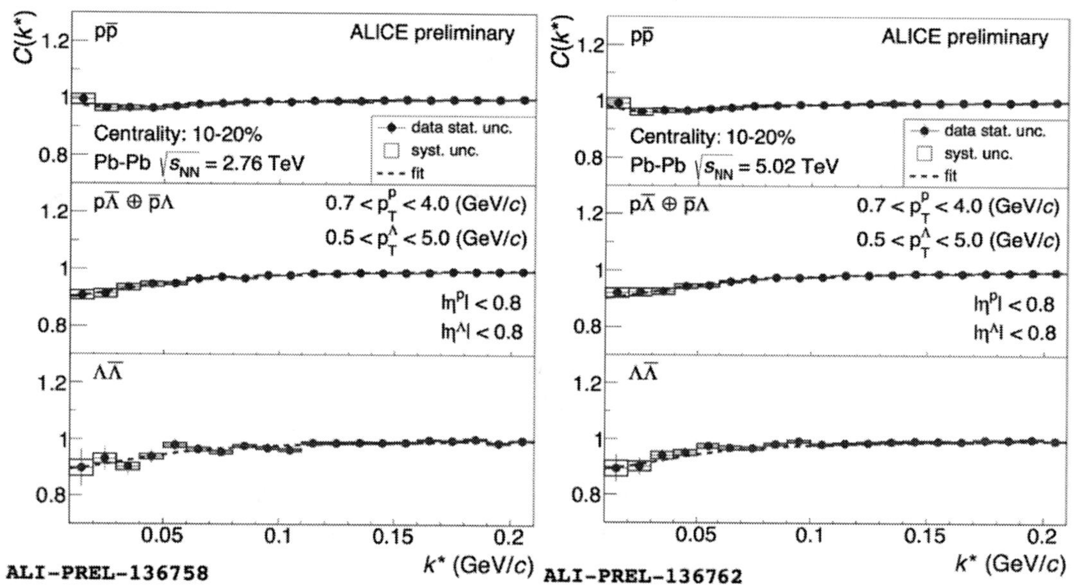

Figure 2. Example of baryon-antibaryon correlation functions measured by ALICE.

2 Description of the fitting procedure

The fitting was performed for all baryon-antibaryon correlation functions simultaneously. An appropriate form of the theoretical function was chosen depending on the type of particles, taking into account strong and Coulomb interactions and Quantum Statistics. Three sets of parameters were introduced, for $p\bar{\Lambda}$, $\Lambda\bar{\Lambda}$ and effective parameters for heavier $b\bar{b}$ pairs. In each set, three parameters were free in the fit: the real and imaginary part of the scattering length and the effective range.

The fitting procedure fully takes into account so-called residual correlations that are the consequence of the fact that a fraction of measured particles comes from decays of heavier ones. This is illustrated in Fig.3 on the example of pp correlation function. In this case, two protons are observed in the experiment and influence the shape of proton-proton correlation functions, but the interaction actually took place between the proton and the Λ baryon, which then decayed into proton and pion. Due to this effect, the final correlation function will be a sum of all residual components:

$$C(k^*) = 1 + \sum_i \lambda_i \left[C_i(k^*) - 1\right] \tag{5}$$

where $C(k^*)$ is the measured correlation function, λ_i are fractions of different residual components and $C_i(k^*)$ is the correlation function of the pair i.

The residual correlations significantly complicate the analysis, but they also allow to link different systems - for example, strong interaction parameters extracted from $\Lambda\bar{\Lambda}$ functions can be used for the residual components of the $p\bar{\Lambda}$ and $p\bar{p}$ functions (see Fig.3).

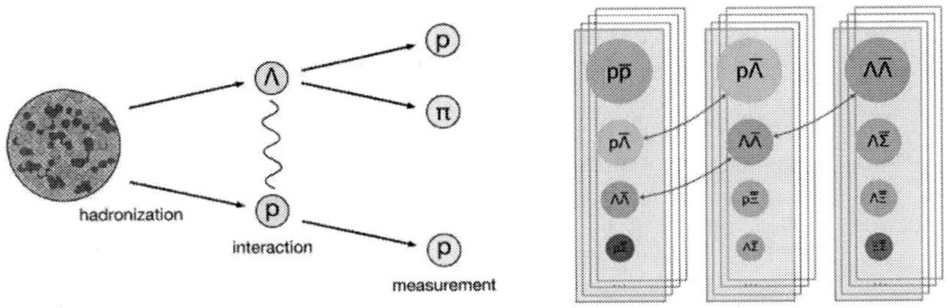

Figure 3. Left: Illustration of the residual correlations mechanism. Right: Illustration of the interlinks between different correlation functions through the residual components.

The source sizes were fixed in the fit. The femtoscopic radii scale as the cube root of multiplicity, which in turn depends on the collision centrality. This allows to link correlation functions for different centralities through the source sizes. An approximate scaling with the transverse mass m_T was also used to determine radii for baryon-antibaryon pairs based on previous measurements for pions, kaons and protons performed by ALICE [6]. In this analysis, the same scaling was used for the source sizes of the residual pairs, while previously the same radius was used for all residual components [7][8]. The functions were suitably normalised and non-femtoscopic background was corrected before fitting. The decay kinematics were taken into account for the residual components. The theoretical correlation functions were modified to take into account the limited momentum resolution of the detector [8].

3 Results of the analysis

Preliminary values of the measured strong interaction parameters for $p\bar{\Lambda}$, $\Lambda\bar{\Lambda}$ and heavier $b\bar{b}$, together with statistical and systematic uncertainties, are presented in Fig.4. A direct comparison with the values obtained from previous experiments and model predictions reveals good relative precision of the measurement. The extracted values of the interaction parameters for all baryon-antibaryon pairs are similar, the real part of the scattering amplitude is negative, the imaginary part is positive and the effective range is non-zero.

The imaginary part of the scattering length is associated with inelastic processes, in which the types of particles change [9]. The obtained results suggest that inelastic channels are present for all baryon-antibaryon pairs (even non-identical ones, such as $p\bar{\Lambda}$) and they have similar magnitude.

The negative real part of the scattering amplitude means that either strong interaction for baryon-antibaryon pairs is repulsive, or that bound states exist (such as $p\bar{\Lambda}$, $\bar{p}\Lambda$, $\Lambda\bar{\Lambda}$, $\Xi\bar{\Sigma}$ etc.). In the low energy limit there is no way to distinguish between these two interpretations. However, existence of the inelastic channels favors the bound state scenario, in which baryon-antibaryon form a bound state (negative $\Re f_0$) and then decay into e.g. three mesons (non-zero $\Im f_0$) [10].

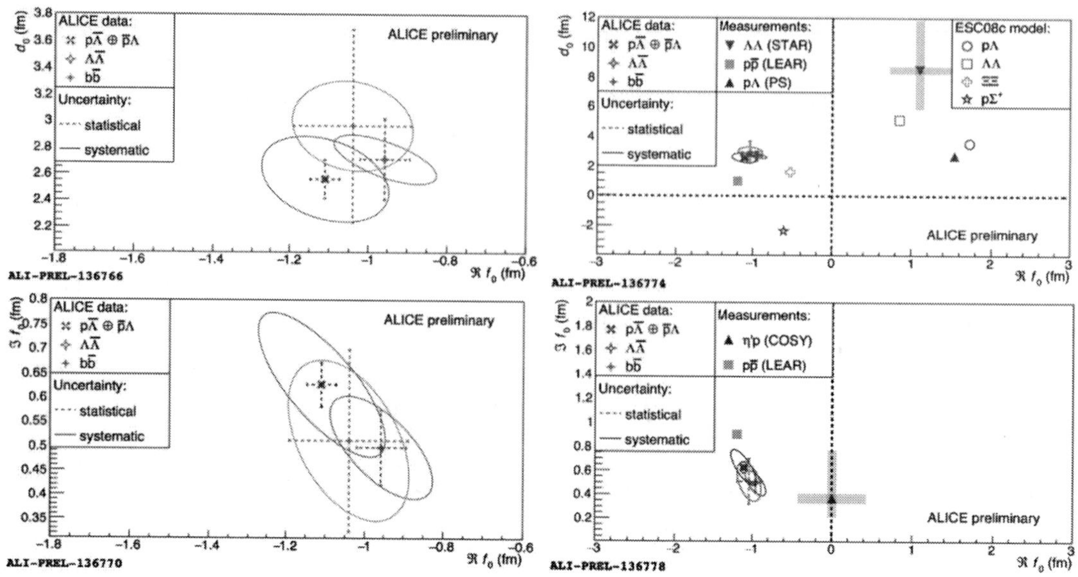

Figure 4. Left: Strong interaction parameters for different baryon-antibaryon pairs measured by ALICE. Right: Comparison of the obtained values with previous measurements and model predictions. Statistical and systematic uncertainties are shown with dashed and continuous lines respectively.

4 Conclusions

The correlation functions for all combinations of (anti)protons and (anti)lambda baryons were measured by ALICE in Pb-Pb collisions at 2.76 TeV and 5.02 TeV per nucleon pair. A new procedure was developed to fit simultaneously several systems, in centrality bins, taking into account residual correlations. The strong interaction parameters were measured for $p\bar{\Lambda}$, $\Lambda\bar{\Lambda}$ and heavier $b\bar{b}$ pairs.

The values of parameters which were obtained suggest that strong interaction for all $b\bar{b}$ pairs has similar properties and it is either repulsive, or it causes the formation of bound states. Inelastic channels are present for all $b\bar{b}$ pairs, including non-identical ones.

The presented results and their interpretation open further research paths in the direction of the bound states searching, or possible subtle spatial separation of baryons and antibaryons in heavy-ion collisions caused by a repulsive strong interaction.

References

[1] C. Patrignani et al. (Particle Data Group), Chin. Phys. C **40**, 100001 (2016) and 2017 update
[2] S. E. Koonin, Physics Letters B **70**, 43-47 (1977)
[3] A. Kisiel, Nuclear Physics A **956**, 385-388 (2016)
[4] R.Lednicky, V. L. Lyuboshits, Sov. J. Nucl. Phys., **35**, 770 (1982)
[5] ALICE Collaboration, Int. J. Mod. Phys., **A29**, 1430044 (2014)
[6] ALICE Collaboration, Phys. Rev. **C92**, 054908 (2015)
[7] A. Kisiel, H. Zbroszczyk, and M. Szymanski,, Phys. Rev. C, **89**, 054916 (2014)
[8] M. Szymański, CERN-THESIS-2016-172, (2016)
[9] J. Niedziela, CERN-THESIS-2017-274, (2017)
[10] C. Hategan, R. Ionescu, Rom. Rep. Phys., **60**, 515-528 (2008)

EPJ Web of Conferences **177**, 04009 (2018) https://doi.org/10.1051/epjconf/201817704009
AYSS-2017

Processing of higher count rates in Troitsk nu-mass experiment

Alexander Nozik[1,2,⋆] and *Vaslily* Chernov[1]

[1] *Institute for Nuclear Research RAS, prospekt 60-letiya Oktyabrya 7a, Moscow 117312*
[2] *Moscow Institute of Physics and Technology, 9 Institutskiy per., Dolgoprudny, Moscow Region, 141700, Russian Federation*

Abstract. In this article we give a short outline of current status of search for sterile neutrinos with masses up to 4 keV in "Troitsk nu-mass experiment". We also discuss major sources of systematic uncertainties and methods to lower them.

1 Introduction

Neutrino studies are one of the most motivating research areas in modern particle physics. While the particle itself was theoretically proposed in 1930-s, neutrino exact properties are still not known. Due to very small interaction cross-section, neutrino proved to be very hard to study experimentally. For now, one of the most interesting question is the absolute mass scale of neutrinos. The discovery of neutrino oscillations points to the fact that neutrino mass is not zero, but direct measurements still provide only upper limits for its mass of about 2 eV ([1]).

Additional point of interest is the search for so-called sterile neutrinos (special type of neutrino that does not participate in weak interaction). There is a number of indirect indications that such sterile neutrinos could exist ([2]), but there is almost no direct knowledge about specific region of masses or mixing parameters.

2 Troitsk nu-mass experiment

Troitsk nu-mass experiment detailed description as well as sterile neutrino program proposal could be found in [3]. The recent results for sterile neutrino with mass up to 1 keV are published in [4].

The main complication of search for sterile neutrino compared to search for electron neutrino mass performed previously in Troitsk nu-mass experiment ([1][5]) is the broader energy region. It produces two additional problems:

- The uncertainty caused by imprecise knowledge of electronics dead time.
- The uncertainty due to different number of events under the detector threshold for different retarding potentials.

The second problem is caused by back scattering effects discussed in [6] and could be solved only with precise simulations. In this article we will discuss the first problem and ways to solve it.

⋆e-mail: nozik@inr.ru

3 Reducing dead time uncertainty

The current DAQ hardware, used in the Troitsk nu-mass has a dead time of approximately 6.5 μs. It was sufficient for count rates of about 100 Hz used when the spectrum endpoint was investigated in search for electron neutrino mass ([1]). However, at count rates required to get sufficient statistics for sterile neutrino search (20 – 40 kHz) even small deviations of estimated dead time could produce huge differences in count rate (see systematic uncertainties estimation in [7]). Furthermore, since dead time was determined by analog readout system, it could be precisely estimated only experimentally.

3.1 Continuous digitization

One of possible solutions to reduce the impact of dead time uncertainty is to reduce dead time itself. It could be done either by using faster electronics, or by using advanced signal shape analysis to designate pileup events in off-line analysis. The first approach is currently impossible to implement on Troitsk nu-mass because shorter shaping times introduce additional systematic error caused by additional events from detector back scattering (discussed in [6]). Therefore it was decided to switch to continuous digitization without changing shaping time.

There were two candidates for digitization hardware: CAEN DT-5270 and Rudnev-Shilaev Lan10-12PCI. DT-5270 allowed for much faster sampling rate (31.25 MHz and higher), but it proved to be unsuitable for continuous digitization due to limitations of computer interface (USB channel bandwidth limit of 30 mb/s). Lan10-12PCI, on the other hand, provides much lower sampling rate and have direct PCI-express interface with the computer bus. It was discovered that optimal Lan10-12PCI performance could be obtained with sampling rate of 3.125 MHz. In that case live time losses due to data download to computer amounts only to about 13%.

Additional benefit of continuous digitization is the ability to simulate exact pile up effect using known signal shape. In order to do so one need to introduce an artificial signal and background noise distribution indistinguishable from real experimental data. In order to check the quality of analytic signal shape as well as background distribution, we plotted the distribution of χ^2 over degree of freedom for experimental and simulated data versus analytic signal shape in Fig. 1.

It could be clearly seen, that both shapes are very similar and produce very good average quality of fit.

Three different algorithms were considered for off-line analysis of the signal shape:

1. **Direct search for local peaks**. This is the simplest and fastest method. However, it has rather high miss rate for close events because one needs a clear local minimum between events to separate them.

2. **Search for local peaks with compensation**. This method is a modification of the first one. In this case when the peak is found, it is reconstructed using analytic signal shape and subtracted from the initial signal to reduce influence on subsequent peaks.

3. **Signal front fit**. In this case reconstruction of signal performed not by finding its peak, but by fitting its front with analytical shape.

The results of simulations of different methods effectiveness for count rate about 40 kHz are shown in table 1. Front fit method, as expected, gives better results for pileup recognition, but it produces a lot of false positive signals. Also it gives much worse performance then other two. Both methods 1 and 2 give similar effectiveness and performance. Further tests showed that using either of these methods produce an effective dead time caused by indistinguishable pile-up events of about 2.1 μs

https://doi.org/10.1051/epjconf/201817704009

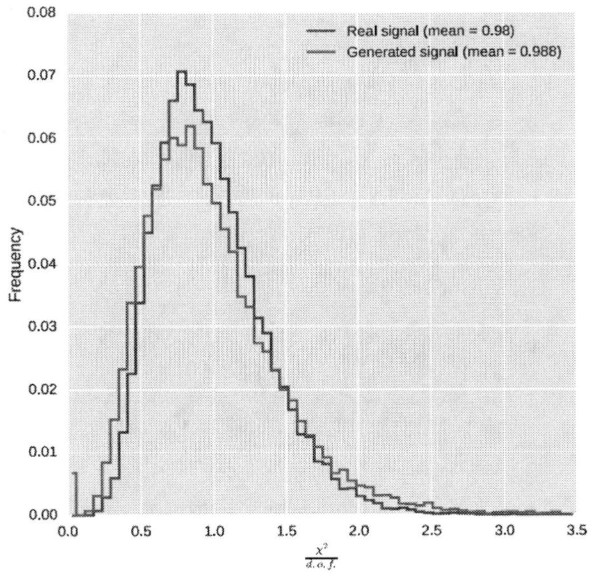

Figure 1: Distributions of χ^2 over degree of freedom for simulated and experimental events versus analytic signal shape.

	Direct	Direct with correction	Front fit
Computation time, s	1.222999	3.176033	195.431107
Total events	80526	80526	80526
Recognized events	72174	72244	79067
False negatives	1358	1269	467
False positives	6	4	1767
Pileup	6467	6470	2607

Table 1: Comparison of different peak recognition methods

which is almost 3 times better than the dead time of analog system. Furthermore, computation time of 1-3 s for 2 s measurement after optimization will allow to use these algorithms on-line to reduce stored data size.

3.2 Statistical approach

Another approach to dead time problem is to use information on events time distribution. The distribution of time delays between signals follows exponential distribution:

$$f(t) = \frac{1}{\tau}e^{-t/\tau}, \tag{1}$$

where τ is the average time between events and $1/\tau$ is the count rate.

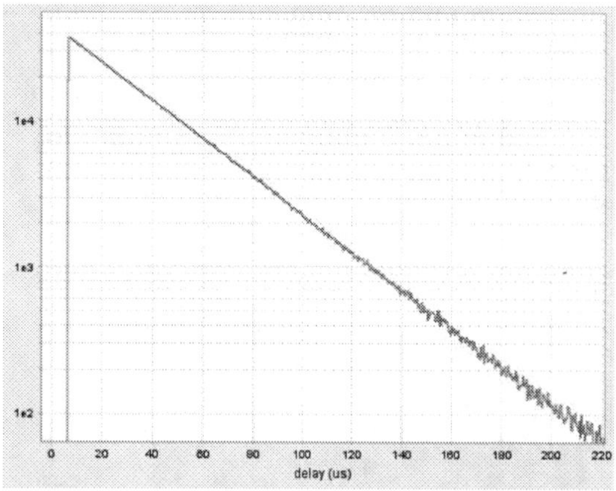

(a) Histogram of delays between events for dead time of 6.5 μs.

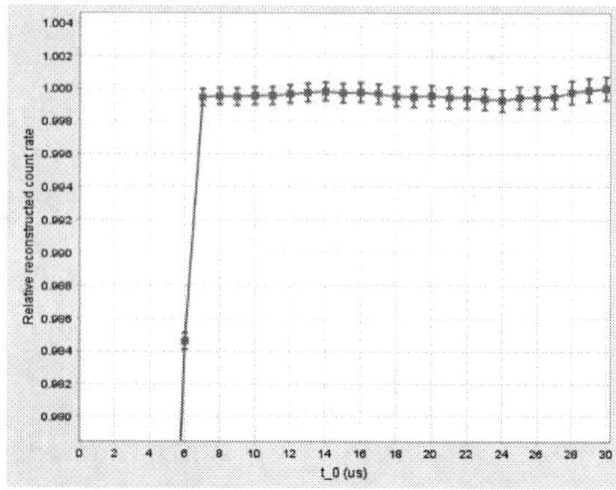

(b) Reconstructed count rate with error for different cut-off times t_0 (in μs.)

Figure 2: Statistical count rate reconstruction of simulated data.

Let us consider that there is sharp dead time margin t_0 that does not allow two events to be closer in time than this t_0. In this case the shape of time distribution above t_0 wont change, but there will be change in distribution normalization:

$$f^*(t) = \frac{e^{t_0/\tau}}{\tau} \begin{cases} e^{-t/\tau} & t \geq t_0 \\ 0 & t < t_0 \end{cases} \tag{2}$$

In order to avoid uncertainties from electronics dead time, one could select a t_0 slightly above the apparatus cutoff and estimate τ using modified distribution $f^*(t)$. Since t_0 is artificial, it could be selected with any needed precision and does not produce additional systematic error. Of course statistical error will be slightly increased because some events with delay below t_0 are excluded from analysis.

To get estimate of τ one can use the maximum likelihood method. The likelihood function for modified distribution could be written in a following way:

$$L(\tau) = \prod_{i=1}^{N} f(t_i) = \left(\frac{e^{t_0/\tau}}{\tau}\right)^N \exp\left(-\frac{1}{\tau}\sum_{i=1}^{N} t_i\right), \tag{3}$$

where t_i are experimental times between events greater then t_0 and N is total number of events with $t > t_0$. Designating $\sum_{i=1}^{N} t_i$ as T, the likelihood logarithm could be written in a following way:

$$\ln L(\tau) \sim -N \ln \tau - \frac{1}{\tau}(T - Nt_0) \tag{4}$$

The maximum of $L(\tau)$ corresponds to

$$\tau_0 = \frac{T}{N} - t_0. \tag{5}$$

The difference between regular exponential distribution and modified one is additional shift t_0. The statistical uncertainty for τ_0 is defined by the same formula as for regular one $\sigma_{\tau_0} = \tau_0/\sqrt{N}$. The ratio between statistical error for the whole data set and for cut one equals the square root of ratio between total number of events before and after the cut. For count rates up to 30 - 40 kHz and t_0 about 5-10 μs one can completely eradicate systematic error from the dead time, sacrificing only 10-15% of statistic sensitivity.

Fig. 2a shows the histogram of delays between subsequent events for a count rate of 30 kHz with a dead time of 6.5 μs. Fig. 2b shows count rate reconstructed via statistical procedure for different values of t_0. It could be clearly seen, that for t_0 above real dead time, the value of reconstructed rate suffers only slight fluctuations well within statistical errors and for higher values of t_0 errors are actually larger.

4 Conclusion

The combination of continuous digitization and statistical count rate estimation allow to significantly decrease the systematic error from dead time. The second primary source of uncertainty: events under the threshold, requires additional study and simulations. Algorithms used to distinguish pileup events could be optimized to be used on-line, thus allowing to use continuous digitization not only for one-pixel detector, but also for detectors with a lot of pixels without developing complicated storage system.

The work was supported by RFBR grant 17-02-00361 A. Authors would like to thank the whole Troitsk nu-mass group and particularly its leader Vladislav Pantuev.

References

[1] V.N. Aseev et al. (Troitsk), Phys. Rev. **D84**, 112003 (2011), 1108.5034
[2] R. Adhikari et al., Submitted to: White paper (2016), 1602.04816
[3] D. Abdurashitov, A. Belesev, A. Berlev, V. Chernov, E. Geraskin, A. Golubev, G. Koroteev, N. Likhovid, A. Lokhov, A. Markin et al., arXiv preprint arXiv:1504.00544 (2015)
[4] J.N. Abdurashitov et al., JETP Lett. **105**, 753 (2017), 1703.10779
[5] A.I. Belesev, A.I. Berlev, E.V. Geraskin, A.A. Golubev, N.A. Likhovid, A.A. Nozik, V.S. Pantuev, V.I. Parfenov, A.K. Skasyrskaya, J. Phys. **G41**, 015001 (2014), 1307.5687
[6] P.V. Grigorieva, A.A. Nozik, V.S. Pantuev, A.K. Skasyrskaya (2015), 1511.06129
[7] D.N. Abdurashitov et al., JINST **10**, T10005 (2015), 1504.00544

EPJ Web of Conferences **177**, 04010 (2018) https://doi.org/10.1051/epjconf/201817704010
AYSS-2017

Measurement of the Forward-Backward Asymmetry in the Drell-Yan Dilepton Production in Proton-Proton Collisions at the CMS Experiment at the LHC

V. Shalaev[1,★], *I.* Gorbunov[1,★★], and *S.* Shmatov[1,★★★]

[1] *JINR, Dubna, Russia*

Abstract. In this paper we review the results of a measurement of the forward–backward asymmetry of oppositely charged lepton pairs produced via $Z/\gamma^\star$ boson in pp collisions during the LHC Run 1 at $\sqrt{s} = 8$ TeV with integrated luminosity $19.1 fb^{-1}$ (2012). We also present our preliminary results obtained with Monte Carlo samples at $\sqrt{s} = 13$ TeV.

1 Introduction

The study of the lepton pairs production in the Drell-Yan processes [1], $q\bar{q} \to \gamma^*/Z \to l^+l^-$ (Fig. 1a), is of extreme importance: these processes allow us to verify the Standard Model (SM) predictions for the elementary particles interaction at the new energy region and to search for new physics beyond the SM [2, 3]. The presence in the Drell-Yan processes both the vector (g_V) and axial-vector (g_A) current results in the appearance of an asymmetry in the lepton angular distributions due to P-parity violation:

$$\frac{d\sigma}{d(cos\theta^*)} = \frac{4\pi\alpha^2}{3s}\left[\frac{3}{8}A\left(1 + cos^2\theta^\star\right) + Bcos\theta^\star\right], \tag{1}$$

where $\theta^\star$ is the angle of lepton emission with respect to the direction of the quark momentum in the center mass of the lepton pair system (Fig. 1b), A and B – the coefficients dependent on weak isospin and a charge of fermions in the initial states. In general the value of the asymmetry given by the coefficient B depends on the values of vector (g_V) and axial-vector (g_A) couplings which makes B sensitive to the presence of additional contributions to the SM processes. Therefore, the possible difference between the asymmetry value and the SM predictions can indicate the existence of processes beyond the SM namely new neutral gauge bosons [4], the internal structure of quarks and leptons [5], supersymmetric particles and extra dimension [6]. Moreover measurement of the exact value of the asymmetry could shed light on the new phenomena nature such as the origin of a new particle and which theoretical model predicts the observed effect. For example, one of the distinguishing features of the asymmetry is its dependence on the pseudorapidity of the new neutral gauge boson predicted by various models with extended gauge sector [7]. This will allow us to choose between different

★e-mail: Vladislav.Shalaev@cern.ch
★★e-mail: Ilya.Gorbunov@cern.ch
★★★e-mail: Sergei.Shmatov@cern.ch

theoretical scenarios even for particles having the same spin. Also, a measurement of the asymmetry value may provide information indicative of the correctness of our understanding of the quark and gluons distribution functions. Moreover, the measurement of the asymmetry in the Z–pole allows us to measure the effective weak mixing angle $sin^2\theta_W$ values [9].

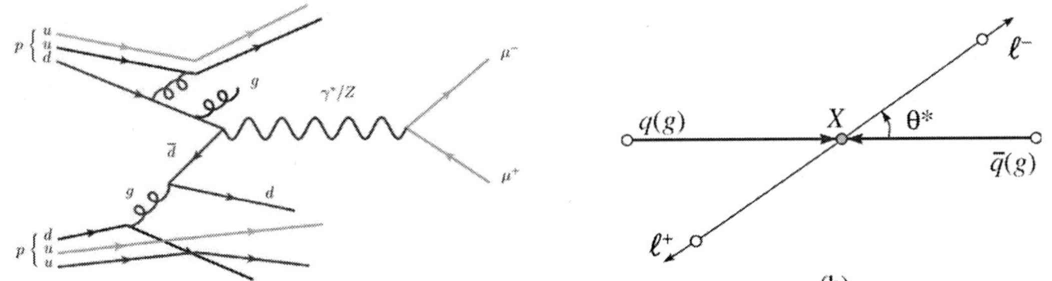

(a) (b)

Figure 1. (a) Dilepton production in the annihilation of a quark–antiquark pair, $q\bar{q} \rightarrow \gamma^*/Z \rightarrow l^+l^-$. (b) Definition of the angle θ^* in the decay of the exchanged particle X.

In this paper we review the results of a measurement of the forward–backward asymmetry of oppositely charged lepton pairs (both $\mu^+\mu^-$ and e^+e^-) in pp collisions during the LHC Run 1 at $\sqrt{s}$ = 8 TeV [10]. The data sample corresponds to an integrated luminosity of 19.7 fb^{-1} collected by the CMS detector(2012). We also present our preliminary results obtained with Monte Carlo samples at $\sqrt{s}$ = 13 TeV.

2 CMS Experiment

The central part of the experimental setup [8] is superconductive magnet solenoid. It is 13 m long and 6 m internal diameter, providing magnetic field of 3.8 T. The magnetic system contains a tracker and calorimetric systems. The tracker system covers the pseudorapidity range $|\eta| \leqslant 2.5$ and includes two types of silicon detectors: a pixel detector and a silicon strip detector. The pixel detector is located near by the beam interaction point and consists of 65 million Individual sensitive elements namely pixel size of 100×150 μm grouped by in three layers with radii of 4, 7 and 11 cm. The strip detector is assembled from 10 million silicon strips which are located at a distance up to 130 cm from the beam axis.

The solenoid is surrounded by detector consisted of three types of muon system such as drift tubes in the central part of the CMS ($|\eta| \leqslant 1.2$), cathode strip chambers in the front part ($1.9 \leqslant |\eta| \leqslant 2.4$) and the resistive plates chambers in the entire range of pseudorapidity of a muon system ($|\eta| \leqslant 2.4$). Joint use of the muon system and the tracker system allows us to measure the transverse muon momentum with a resolution between 1 and 5% for values of transverse momentum up to 1 TeV.

3 A_{FB} measurements with 8 TeV Data

In the experiment the asymmetry can be obtained by means of an independent measurements of the number of dilepton events for the different invariant mass bins in the forward and backward regions defined by the polar angle θ of the lepton with respect to the quark momentum direction:

$$A_{FB} = \frac{\sigma_F - \sigma_B}{\sigma_F + \sigma_B}, \quad \sigma_F = \int_0^1 \frac{d\sigma}{d(cos\theta^\star)} d(cos\theta^\star), \quad \sigma_B = \int_{-1}^0 \frac{d\sigma}{d(cos\theta^\star)} d(cos\theta^\star), \qquad (2)$$

where σ_F and σ_B are the total cross section for the events in the forward and backward directions. The direction "forward" ("backward") is given by a hemisphere with $cos\theta^\star \geqslant (<0)$, where θ is the angle between the outgoing lepton l^- and quark q in the lepton pair rest system. This definition suggests that the initial direction of the quark is known, but this condition is not valid in the case of experiments on pp-collisions. However, it is possible to fix the direction of the quark momentum by the two-lepton system emission direction since the pair is produced as a result of a predominantly valence quark and a sea antiquark annihilation and the momentum of the valence quark is on average larger than the momentum of the sea antiquark. The asymmetry defined this way allow us to study the spin structure of a muon pair produced in symmetric proton beams. We note that this approximation causes an increase in systematic uncertainties due to the errors in quark direction determination that must be taken into account while calculating asymmetry values.

To reduce uncertainties due to a nonzero transverse momentum of the quarks the Collins-Soper [11] coordinate system was used. In this system $\theta^\star_{CS}$ is the angle between the muon momentum and the z-axis which halves the angle between the quark and antiquark momentum. The A_{FB} was measured by analyzing the leptons angular distributions for each invariant mass bin. By using quantities written in the laboratory frame, one can represent the angle $\theta^\star_{CS}$ in the form

$$cos\theta^\star_{CS} = \frac{2\left(P_1^+ P_2^- - P_1^- P_2^+\right)}{\sqrt{Q^2\left(Q^2 + Q_T^2\right)}}, \tag{3}$$

where Q and Q_T are 4-momenta and transverse momentum of dilepton system, respectively; $P^i_{1,2}$ - the i component of lepton and antilepton 4-momenta; and $P_i^\pm = 2^{-1/2}\left(P_i^0 \pm P_i^3\right)$. In this case, the direction "forward" ("backward") is determined by the cosine sign: $cos\theta^\star_{CS} \geqslant 0 (<0)$.

The $cos\theta^\star_{CS}$ distributions for $\mu\mu$ events at $\sqrt{s} = 8$ TeV are presented in the Fig. 2. Only statistical uncertainties are shown. The stacked histograms represent the sum of the background contribution and the signal. The plots on the left (right) panels correspond to events with dilepton invariant mass 50 $< m <$ 60 GeV (133 $< m <$ 150 GeV). The Monte Carlo predictions are normalized to the integrated luminosity of the data.

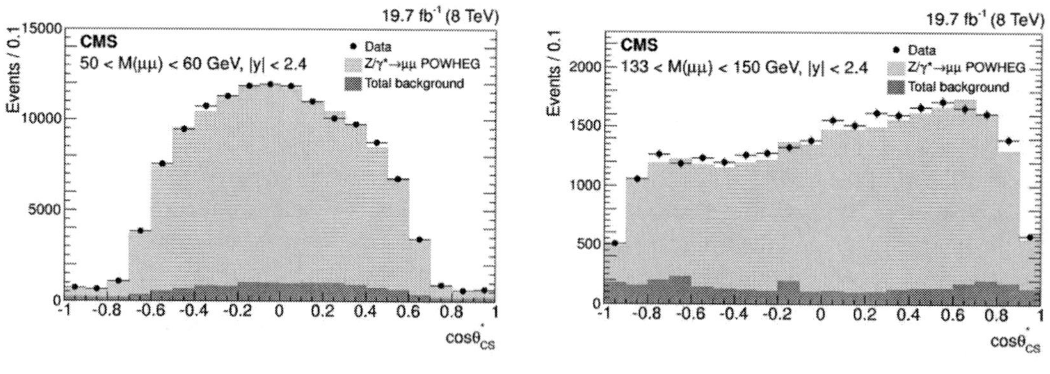

Figure 2. The $cos^\star\theta_{CS}$ distributions for $\mu\mu$ events at $\sqrt{s} = 8$ TeV [10].

The measurement of A_{FB} is performed for dilepton masses between 40 GeV and 2 TeV in five regions of absolute rapidity: 0-1, 1-1.25, 1.25-1.5, 1.5–2.4, and 2.4–5. The Fig. 3 shows the combined ($\mu^+\mu^-$ and e^+e^-) unfolded A_{FB} distributions in the four central rapidity regions. The statistical (thick vertical bar) and statistical plus systematics (thin vertical bar) uncertainties are presented. The measurements are compared with the next-leading-order prediction of Standard Model (POWHEG Monte Carlo generator [12]). The total uncertainties (considering the statistical, PDF, and scale uncertainties) in the powheg prediction are shown as shaded bands. The lower panel in each plot shows the difference of A_{FB} in data and prediction divided by the total uncertainty of data and prediction.

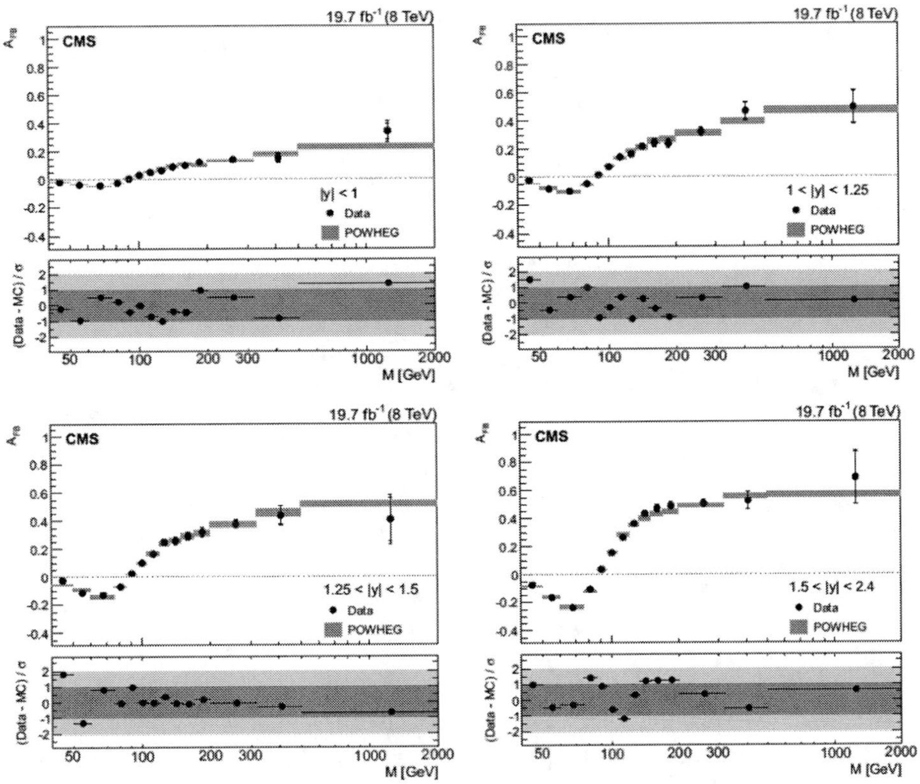

Figure 3. A_{FB} distributions for different invariant mass and rapidity at $\sqrt{s}$ = 8 TeV [10].

4 Monte Carlo Studies at 13 TeV

The LHC operation at $\sqrt{s}$ = 13 TeV, referred to as the LHC Run 2, started in 2015, and by the end of 2017 year the integrated luminosity collected by the CMS experiment reached 37.8 and 45.1 fb^{-1} for 2016 and 2017, respectively.

The first study of the forward-backward asymmetry A_{FB} at $\sqrt{s}$ = 13 TeV was performed with the Monte Carlo data. To compare data collected during the Run 2 with the SM predictions, a simulation of both signal and background events with the PYTHIA 8.0 Monte Carlo events generator was completed [13].

For example, the $cos\theta^{\star}_{CS}$ distribution of dimuon events is given at Fig. 4a. Fig.4b shows the dependence of the simulated asymmetry A_{FB} on the dimuon invariant mass. Presented distributions were obtained with the Monte Carlo generator without simulation of the particle response in the CMS detecting system.

A simulation of events signals obtained with PYTHIA 8 and passed to a full Geant4 simulation of the particle response in the CMS detecting systems, with using vital selection criteria and event reconstruction algorithms was performed but this results are not presented in this paper.

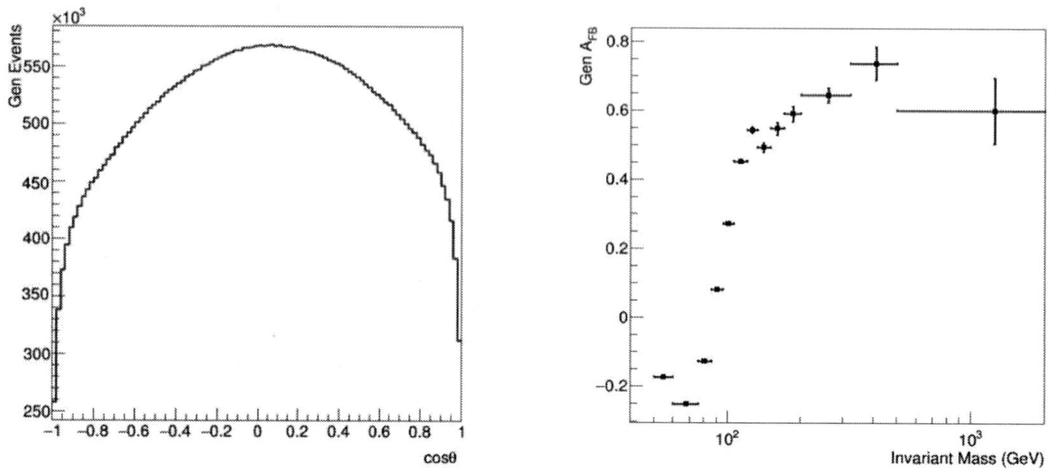

(a) (b)

Figure 4. The results of simulation within the CMS at $\sqrt{s}$ = 13 TeV (without detector response): (a) the $cos\theta^{\star}_{CS}$ distribution of dimuon events; (b) A_{FB} distribution for different dimuon invariant mass.

We would like to thank the management of the CMS theme at JINR for support.

References

[1] S. D. Drell and T.-M. Yan, Phys. Rev. Lett. **25**, 316 (1970).
[2] CMS Collab. (G. L. Bayatian et al.), J. Phys. G **34**, 995 (2007).
[3] V. A. Rubakov, Phys. Usp. **55**, 949 (2012); A. V. Lanev, Phys. Usp. **57**, 923 (2014); Phys. Atom. Nucl. **78**, 521 (2015); M. V. Savina, Phys. Atom. Nucl. **78**, 532 (2015); S. V. Shmatov, Phys. Atom. Nucl. **76**, 1106 (2013); Phys. Atom. Nucl. **78**, 509 (2015); Phys. Atom. Nucl. **79**, 266 (2016); in Proc. Joint RDMS CMS Seminar on Physics at LHC (2016), vol. 4, p. 13.
[4] D. London and J. L. Rosner, Phys. Rev. D **34**, 1530 (1986); J. L. Rosner, Phys. Rev. D **35**, 2244 (1987); Phys. Rev. D **54**, 1078 (1996); A. Bodek and U. Baur, hep-ph/0102160; F. Abe et al., Phys. Rev. Lett. **79**,2192 (1997).
[5] F. Abe et al., Phys. Rev. Lett. **79**, 2198 (1997).
[6] H. Davoudiasl, J. L. Hewett, and T. G. Rizzo, Phys. Rev. Lett. **84**, 2080 (2000).
[7] I. A. Golutvin, V. V. Pal'chik, M. V. Savina, and S. V. Shmatov, Phys. At. Nucl. **70**, 56 (2007).
[8] CMS Collaboration (S. Chatrchyan et al), JINST **3**, S08004 (2008).
[9] CMS Collaboration (S. Chatrchyan et al). Phys. Rev. D **84**, 112002 (2011).
[10] CMS Collaboration (V. Khachatryan et al), Eur. Phys. J. C **76**, 325 (2016).
[11] J. C. Collins and D. E. Soper, Phys. Rev. D **16**, 2219 (1977).
[12] S. Frixione, P. Nason, C. Oleari, JHEP **11**, 07 (2007); S. Alioli, P. Nason, C. Oleari, E. Re, JHEP **07**, 060 (2008); JHEP **06**, 043 (2010).
[13] T. Sjostrand et al, Comp. Phys. Comm. 191, 159 (2015)

EPJ Web of Conferences **177**, 04011 (2018) https://doi.org/10.1051/epjconf/201817704011
AYSS-2017

Test bench for measurements of NOvA scintillator properties at JINR

D.S. Velikanova,[1*] A.I. Antoshkin,[1] N.V. Anfimov,[1] O.B. Samoylov [1]
[1] DLNP, JINR, Dubna, Russia

ABSTRACT

The NOvA experiment was built to study oscillation parameters, mass hierarchy, CP-violation phase in the lepton sector and θ_{23} octant, via ν_e appearance and ν_μ disappearance modes in both neutrino and antineutrino beams. These scientific goals require good knowledge about NOvA scintillator basic properties.

The new test bench was constructed and upgraded at JINR. The main goal of this bench is to measure scintillator properties (for solid and liquid scintillators), namely α/β discrimination and Birk's coefficients for protons and other hadrons (quenching factors). This knowledge will be crucial for recovering the energy of the hadronic part of neutrino interactions with scintillator nuclei.

α/β discrimination was performed on the first version of the bench for LAB-based and NOvA scintillators. It was performed again on the upgraded version of the bench with higher statistic and precision level. Preliminary result of quenching factors for protons was obtained. A technical description of both versions of the bench and current results of the measurements and analysis are presented in this work.

INTRODUCTION

NOvA is a long baseline (810 km) neutrino experiment that studies various parameters of neutrinos. It consists of two main elements: the Fermilab accelerator complex (NuMI – Neutrinos at the Main Injector) as the neutrino source, and two detectors. For more details see [1].

The NOvA experiment uses two detectors: a 330 metric-ton near detector at Fermilab (near Chicago) and a much larger 14 metric-kiloton far detector in Minnesota just south of the U.S.-Canada border. The NOvA detectors are assembled with modules of extruded PVC (**P**oly**v**inyl **C**hloride) which is coated with titanium dioxide to enhance reflectivity. The technology is similar to that used commercially for garage doors and fencing. Modules of 32 cells are assembled by gluing two 16-cell extrusion together (See Fig. 1). Each cell has an interior size of 3.8 cm transverse to the beam direction and 5.9 cm along the direction. The length of the modules ranges from 15.6 m for the far detector to 4.2 m for the NOvA near detector.

What is a scintillator?

One of the methods of detecting ionizing radiation in experimental physics is to use scintillators. A charged particle passing through a substance deposits its energy within. Part of this energy goes to photon production. For some substances (scintillators) this portion is significant, so that generated light can be detected and measured by photosensors or photodetectors. The spectrum and intensity of the light signal depends on the intensity of the energy release, and the type of passing particle and attributes of the scintillator, see [2] and [3]. Many scintillators, depending on radiation length, are sensitive not only to charged particles, but also to gamma-radiation and neutrons. These are the main attributes

*darya-velikanova@mail.ru, Work supported by RFBR project No. 14-22-03090

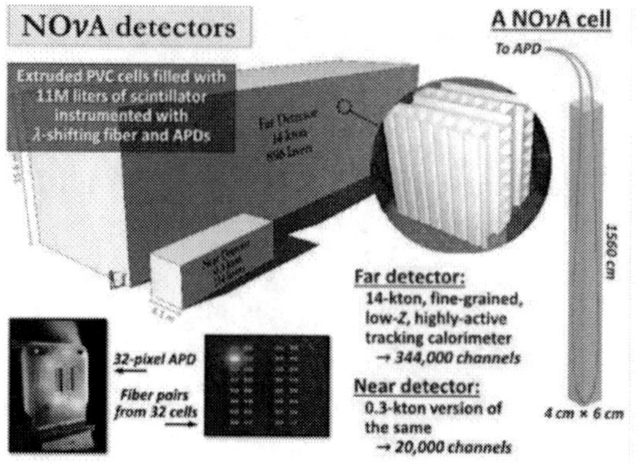

Figure 1. General design of NOvA detectors

of scintillators: light yield, the spectral composition of radiation, energy resolution, decay time, radiation resistance, radiation length and quenching factors – Birks Law.

Birks Law

For our work the most important attribute is the Quenching factor.

The scintillation response, S, of organic crystals depends on the nature and energy, E, of the incident ionizing particle, of residual range r. The specific fluorescence, $\frac{dS}{dr}$, is not in general proportional to the specific energy loss $\frac{dE}{dr}$. Quenching effect means that molecules are damaged by the particle and by the "excitons" produced by it. By considering this effect it is possible to show that:

$$\frac{dS}{dr} = \frac{A\frac{dE}{dr}}{1 + kB\frac{dE}{dr}} \tag{1}$$

Where A and kB are constants, which have been evaluated for scintillator from observations of S and E, and the range-energy data. The method used for evaluating the relative response is applicable to ionizing particles of any nature or energy, and also to the different organic scintillation crystals or liquids.

The quenching factor is a very important attribute of any scintillator. If we do not know the quenching factor, we cannot say what kind of particle passed through the scintillator and what primary energy it had. So, for example, in large neutrino experiments (Daya-Bay, JUNO, NOvA, DANSS) it is very important to know the precise value of the quenching factor for the scintillators.

METHODICAL ASPECT OF THE MEASUREMENT AND HARDWARE SETUP (SEE FIG. 3)

The first version of the test bench consists of a few components (see Fig. 2 and 3). First is the Black Box with three sections. The top and bottom sections are used like a muon telescope. A muon telescope helps one to divide the light from muons and radioactive sources. Light is measured by PMTs (Photomultiplier tubes) and digitized by ADCs

(Analog-to-digital converters). Inside the middle section it is possible to place scintillator samples and all necessary radioactive sources (for α/β ratio measurements). Second is the neutron source (Am-Be) with the NaI crystal. And the final component is scintillators and different radioactive sources (see Fig. 3).

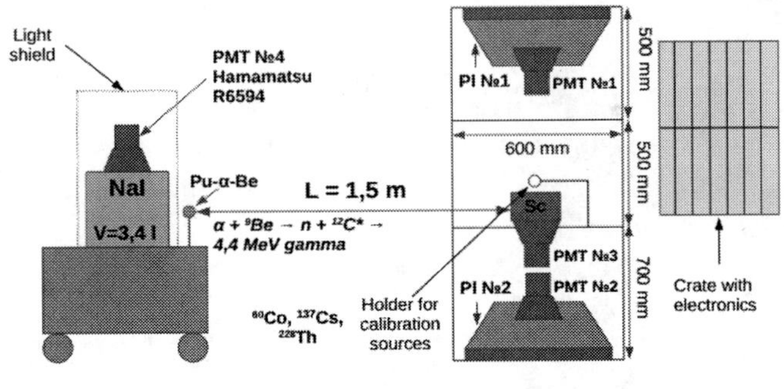

Figure 2. First version of the test bench

Firstly we have to measure the α/β ratio for a scintillator (see Fig. 3). This procedure is necessary for the calibration of our ADC and testing of our method. After that it is possible to measure the quenching factors for different hadrons. In the case of NOvA we want to measure Birk's coefficients for protons with energy in the range of 1-14 MeV. But it is not so easy to deliver protons with this energy inside the scintillator. We have to use a neutron source and measure the energy of recoil protons. Recoil protons are produced in the liquid scintillator by neutron-proton scattering events using a neutron generator ING-27 with a monochromatic energy distribution or a Am-Be source with continuous spectrum. Hence, the proton light output function can be determined from the position of the recoil proton edge in the pulse-height spectra produced by mono-energetic neutrons (see Fig. 5). In the case of a continuous spectrum it is possible to use the time-of-flight method for speed/energy determination. It is rather crucial to have a GEANT4 model for our bench and measurement process. With its help we can compare theoretical and measured spectra, which allows one to extract all necessary parameters.

EXPERIMENTAL SETUP OF THE UPGRADED TEST BENCH

With this hardware configuration (see Fig. 2) we had a problem with precision level. It is mostly connected with the electronics. So we decided to upgrade the test bench.

In addition to the existing part we added several components. The first is a Black metal box with sealing rubber and all necessary connectors. It is used to isolate the inner structure of the box from light. The second is a PMT with divider and teflon cuvette. Additional cuvette and PMT increases the statistic and allows us to cross-check the measurement process and result. The PMT and cuvette are put inside the box in a vertical orientation (see Fig. 6). The third is all support electronics like ADC (with higher sampling – it gives us required precision level), High-Voltage source and laptop with required software.

MEASUREMENTS

We repeated α/β ratio measurements using the same method (see Fig. 3) but on the new version of the bench with higher statistic and precision level. Also we performed a

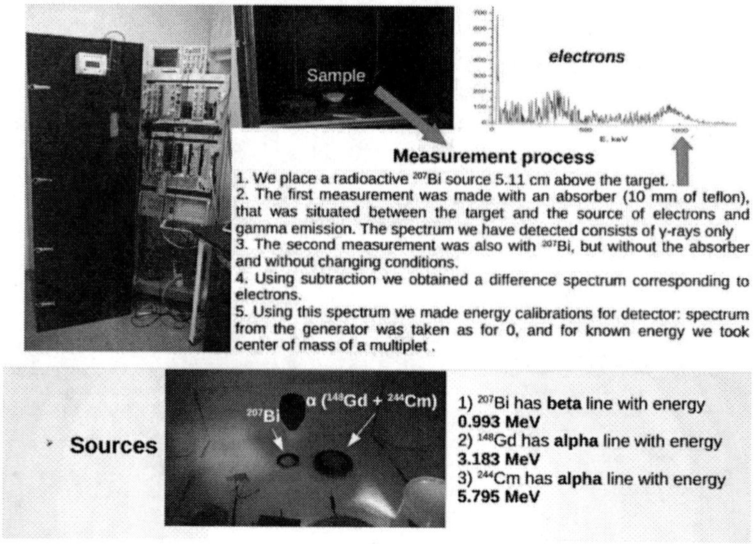

Figure 3. Radio-Chemical Lab bench and measurements. Radioactive sources

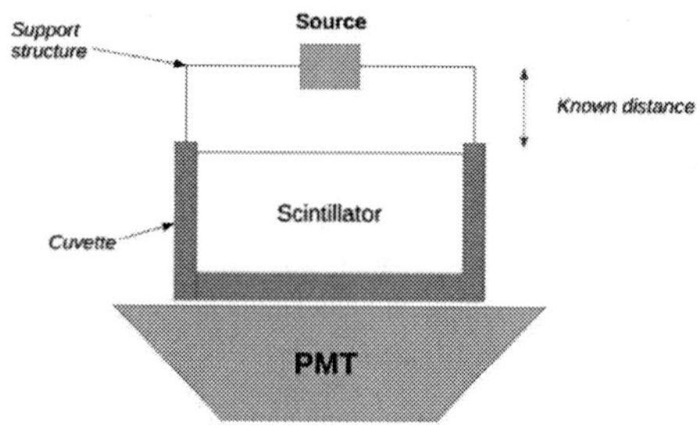

Figure 4. α/β **ratio (evaluation of the quenching factor)**

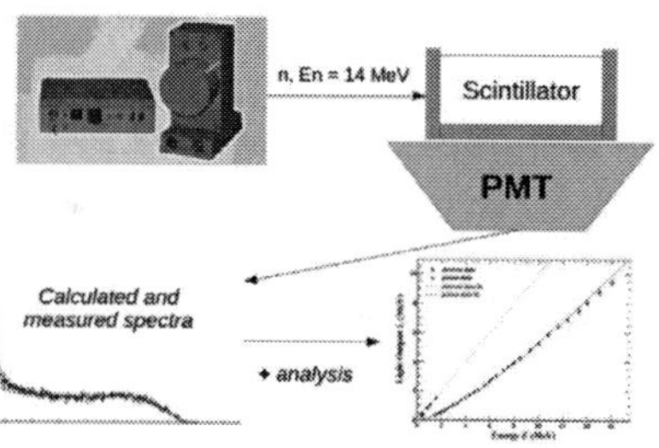

Figure 5. Measurement with a neutron source ING-27

measurement of the quenching factor for recoil protons using a Am-Be neutron source (see Fig. 2). The resulting curve is pretty well described by Birks Law (1 and Fig. 8).

Results

Preliminary results are as follows (see 7 and 8):

For the Gd+Cm source quenching factors for LAB-based scintillator (α) are equal to **22,2** for Gd (energy 3,183 MeV) and to **16,5** for Cm (energy 5,795 MeV). For the NOvA scintillator they are equal to **23,58** and **17,6**. Using these numbers and (1) one can evaluate the real energy of an α particle.

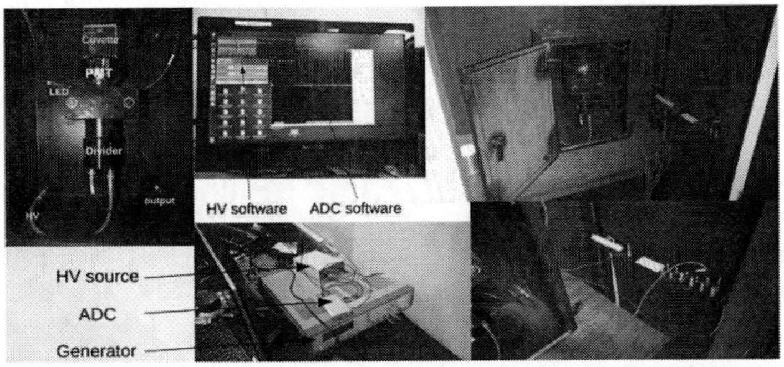

Figure 6. General design of new version of the bench

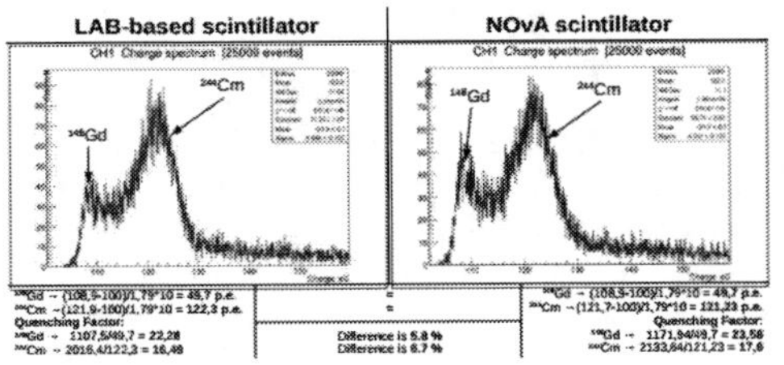

Figure 7. Resulting spectra for α sources

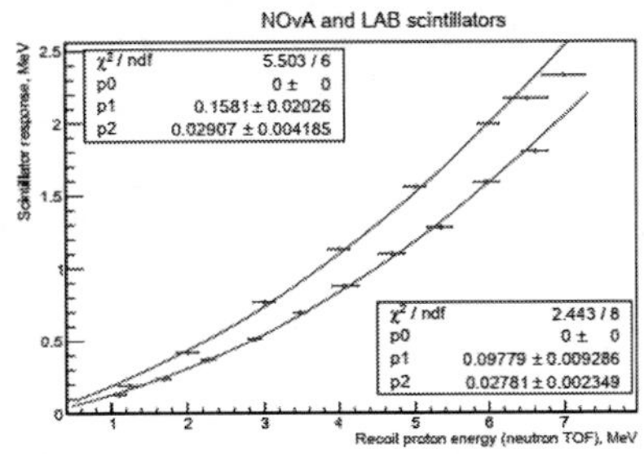

Figure 8. LAB-based(blue) and NOvA(red) light response for recoil protons

CONCLUSION

We performed measurements with α and β sources on the test bench at the Radio-Chemical Lab. Due to a problem with precision level we decided to upgrade the test bench. In addition to the existing part we added a new sensitive part and better electronics. The resolution ($\frac{\delta E}{E}$) of the upgraded bench has been increased by 30%. We performed the measurement (α/β evaluation) for LAB-based scintillator and NOvA scintillator on the upgraded bench. We performed measurements with a Am-Be neutron source and got the first resulting curve. This curve is pretty well described by Birks Law.

ACKNOWLEDGMENTS

We would like to thank the NOvA Collaboration for providing us with unpublished and public information.

REFERENCES

[1] Site of the NOvA experiment. URL: https://www-nova.fnal.gov/

[2] Birks J. B. The Theory and Practice of Scintillation. Counting, Pergamon Press, 1964.

[3] Kharzheev Yu. N. Scintillation counters in modern experiments at high-energy physics. Physics of elementary particles and atomic nuclei. Issue 4. Dubna : JINR, 2015.

The Unified Database for BM@N experiment data handling

Konstantin Gertsenberger[1,*,#], *Oleg* Rogachevsky[1]

[1]Laboratory of High Energy Physics, Joint Institute for Nuclear Research, Russia, Dubna

Abstract. The article describes the developed Unified Database designed as a comprehensive relational data storage for the BM@N experiment at the Joint Institute for Nuclear Research in Dubna. The BM@N experiment, which is one of the main elements of the first stage of the NICA project, is a fixed target experiment at extracted Nuclotron beams of the Laboratory of High Energy Physics (LHEP JINR). The structure and purposes of the BM@N setup are briefly presented. The article considers the scheme of the Unified Database, its attributes and implemented features in detail. The use of the developed BM@N database provides correct multi-user access to actual information of the experiment for data processing. It stores information on the experiment runs, detectors and their geometries, different configuration, calibration and algorithm parameters used in offline data processing. An important part of any database – user interfaces are presented.

1 The BM@N experiment at NICA

Relativistic heavy ion collisions provide the unique opportunity to study nuclear matter under extreme density and temperature. At these extreme conditions, the following features of strongly interacting matter can be studied: the equation-of-state (EoS) of strongly interacting matter at high temperatures and high baryon densities; the microscopic structure of strongly interacting matter as a function of temperature and baryon density; the in-medium modifications of hadrons, which might provide information on the onset of chiral symmetry restoration. Theoretical models, however, suggest different possible scenarios to describe features of strongly interacting matter, so that new experimental data are needed in order to disentangle different theoretical predictions [1].

The research program on heavy-ion collisions at the Nuclotron of the Joint Institute for Nuclear Research includes the following topics: investigation of the reaction dynamics and nuclear EoS, study of the in-medium properties of hadrons, production of (multi)-strange hyperons at the threshold and search for hyper-nuclei. According to the program, the Nuclotron-based ion collider facility (NICA) is being constructed as an ion accelerator complex for the range of collision energy from 4 to 11 GeV (for Au^{79+}).

[*] email: gertsen@jinr.ru

[#] on behalf of the BM@N collaboration

A new generation of heavy-ion physics experiments is expected to emerge at NICA. The BM@N experiment is a fixed target experiment at extracted Nuclotron beams to study collisions of particles and ions with fixed targets at the energies up to 6 GeV per nucleon. It is proposed to study elementary reactions and cold nuclear matter, properties of dense baryonic matter in heavy ion collisions. This information could be obtained from the measurement of particle yields and ratios, transverse momentum spectra, rapidity and angular distributions, as well as fluctuations of event properties and correlations between particles. Production of (multi)-strange hyperons and search for hyper-nuclei will be studied also.

The detector should measure particles produced in collisions with high efficiency and estimate the parameters with high precision for the full study of hot matter that is why the BM@N is constructed as a set of different types of subdetectors. It combines high precision track measurements with time-of-flight information for particle identification and uses total energy measurements for the analysis of the collision centrality. The target is located inside the large-acceptance dipole magnet with the magnetic field up to 1.2 T. The BM@N setup divides detectors into inner tracker, such as Silicon detectors and GEMs, and outer tracker, such as Drift Chambers and Time-of-Flight detectors, to measure particles with both low and high momentum. The conceptual structure of the detector is shown in figure 1.

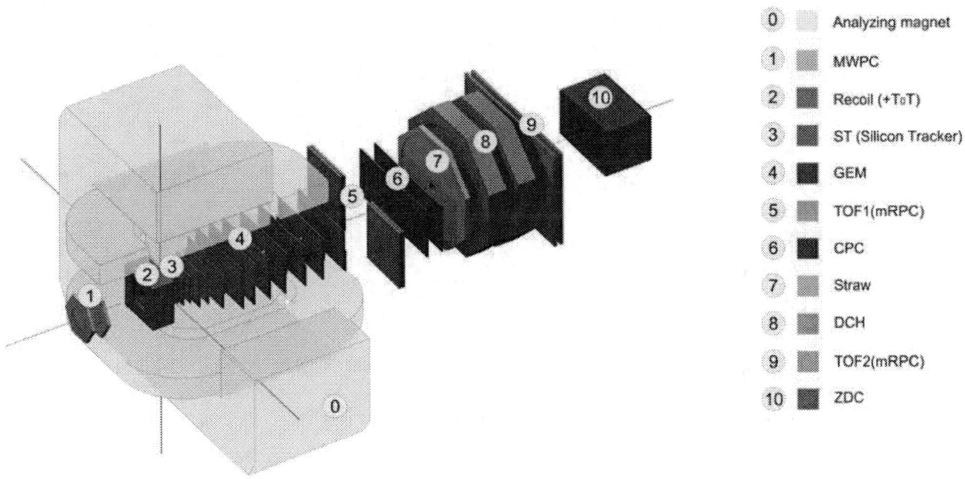

Fig. 1. Three-dimensional scheme of the BM@N facility.

2 Purposes of the BM@N database development

Since 2015, four technical runs of the BM@N experiment have been performed with deuteron and carbon beams collided with various targets, such as carbon, copper and lead, at the energies 3.5 – 5 GeV per nucleon. About 1000 raw data files have been obtained and converted to the ROOT format [2]. Multiple parameter data of the BM@N experiment used in offline data processing have been stored in files of different formats, such as binary data, text, CSV and XML.

The first runs of the BM@N experiment showed the following main problems concerned with the BM@N file storing approach previously used:

• The usage of multiple files and different formats, duplication of information in different files lead to data isolation, redundancy and inconsistency.
• Sequential, non-indexed file access and search are not efficient.
• No mechanism exists for relating data between these files.
• It is difficult to access and manipulate the data.

• Uncontrolled concurrent access by multiple users also often leads to inconsistencies.

Today the use of databases is a prerequisite for qualitative management and unified access to data of modern high-energy physics experiments. The Unified Database designed as a central data storage of the BM@N experiment for offline data analysis offers solutions to the above problems and provides unified access and data management for collaboration members, correct multi-user data processing, ensuring the actuality of the information being accessed, data consistency and integrity, excluding the multiple duplication and use of outdated data. Furthermore, automatic daily backup of the stored data is provided to ensure that data of the experiment will not be lost due to software or hardware failures.

3 The Unified Database of the BM@N experiment for offline data processing

The entity-relationship diagram of the Unified Database implemented on PostgreSQL Database Management System is shown in figure 2.

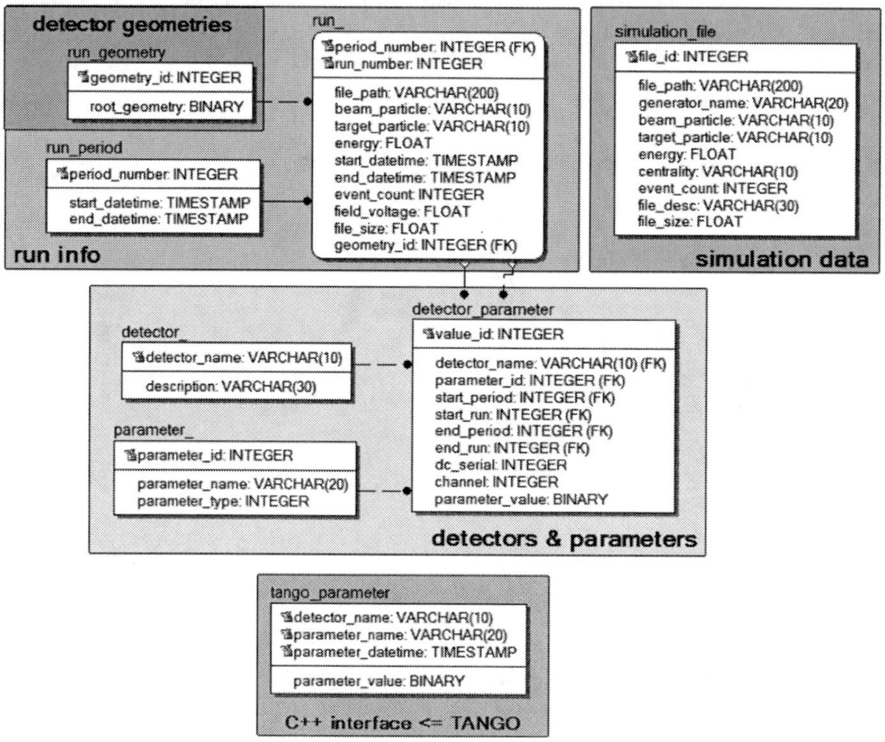

Fig. 2. The Unified Database diagram.

The following parts of the database can be highlighted: storing simulation data, detector geometries, information on experiment runs and experimental data files, detectors and parameters with their values. Hardware parameters of the BM@N Slow Control System "Tango" do not correspond to any table of the Unified Database, but a special C++ interface was implemented. During the design, the structure of the Unified Database was simplified as much as possible.

The parameters being stored in the Unified Database can be classified into 4 groups. Configuration data is concerned with the detector running mode, as well as some programmable parameters for front-end electronics. Calibration data describes the calibration and alignment of the subdetectors. Usually these quantities are evaluated by running offline algorithms. Parameter data presents the state of the detector subsystems. Algorithm data is used to control the way the offline algorithms operate. It includes, for example, cuts for selection and noise channels excluded from the data processing.

To support the BM@N experiment, the BmnRoot [3] software is developed for event simulation, reconstruction of experimental or simulated data and following physical analysis of particle collisions with a fixed target. The developed database integrates with the BmnRoot framework to use the database in offline raw data processing, reconstruction and physical analysis tasks. Also, it can be noted that BmnRoot converts BM@N *.geo* and *.root* subdetector geometry files to a single ROOT geometry file used to visualize BM@N events on the screen and stored in the developed database.

To process the experiment data via the BmnRoot environment, a C++ database interface was implemented as a set of C++ class wrappers for all database tables (figure 2) with many specific functions. It allows for the access to the data of the Unified Database without any SQL statements. The developed classes correspond to the database tables and manage information on the experiment runs, detector geometries, detectors, their parameters and parameters' values, and generated simulation files. Among functions for creating and changing the objects and their attributes, *Search* function should be noted which was developed to select runs and detector parameters for analysis.

Some additional classes of the C++ database interface were developed. *UniDbConnection* serves to open and close database connections, *UniDbGenerateClasses* – to generate automatically skeletons of the class wrappers for all tables of the database, *UniDbParser* – to parse existing data of the experiment from HTML, Excel and text files and to write them to the Unified Database. Also, the interface includes ROOT macros to execute tasks presented by the above classes, numerous examples to work with the BM@N database and documentation. *UniDbTangoData* interface is used to get hardware data from the Tango control system. Figure 3 presents an example of using the interface to get BM@N magnet field values for selected runs.

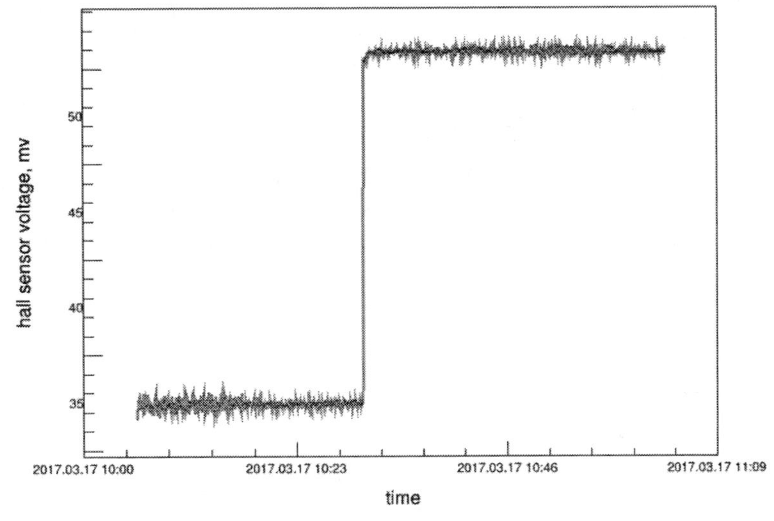

Fig. 3. BM@N magnet field from Run 1886 to Run 1888.

The overview scheme of the Unified Database used in offline processing is shown in figure 4. The BmnRoot environment uses the C++ database interface to get data for detector simulation, raw data processing, event reconstruction and physics analysis tasks. Also, the BmnRoot has an access to the Tango hardware data by the developed interface. The Web interface of the BM@N database is under development now to simplify reading and changing the experiment data for users over the Web.

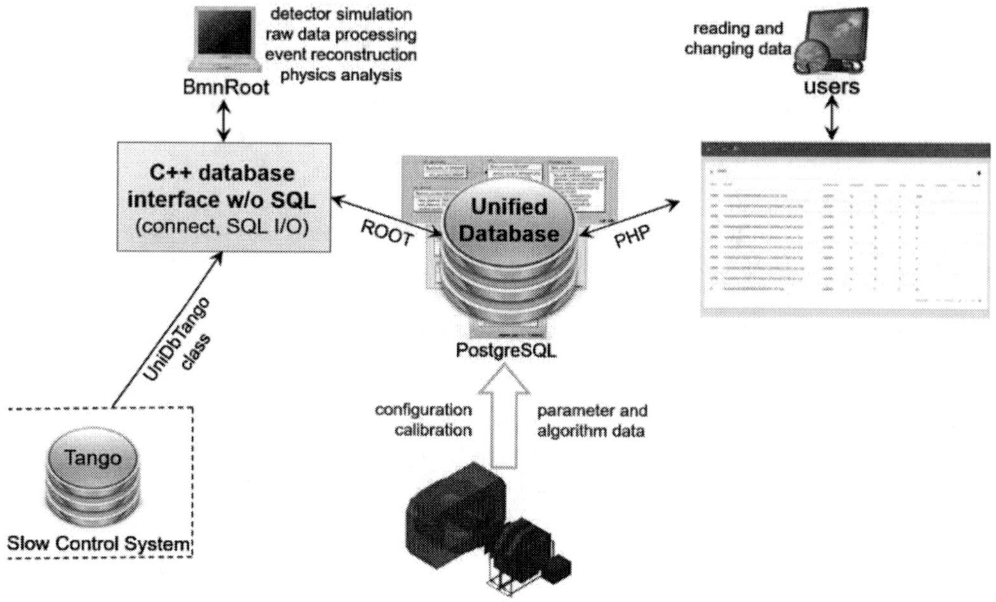

Fig. 4. The overview scheme of the BM@N database usage.

4 Conclusions

The BM@N database has been developed based on PostgreSQL and deployed in the cloud of the Laboratory of Information Technologies, JINR. The Unified Database includes existing data of conducted BM@N runs from multiple files (binary, text, CSV, HTML and XML). The developed database is a comprehensive data storage of the BM@N experiment, which provides unified user access to actual information of the experiment: experiment runs, detectors and their geometries, configuration, calibration and algorithm parameters. The C++ database interface has been implemented as a part of the BmnRoot environment to use it for offline raw data processing, reconstruction and physical analysis tasks. The C++ Tango interface has been developed to get hardware data from the BM@N Slow Control System. The Web interface is under development now to simplify reading and changing the experiment data for users. The detailed information on the Unified Database and user interfaces is presented in the «Databases» section on our technical Web-site *mpd.jinr.ru*. A future version of the Unified Database can be used for the MPD/NICA experiment.

References

1. BM@N Collaboration, *BM@N Project. Studies of Baryonic Matter at the Nuclotron* (http://bmnshift.jinr.ru/wiki/lib/exe/fetch.php?media=bmnproject_2016.pdf, 2016)
2. CERN, *ROOT – An Object-Oriented Data Analysis Framework* (CERN, 2013)
3. K. Gertsenberger, S. Merts, O. Rogachevsky, A. Zinchenko, Eur. Phys. J. A. 52, 8 (2016)

Integrated cloud infrastructure of the LIT JINR, PE "NULITS" and INP's Astana branch

Yelena Mazhitova[1,2,*], *Nikita* Balashov[1], *Aleksandr* Baranov[1], *Nikolay* Kutovskiy[1], and *Roman* Semenov[1,3]

[1] Laboratory of Information Technologies, Joint Institute for Nuclear Research, 141980, 6 Joliot-Curie St., Dubna, Moscow region, Russia
[2] Institute of Nuclear Physics, 050032, 1 Ibragimova street, Almaty, Kazakhstan
[3] Plekhanov Russian University of Economics, 117997, 36 Stremyanny per., Moscow, Russia

Abstract. The article describes the distributed cloud infrastructure deployed on the basis of the resources of the Laboratory of Information Technologies of the Joint Institute for Nuclear Research (LIT JINR) and some JINR Member State organizations. It explains a motivation of that work, an approach it is based on, lists of its participants among which there are private entity "Nazarbayev University Library and IT services" (PE "NULITS") Autonomous Education Organization "Nazarbayev University" (AO NU) and The Institute of Nuclear Physics' (INP's) Astana branch.

1 Service Implementation

The JINR cloud service [1] is based on an open-source IaaS system OpenNebula [2]. Two main components of the system can be marked out:

- front-end node (FN) which contains the system core and user interfaces to interact with the service;
- cluster nodes (CNs): the physical servers which host the users' virtual machines (VMs) and containers (CTs).

Two user interfaces are available to access the service:

- command line interface (CLI);
- web-based graphical user interface "Sunstone".

Figure 1 shows the interactions between the cloud service components.

* Corresponding author: emazhitova@jinr.ru

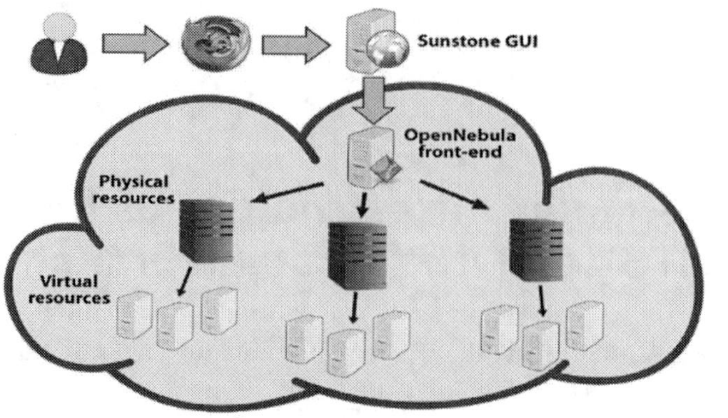

Fig.1. JINR cloud service structure scheme showing interactions between its components.

Currently the service uses two virtualization technologies to provide virtual instances (VIs):

- OpenVZ containers [3] (an operating system-level virtualization);
- KVM virtual machines [4] (full hardware virtualization).

The reason why two different virtualization technologies are used is to better fit the variety of the tasks: OpenVZ CTs are lightweight and fast but they are bound to use the hosts operating system kernel, while KVM VMs support any type of operating systems inside but have higher overhead. Originally OpenNebula had no OpenVZ containers support but extensible and modular architecture allowed us to add such support by implementing the custom driver.

The JINR cloud has two types of CNs:

- servers with two mirrored disk drives (RAID1) used to host highly reliable VIs;
- servers with one disk used for educational, research or test VIs.

The virtual instances can be accessed either with the use of an rsa/dsa-key or a password. A plugin implementing Kerberos authentication was developed for user authentication in Sunstone. To increase a security of the transmissions between the Sunstone and user's browsers an SSL encryption is used.

The architecture of the JINR cloud is shown on the Figure 2.

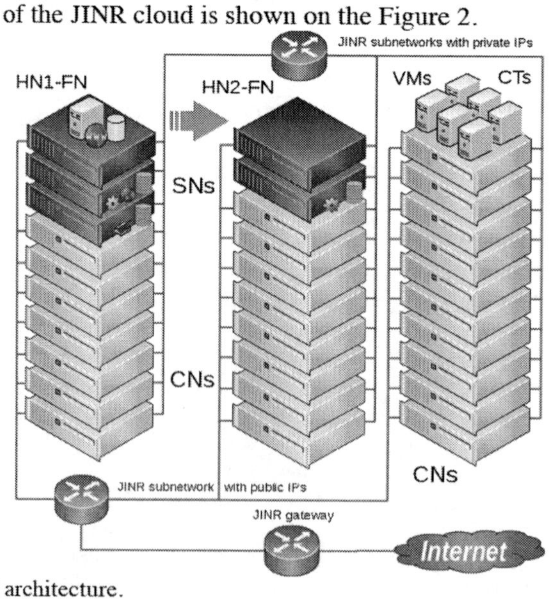

Fig.2. The JINR cloud architecture.

OpenNebula front-end (FN) consists of the following components: a core, a scheduler, a MySQL backend, a User (Web-GUI, CLI) and API (rOCCI) interfaces, OneGate and OneFlow services. The FN is deployed in OpenVZ container (CT) running on distributed replicated block device (DRBD) partition which is shared between two physical servers (marked on the Fig. 2 as HN1-FN, HN2-FN). They track each other's status with help of Heartbeat service. HN1-FN and HN2-FN are connected to different UPSes to increase the reliability. All VI images and data are stored on the distributed network file system (DNFS) based on LizardFS with automatic data replication. To split external and internal traffic all CNs are connected to public and private networks whereas storage nodes (SNs) – to the private one only.

2 Clouds integration

To join resources of partner organizations from the JINR Member States for solving common tasks as well as to distribute a peak load across them, a cloud bursting driver [5] was developed by the JINR cloud team. It allows one to integrate the JINR cloud with partner clouds either OpenNebula-based one (and in that case it is possible to enable real time external cloud resources monitoring) or any other cloud platform which supports Open Cloud Computing Interface (OCCI), but then statically defined values of external cloud resources are only possible. The driver is based on ruby implementation of OCCI (rOCCI) and OpenNebula XML-RPC interface. The reason why both interfaces are used at the same time is that OCCI allows one to better reuse the code, since it is provided by other popular cloud platforms like OpenStack, while OpenNebula XML-RPC is OpenNebula-specific and cannot be used for integration with other platforms. At the same time, OCCI does not provide monitoring operations, while OpenNebula XML-RPC does. It was decided to use rOCCI for management operations and XML-RPC for monitoring. The driver's scheme is shown in the figure 3.

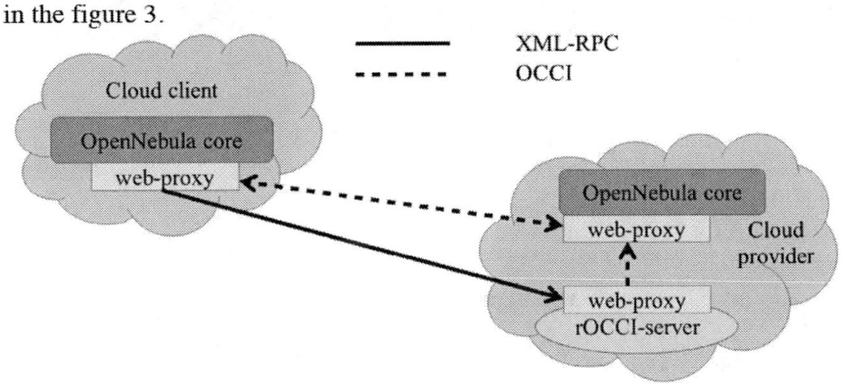

Fig.3. Cloud bursting driver.

The JINR cloud is a core of the distributed cloud infrastructure. At the moment, the integration of the clouds of the JINR Member State organizations is at different stages, in particular:

- Institute of Physics of Azerbaijan National Academy of Sciences – IP (Baku, Azerbaijan, integrated);
- Bogolyubov Institute for Theoretical Physics of the National Academy of Sciences of Ukraine – BITP (Kiev, Ukraine, integrated);
- Plekhanov Russian University of Economics – PRUE (Moscow, Russia, integrated);

- Research Institute for Nuclear Problems of Belarusian State University (Minks, Belarus, integrated);
- Nazarbaev University (Astana, Kazakhstan, deployment);
- Institute of Nuclear Physics (Astana, Kazakhstan, deployment);
- Institute for Nuclear Research and Nuclear Energy (Sofia, Bulgaria, deployment);
- Sofia University (Sofia, Bulgaria, deployment);
- Yerevan Physics Institute (Yerevan, Armenia, deployment);
- Georgian Technical University (Tbilisi, Georgia, deployment);
- University "St. Kliment Ohridski" (Bitola, Macedonia, negotiations).

Figure 4 shows on the map the locations of such distributed cloud infrastructure participants.

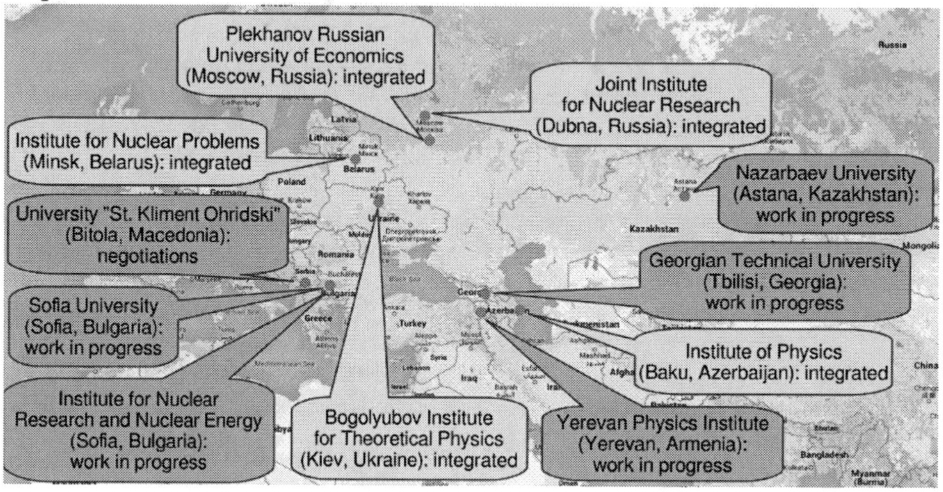

Fig.4. Map with the distributed cloud infrastructure participants.

An activity related to the integration of PE "NULITS" and INP's Astana branch resources into the distributed cloud infrastructure is ongoing within framework of 3 years project with priority funding in the period 2017-2019.

Working plan for the integration of the JINR, PE "NULITS" and INP's Astana branch clouds is presented below:

1. a research of possible solutions for joining resources of LIT JINR and AO NU (2017);
2. purchase of an equipment and building of a testbed based on the selected solution (2017);
3. conducting training events for Kazakh specialists (2017);
4. integration the resources of LIT JINR, PE "NULITS" and INP's Astana branch into a single distributed information and computing environment (DICE) (2018);
5. testing of the DICE (2018);
6. involvement of potential participating organizations in the project (L.N. Gumilyov ENU, E. Buketov State University, Al-Farabi Kazakh National University, private companies) (2019).

3 Conclusion

Studies were carried out on the possible ways of integrating the LIT JINR, PE "NULITS", INP's Astana branch resources. The cloud bursting approach was considered as an optimal one. A set of training events for specialists from Kazakhstan on the OpenNebula deployment and DICE resources usage was organized and conducted. As a prototype a testbed at PE "NULITS" was installed on the currently available equipment. It is expected that resources integration will help to speed up a new scientific results and achievements in various domains the JINR and its Member States in common and Kazakhstan in particular are involved in as well as it will become a basis for conducting training events for students, researches and IT-specialists to use modern information technologies for storing, processing and analyzing large volumes of scientific data.

References

1. Baranov A.V. et al. JINR cloud infrastructure evolution, Particles and Nuclei Letters, **13**, 5, 672-675, DOI: 10.1134/S1547477116050071 (2016).
2. OpenNebula web-portal. URL: http://opennebula.org.
3. OpenVZ Virtuozzo Containers Wiki. URL: https://openvz.org.
4. Kernel Virtual Machine web-portal. URL: //www.linux-kvm.org.
5. Baranov A.V. et al. Approaches to cloud infrastructures integration, Computer Research and Modeling, **8**, 3, 583-590 (2016).

EPJ Web of Conferences **177**, 05003 (2018) https://doi.org/10.1051/epjconf/201817705003
AYSS-2017

DataForge: Modular platform for data storage and analysis

Alexander Nozik*

[1] *Institute for Nuclear Research RAS, prospekt 60-letiya Oktyabrya 7a, Moscow 117312*
[2] *Moscow Institute of Physics and Technology, 9 Institutskiy per., Dolgoprudny, Moscow Region, 141700, Russian Federation*

Abstract. DataForge is a framework for automated data acquisition, storage and analysis based on modern achievements of applied programming. The aim of the DataForge is to automate some standard tasks like parallel data processing, logging, output sorting and distributed computing. Also the framework extensively uses declarative programming principles via meta-data concept which allows a certain degree of meta-programming and improves results reproducibility.

1 Introduction

Modern experimental research, both in the field of particle physics, and in other fundamental and applied problems, is usually associated with the acquisition, storage and analysis of large amounts of data. In all three stages of working with data (obtaining, storing and processing) in modern conditions, the maximum level of automation is required in order to both save the working time of highly qualified employees, and reduce errors caused by manual work with data. To provide automation, in turn, it is necessary to have specialized software that would allow, on the one hand, to solve the entire set of tasks for working with data, on the other - it would not require high programmer skills from users.At present, there are a few data analysis platforms, but most of them were created in the early 2000s or even earlier, and as a consequence, they do not use all the variety of software tools that have emerged since then. In addition, almost all existing tools were created for a highly specialized task and as a result are poorly suited to work with the entire chain: acquisition, storage and analysis.

In particle physics the most used framework for data storage and analysis is ROOT ([1]). The system is developed and supported by CERN and is used in most of modern large-scale experiments. The framework is monolithic, implements obsolete architecture and also has a lot of minor and major flaws, discussed in the literature. There were a few attempts to create a better alternative for ROOT like JAS3([2]) or DataMelt, but none managed to solve all of the problems. Software solutions for data acquisition are usually commercial projects with closed code, which makes it difficult to create a more or less universal system. The DataForge is a project intended to address these problems, namely provide modern lightweight framework to perform the full cycle of data processing without a lot of programming expertise.

*e-mail: nozik@inr.ru

2 Basic concepts

DataForge ([3]) introduces a several major principles which serve both as ideological guide for developers and design hints for users.

The main one is the principle of **metadata processor**. According to this principle, during the processing of data, only data (immutable initial data in any textual or binary format) and metadata (user-defined tree of values) are allowed to be used as input, meaning no user instructions (scripts) or manually managed intermediate states are possible. This means that user is forced to use declarative approach to describe what should be done instead of imperative one. Of course such restriction limits user capabilities, but on the other hand, processing description in the form of metadata could be a subject of programming itself. One can automatically transform metadata, use it to cache analysis results or even send instructions to remote computing node. All these features are hard to implement using scripting approach. Automatic script transformation is unreliable and requires sophisticated code processing tools like parsers and syntax analyzers. While remote procedure call is possible via scripts, in most cases it is restricted to simple imperative commands. Here we give only three examples of the usefulness of the metadata processor:

1. The results of the analysis are completely determined by metadata. This means that each result is mapped to a fixed configuration. Configurations, unlike scripts, can be compared. If the configuration of some part of the analysis has not changed, it can be considered guaranteed that the result of this part has not changed. This result can be calculated once, placed in the cache and then used ready, significantly saving computational time.

2. It is frequently necessary to do some analysis for a list of different values of some parameter. In most cases, this is done with the help of macros recording a sequence of user actions. Macros are generally cumbersome and not reliable. In the case of a metadata processor, it is sufficient to make a metadata generator that will change one or more values in the source tree and run an analysis for all these configurations.

3. Performing distributed computing, one of the biggest challenges is ensuring the identity of the execution environment and the set of instructions on all nodes of the network. Metadata automatically solve the problem of instruction identity (it is just a tree of values that can be easily compared). For the identity of the environment, the principle of context encapsulation is used (see below).

Separately, it is worth noting the mechanism of layering: two or more metadata sets could be automatically combined without changing individual configurations. Metadata of a higher level overlap certain values of the metadata of a lower level, without changing the lower level itself. The layering system is also very convenient for selective configuration of individual data sets. For example, if there is a general configuration for the entire process, but additional instructions are required for a specific element, rather than overwriting the entire configuration and processing this element separately, a small layering is created that affects only this element. At the same time, all analysis is done, as before, and can be processed in parallel if it is possible.

The second principle is **context encapsulation**. One of the main difficulties in the development of parallel and distributed computing is the influence of the environment (connected libraries, global variables, object states) on the results. DataForge solves this problem by introducing the concept of context. The context contains all I / O tools and connected modules. The context is organized in such a way that the identity of the two contexts can be easily verified (the context description can be transformed into metadata). Neither the context, nor any connected modules can be changed during operation. Thus, processes can not affect the results of other processes. Also, any acquisition, storage or analysis parameter could be accessed from the context using its name. A context encapsulation is extensively used in Android SDK.

Finally, a very important feature of DataForge is its **modularity**. The core of the system contains only basic functionality for working with metadata and organizing the data processing, all specific functions, such as the graphical environment or tools for statistical analysis, are placed in separate plug-ins. It is planned that modules can be dynamically loaded from a remote repository as needed (such a system is implemented, for example, in JAS3 - [2]). Modularity is the standard for software in Computer science, but monolithic systems are still very often used in experimental physics. Monolithic systems are mostly very complex in terms of support and upgrades.

3 Data flow

Metadata processor approach requires a lot of attention to be paid to data flow models. DataForge currently considers two of them:

- **Push-model**. This model actually implements the classical concept of map-reduce. Processing consists of a sequence of actions on the data tree, in each of which a certain transformation of the data takes place. If individual tree elements can be handled independently, then the process automatically (without additional actions from the user) is performed in parallel. The fundamental innovation of DataForge in this data flow model is the process of configuring the actions. The user specifies some configuration of each of the actions, but the metadata of the elements of the data tree themselves overlap the configuration of the action. Thus, if the metadata of an element contains information related to the action, then this information will be automatically taken into account without writing a special processing instruction. This feature significantly simplifies the analysis of large amounts of data when additional external information is required (for example, calibration in physics experiments).
- **Pull-model**. This model is fundamentally new in data processing. Its essence lies in the fact that the user provides some configuration of the result that he wants to see, and the system itself determines what actions and what data it must request to obtain the result. The relations between processing blocks in pull model are called dependencies. When some pull block is called, it first calculates a set of its dependencies based on its metadata and then requests appropriate data or pulls other blocks, using input metadata to produce configuration for dependencies.

Both models involve the use of lazy computing (where possible). The usual procedure is to use pull model globally to define global data processing blocks - tasks and their dependencies and then use push model inside single task. Usually all results of intermediate operations are automatically cached, meaning that calling task or push-action with the same set of metadata does not compute and instead loads results from cache. Since the only thing needed to perform computation is metadata, one can easily delegate computation itself to a remote node or a whole distributed computation network.

Pull data flow model is not so well known as push model and requires additional explanation. An example of existing implementation of the idea is the gradle build system ([4]) used for a wide variety of JVM-based and native projects. In DataForge it is implemented by so-called workspaces and tasks. The workspace is a virtual structure that contains an initial data tree as well as a list of tasks that could be performed. The task execution is performed in three steps:

1. **Task model**. Depending on task definition and input metadata, the system calculates an acyclic dependency graph of tasks and data. Task dependency is represented by the name of task and metadata passed to the task (task could modify metadata passed to dependencies), so each task

calculates its dependencies recursively. Data dependencies consist only of data name or mask. Task and data are resolved by the workspace by name. Task graph is checked for cycles.

2. **Lazy computation model**. A 'Goal' object is created for each specific computation. Goal is basically an elementary lazy computation result which depends on completion of other goals. A task could produce one or many goals for each of data pieces passing through it. There could be pipe goals that transform one input element into one output element, join goals, split goals as well as custom goals with more sophisticated dependency mechanism. Goals are organized in acyclic graph (the complexity of this graph is generally much larger then the one for tasks and user does not have control over it beyond basic canceling). In theory, goal graph could be automatically optimized, but for now there was no practical reason to do it.

3. **Computation**. When the top level goal is triggered, it invokes computation of all goals in chain behind it. If some of goals in cycle is not triggered, it won't be computed thus saving computation time without need for the user to explicitly optimize the data flow. If for some reason, one of the goals is already computed (or just started) when it is triggered, it is not computed again, the previous result is returned. If caching mechanism is enabled and there is a cached goal result with the same medatada, it is restored from cache instead of computation. It is possible to have not only local, but also remote cache, which allows to implement sophisticated distributed network systems.

4 Current state and future plans

Current prototype of the DataForge is written on JVM platform and uses Java 8 and Kotlin languages for core modules. The JVM automatically grants a wide cross-platform portability and by default supports major operating systems like MS Windows, Linux and MacOS. Also JVM is known for its good tooling support, which simplifies development, and has good internal multi-threading support, which is needed for parallel computation. Performance issues experienced in earlier versions of JVM platform were almost completely eliminated in last years and currently JVM just-in-time compiled code has almost the same performance as native C++ code (the statement is bases on multiple case-study articles since direct comparison is not available due to differences between JIT and compiled code operation).

Since the most of DataForge functionality could be accessed via manipulation with metadata trees, it is quite easy to create a client application in any other language which can transform request into metadata, send it to DataForge server and then transform result back into user-friendly representation. It is planned to create thin Python-based client as well as web-based console in the future. Current prototype also supports basic communications with native processes on local computer as well as TCP/IP communication with remote computers or devices and serial port communication (a universal API for most available hardware and software architectures).

Since developing a stand-alone framework for both data acquisition and analysis is a very large scale task, current aim of DataForge is not to completely replace existing tools, but to build additional configuration layer on top of them. Here are examples of such uses for different subsystems:

- **Parallel computations**. Goal API allows to use different implementations for task scheduling and running. DataForge currently utilizes Java 8 fine threading based on fork-join thread pool ([5]) as a default. It also has another implementation using experimental coroutine framework ([6]) developed by JetBrains. It is planned to also implement a back-end for Apache Spark framework ([7]).

- **Caching**. Caching is important part of DataForge since it allows to avoid recalculation of already completed tasks. The caching system utilizes JCACHE specifications ([8]) and thus allows to use different powerful caching packages (though by default it uses internal lightweight implementation).
- **Storage**. The storage API of DataForge is abstract and allows to use different local and remote storage mechanisms like SQL and no-SQL databases. The important distinction is DataForge treats all data as immutable that allows to significantly simplify interaction with storage software.
- **Control**. Device control API uses abstract definitions based on the concept of virtual device and its states. It is planned to introduce a connection to the TANGO slow control system ([9]) as a back-end for large scale projects.
- **Visualization**. DataForge has a small internal tool set for creating graphic user interfaces, but in future it will be mostly replaced by web-based visualization using existing libraries.

Before 2017 DataForge was developed as a part of software package for "Troitsk nu-mass" experiment in search for masses of active and sterile neutrinos ([10]). The experiment data requires a very complicated and multi-step analysis and DataForge allows to greatly simplify the process and avoid making mistakes that would be hard to find. Currently, DataForge is used for the data acquisition, storage and control in this experiment. Currently it is being adopted for simulations and analysis management in center for nuclear physics methods at MIPT ([11]).

5 Conclusion

The DataForge framework is still in early stage of its development, but it demonstrates new approaches to scientific software development (both from ideological and architectural point of view). A metadata processor ideology imposes some limitation on user and developer actions inside the framework, but opens a lot more possibilities like meta-programming, automatic result caching, distributed computing etc.

References

[1] *ROOT homepage*, https://root.cern.ch/ (2017), [Online; accessed 24-January-2018]

[2] M. Donszelmann, T. Johnson, V.V. Serbo, M. Turri, Journal of Physics: Conference Series **119**, 032016 (2008)

[3] *DataForge homepage*, http://npm.mipt.ru/dataforge/ (2017), [Online; accessed 06-November-2017]

[4] *Gradle homepage*, https://gradle.org/ (2017), [Online; accessed 06-November-2017]

[5] *Package java.util.concurrent*, https://docs.oracle.com/javase/9/docs/api/java/util/concurrent/package-summary.html (2017), [Online; accessed 24-January-2018]

[6] *Concurrent Programming in Kotlin: Coroutines*, https://dzone.com/articles/approaching-kotlin-coroutines-concurrent-programmi (2017), [Online; accessed 24-January-2018]

[7] *Apache Spark homepage*, https://spark.apache.org/ (2017), [Online; accessed 24-January-2018]

[8] *JCACHE: JSR-107*, https://www.jcp.org/en/jsr/detail?id=107 (2017), [Online; accessed 24-January-2018]

[9] *TANGO controls*, http://www.tango-controls.org/ (2017), [Online; accessed 24-January-2018]

[10] D.N. Abdurashitov et al., JINST **10**, T10005 (2015), `1504.00544`

[11] *Center for nuclear physics methods at MIPT*, http://npm.mipt.ru/ (2017), [Online; accessed 24-January-2018]

EPJ Web of Conferences **177**, 05004 (2018) https://doi.org/10.1051/epjconf/201817705004
AYSS-2017

Architecture of distributed picture archiving and communication systems for storing and processing high resolution medical images

Victoria Tokareva[1,⋆]

[1] *Laboratory of Nuclear Problems, Joint Institute for Nuclear Research, 6 Joliot-Curie, Dubna, Moscow region, 141980, Russia*

Abstract. New generation medicine demands a better quality of analysis increasing the amount of data collected during checkups, and simultaneously decreasing the invasiveness of a procedure. Thus it becomes urgent not only to develop advanced modern hardware, but also to implement special software infrastructure for using it in everyday clinical practice, so-called Picture Archiving and Communication Systems (PACS). Developing distributed PACS is a challenging task for nowadays medical informatics.
The paper discusses the architecture of distributed PACS server for processing large high-quality medical images, with respect to technical specifications of modern medical imaging hardware, as well as international standards in medical imaging software. The MapReduce paradigm is proposed for image reconstruction by server, and the details of utilizing the Hadoop framework for this task are being discussed in order to provide the design of distributed PACS as ergonomic and adapted to the needs of end users as possible.

1 Introduction

Medical imaging (MI) is a common name representing several technologies (CT, PET, MPI, ultrasound, etc.), which are used to produce images of internal aspect of a human body for the purposes of diagnosis, monitoring and treatment. Also MI is employed for achieving scientific goals in anatomy and physiology by providing a broad database of examinations, which is helpful for the identification of abnormalities. The key benefit of MI is its relative noninvasiveness, i.e. in most of MI techniques no instruments are introduced inside a patient's body to produce images.

During MI we observe a signal (for example, attenuated X-rays) and make a conclusion about the properties of tissues and organs. So, in technical sense, obtaining medical images from the raw data means searching for the solution of mathematical inverse problems.

MARS (Medipix All Resolution System) micro-CT scanner [1], investigated at DLNP JINR, allows obtaining high quality medical images, resulting in large volumes of both raw data and reconstructed images. In order to improve speed and quality of processing, storing, and transfer of medical images for the project, we performed the investigation of relevant standards and industry solutions in the field of big medical data processing, in particular, employing distributed computational systems.

⋆e-mail: tokareva@jinr.ru

2 DICOM standard and PACS systems

In the last few years, DICOM [2] (Digital Imaging and Communications in Medicine), the industry standard by NEMA (National Electrical Manufacturers Association, USA) for creating, storing, transferring and visualizing medical images and documents of patients, has de facto become a world-wide used protocol, that defines a single storage format of medical images and a single protocol for their transmission.

Medical institutions possess systems that control DICOM files and these systems are called Picture archiving and communication systems (PACS). PACS are now designed to handle medical images stored at each hospital or facility.

One of the most striking trends in modern Russian medicine is the widespread informatization, and one of its consequences is a request to ensure that medical data, stored distributedly over various hospitals, at the same time would be available to physicians of a regional network in order to support collaboration in the work of medical institutions. Thus, in the near future several PACS existing in different hospitals or institutions should be able to work together. So, there is a new challenge to develop "meta-PACS" capable of storing and processing region-wide or even nation-wide corpus of medical imaging data.

3 Computed tomography

Both availability and resolution of medical imaging devices are constantly improving, in particular, in the area of computed tomography (CT), which is one of the most popular medical imaging methods. CT is a technique for cross-sections imaging of an object using a series of X-ray measurements taken from different angles around the object [3].

Over the last years, the following trends are being observed in this field.

On the one hand, the improvements in medical image acquisition systems resulted in much better pixel resolutions of detectors and, consequently, in the increase of the amount of medical image data. In recent decades a typical size of a scan has been grown from Kilo- to Gigabytes and is expected to reach Terabytes soon [4]. For example, the new Sky Scan 2011 tomograph with the resolution of 200 nm per pixel could provide 64 Megabyte (MB) per slice.

On the other hand, the popularity of CT as a means of diagnostics is constantly growing: while up until 2010, 5 billion medical imaging studies had been conducted worldwide [5], now more than one billion diagnostic images per year are taken only in the USA, and this number is expected to keep growing [6].

Thus, big data in medical imaging occurs in two ways: first, a large data size from a single scan, and second, thousands and millions of images stored in a PACS server. In practice, both ways multiply.

Not only the increased volumes of data, but also complicated image reconstruction algorithms provide challenges to the future PACS development.

The reconstruction algorithms can be split in two categories. Reconstructing images using *iterative algorithms* could take hours for large scans. *Back-projection algorithms* have computational complexity of $O(n^4)$ [3], while hierarchic back-projection has one of $O(\log(n) \times n^3)$ [7].

At DLNP JINR we investigate the MARS micro-CT scanner [1], employing Medipix electronics based on GaAs semiconductor detectors. It allows obtaining high quality medical images while decreasing the irradiation of an examined object compared to currently used scanner models. Scanner design makes it possible to take up to 720 shadow projections of an object with a spatial resolution up to 55 μm^2. Such accuracy hepls us to obtain high quality resulting images through the reconstruction employing the FDK [8] (Feldkamp-Davis-Kress, 1984) method. The opposite side of such accuracy is a significant growth of both raw data and reconstructed samples, and consequently, slower image processing and network access compared to samples of a smaller size. The solution to these challenges could be a distributed storage and processing data in PACS.

4 Map-reduce based distributed PACS

A popular approach for the implementation of distributed PACS is employing the MapReduce paradigm [9]. The MapReduce is a scheme of distributed computations, which is widely applicable for parallel calculations with big (up to hundreds of Petabytes) data amounts in computing clusters (see Figure 1).

The MapReduce workflow involves two steps: Map and Reduce (called so in reference to high-level functions with the same names, map and reduce). The preprocessing of input data takes place at the Map stage . In order to do this, one of computers in the cluster (so-called NameNode) fetches task input, then divides it into parts and passes to other computers in the network (so-called DataNodes) for preliminary processing. At the Reduce stage the preprocessed data are converged. Then the NameNode receives callbacks from the DataNodes and formulates output on their base. This output is considered to be the solution of the original problem.

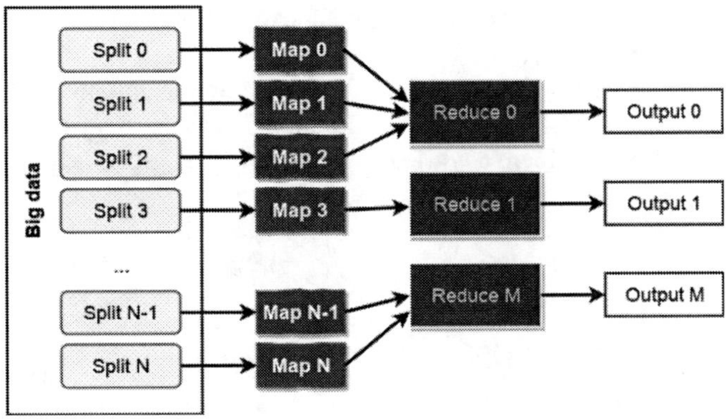

Figure 1. Scheme of MapReduce workflow.

The advantage of MapReduce is based on the fact that it supports distributed execution of preprocessing and data convergence. Although this process happens to be less effective in comparison with more sequential algorithms, the good feature is that MapReduce could be applied to analyze very large amounts of data at several servers simultaneously. Parallel processing also provides us with several ways to recover after node failures and makes our system more fault-tolerant.

Hadoop is a project by the Apache Software Foundation. It involves a set of freely-distributed utilities and libraries, and a framework for development and implementation of distributed programs within the scope of the MapReduce paradigm. Hadoop was developed for running on computing clusters consisting of up to hundreds and thousands of nodes.

So, the implementation of a distributed PACS employing Hadoop framework helps us to benefit from system scalability, cost effectiveness, data replication and a good fault-tolerance level. But, on the other hand, for small clusters image retrieval time grows significantly, so using such Hadoop-based PACS should be discussed for any specific case.

An interesting Hadoop-based PACS implementation was proposed at [10]. Employing MapReduce in a cloud-computing environment authors has implemented scalable CT image reconstruction for cone-beam geometry in high-speed way.

The implementation accelerates the cone-beam FDK reconstruction algorithm by porting it to a MapReduce model. Here authors apply the map functions for filtering and back-projection. Then reduced functions are used to gather parts of back-projected images in full volume. Due to that, the reconstruction time speeded up drastically, so it was found to be in linear correlation with the number of DataNodes employed.

In the study [11], the authors propose an architecture of Medical Image File Access System (MI-FAS) integrating several levels of software platforms: from virtualization infrastructure (OpenNebula+Xen) to virtualization of the fault-tolerance system, then to the Hadoop distributed file system (HDFS), and, finally, several DICOM and Hadoop APIs with user-end web interface. One of the main benefits of this system is its focus on the fault tolerance.

A natural aproach to solve the problem of full volume reconstruction is to decompose the main problem in a sequence of slice reconstruction tasks. But in the case of cone-beam geometry such tasks are not completely data-independent, so decomposition at this level does not provide cost-effective results within the framework of the MapReduce paradigm.

The more convenient aproach in order to gain the benefits of MapReduce-based distributed calculations is employing a lower-level decomposition with smaller tasks like processing a readout of a single pixel string of the detector at a particular angle, consisting of its Fourier transform, filtering and back-projection.

Thus, a MapReduce approach would be effective only when the amount of data exceeds a particular threshold, depending on overheads for creating a subtask in a particular system implementation, and is expected to become more effective while the amount of data grows.

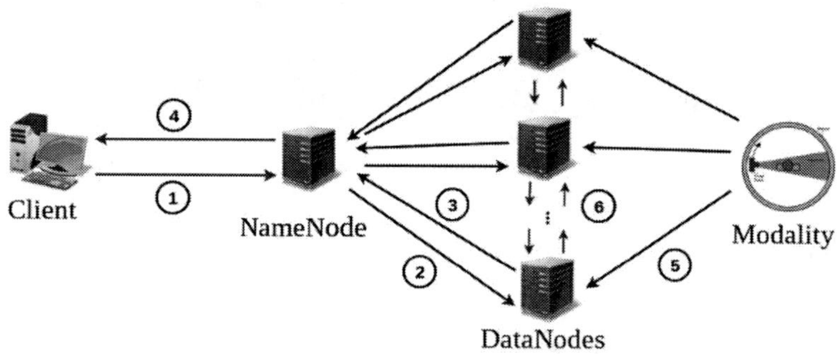

Figure 2. Flows of information in a distributed PACS server. Image reconstruction 1–4: 1) Initial request; 2) Requests to perform the Map step; 3) Gathering data packages for the Reduce step; 4) Providing a final result. Raw data information flow 5–6: 5) Raw data flow from modality; 6) Data exchange between server nodes.

Observing the investigations given before, we propose for our distributed PACS the architecture, shown in Figure 2. Information flow for a reconstruction or data-obtaining request can be described in the following way. A client sends a request to a NameNode of the distributed PACS server. After processing the request, the NameNode sends pieces of code that would process data on DataNodes. The DataNodes send results of data processing back to the NameNode, where the whole reconstruction is gathered using the Reduce algorithm. Optionally, a reconstructed image can be saved in a dedicated file storage connected to the NameNode. Finally, NameNode provides a final result to the client via user API.

The information flow of the raw data could be given as follows. A PACS modality (e.g. a CT scanner) sends the obtained raw data to nodes of the distributed PACS server. Server nodes exchange data according to built-in algorithms of the distributed file system, e.g. HDFS [12].

5 Conclusion

Big data is going to play an essential role in future medicine. The availability of medical devices and their resolution are growing rapidly, resulting in a fast increase of the size of medical data, which requests innovative solutions to process these data in reasonable timescales. One of these solutions could be new-generation PACS using distributed data storage and processing.

The MapReduce programming model could be especially useful for medical imaging, since this paradigm implies moving code to data instead of vice versa, and thus eliminates the necessity to transfer large data volumes through the storage system. Besides, it was designed for being deployed on inexpensive clusters using commodity hardware. The practical implementation of distributed PACS based on the MapReduce model could be done using recent developments in open-source software, namely, the Hadoop project and the associated frameworks. Such systems could provide means for storing Terabytes and Petabytes of medical imaging data and for parallel fault-tolerant processing of these data. These tools will form the core of future medical informatics software as the quantity and quality of medical data are continuously increasing.

References

[1] R. Zainon, A.P.H. Butler *et al.*, Internetworking Indonesia Journal **2(1)**, 3–10 (2010).
[2] DICOM standard official site, available at http://dicom.nema.org/standard.html.
[3] H. Turbell, *Cone-beam reconstruction using filtered back-projection*, Linköping University Electronic Press, 189 (2001).
[4] I. Scholl, T. Aach, T.M. Deserno *et al.*, Comput. Sci. Res. Dev. **26**, 5–13 (2011).
[5] C.A. Roobottom, G. Mitchell, G. Morgan-Hughes, Clin. Radiol. **65 (11)**, 859–67 (2010).
[6] T. Hiroyasu, Y. Minamitani, M. Yoshimi, M. Miki, WORLDCOMP'11 **2011-MRS-84 (5)** (2011).
[7] S. Xiao, Y. Bresler, D.C. Munson, Proceedings 2003 International Conference on Image Processing **2**, II–819–22 (2003).
[8] L. Feldkamp, L. Davis, J. Kress, Journal of the Optical Society of America A **1(6)**, 612–619 (1984).
[9] A. Belle, BioMed Research International **2015**, 16 (2015).
[10] B. Meng, G. Pratx, L. Xing, Med. Phys. **38(12)** 6603–6609 (2011).
[11] C.-T. Yang, C.-T. Kuo, W.-H. Hsu, W.-C. Shih, Proceedings of the 7th international conference on Advances in Grid and Pervasive Computing **7296**, 338–349 (2012).
[12] HDFS documentation, available at http://hadoop.apache.org/docs/current/hadoop-project-dist/hadoop-hdfs/HdfsDesign.html.

EPJ Web of Conferences **177**, 06001 (2018) https://doi.org/10.1051/epjconf/201817706001
AYSS-2017

Induction and repair of DNA double-strand breaks in hippocampal neurons of mice of different age after exposure to ^{60}Co γ-rays *in vivo* and *in vitro*

R.A. Kozhina[1, 2], *V.N.* Chausov[2], *E.A.* Kuzmina[1, 2], and *A.V.* Boreyko[1, 2]

[1]Dubna State University, Department of biophysics, 141982 Dubna, Russia
[2]Joint Institute for Nuclear Research, 141980 Dubna, Russia

Abstract. One of the central problems of modern radiobiology is the study of DNA damage induction and repair mechanisms in central nervous system cells, in particular, in hippocampal cells. The study of the regularities of molecular damage formation and repair in the hippocampus cells is of special interest, because these cells, unlike most cells of the central nervous system (CNS), keep proliferative activity, i.e. ability to neurogenesis. Age-related changes in hippocampus play an important role, which could lead to radiosensitivity changes in neurons to the ionizing radiation exposure. Regularities in DNA double-strand breaks (DSB) induction and repair in different aged mice hippocampal cells *in vivo* and *in vitro* under the action of γ-rays ^{60}Co were studied with DNA comet-assay. The obtained dose dependences of DNA DSB induction are linear both *in vivo* and *in vitro*. It is established that in young animals' cells, the degree of DNA damage is higher than in older animals. It is shown that repair kinetics is basically different for exposure *in vivo* and *in vitro*.

1 Introduction

Space journey will imminently affect the astronauts' health, because they will be exposed to Solar Cosmic Rays and Galactic Cosmic Rays in deep space, which include a wide spectrum of different quality radiation: from protons and gamma-rays to heavy charged particles [1].

Studies in recent years have shown that a lot of radiation-induced physiological impairment to CNS functions is due to damage formed in the hippocampus, which plays a key role in the formation of short-term memory and further transition into long-term memory, as well as spatial memory [2]. Moreover, the formation of new neuronal elements is carried out in hippocampus region, which is called dentate gyrus. Neurogenesis also plays a pivotal role in hippocampus function. So, the structure of the hippocampus is most radiosensitive compared to the neurons of other parts of the brain [3]. The majority of adult brain neurons are highly differentiated cells, which have lost the ability to divide and, therefore, are adequate resistant to radiation (the Bergonie and Tribondeau law, 1906).

DNA DSBs are the hardest damage by ionizing radiation, which could cause neurons death and involve disorders of the higher integrative functions of the CNS later [4-5]. Also not repaired DNA damage lead to a loss in the fidelity of information transferred from DNA to proteins [6].

The purpose of this research is to study the regularities of formation and repair of DNA DSB in hippocampal cells of different aged mice under the action of ^{60}Co γ-rays *in vivo* and *in vitro*. Because age plays a major role in susceptibility to radiation, and as it established, young aged mice brains are more radiosensitive [3, 7]. So, hippocampus is one of the brain regions most susceptible to aging [8].

2 Materials and methods

Female CBAxC57Bl, F1 mice were purchased from nursery of laboratory animals "Pushchino" (Pushchino, Russian Federation). The age of the animals in *in vivo* experiments was 8 weeks; the body weight was from 26 to 35 grams. Mice were exposed totally to Co^{60} doses from 1 to 5 Gy of γ-rays. During the experiments animals were kept in individual cages (maximum 6 animals/cage) and were provided with standard food and water *ad libitum*. At the end of experiments, animals were sacrificed by cervical dislocation. After cervical dislocation the hippocampus was dissected from the whole brains.

The cell suspension was isolated from the hippocampus of mature female hybrid mice CBAxC57Bl, F1 for irradiation with Co^{60} γ-rays *in vitro*. The age of the animals was 8 and 48 weeks; the body weight was from 26 to 35 grams. Irradiation of hippocampal cells was carried out in 0.5 ml microtubes (Eppendorf); the volume of the resulting suspension was 200 μl per dose. For irradiation, the Rocus-M radiation therapy facility (LNP, JINR) was used, delivering 0.01 Gy/s. Upon irradiation, samples were placed in an ice bath (0–4°C) for the inhibition of repair.

The procedure isolation and homogenization of mouse hippocampal cells of Seibenhener [9] was followed, with some modifications. All procedures were performed at room temperature. The extracted hippocampus was placed in a balanced HBSS (Mg [-] and Ca [-]) (Gibco). For making single-cell suspension the hippocampus was cut into small pieces (1-2 mm^3) and placed in a collagenase solution (340 U/ml) (Paneco) in HBSS(Ca [+], Mg [+]), after which the tissue samples were incubated for 15 minutes at 37°C. After dissociation, the tissue pieces were carefully pipetted until a homogeneous cell suspension was obtained. The resulting suspension was washed several times by centrifugation in HBSS (200 g), to which veal embryonic serum (10% of the volume) was added. After washing, the pellet was resuspended in 1 ml into L-15 Medium (Leibovitz) (Sigma) culture medium. Comet-assay was performed by the method described earlier [10, 11].

An Axiolab fluorescent microscope (Carl Zeiss) with an AxioCam ICc3 was used for comet image registration. 100 comets were registered for each slide. The scanning of each comet provided its fluorescence intensity profile; the comet head and tail were easily identified on this profile. The comet image was processed using CASP code [12]. The results were analyzed and the plots were constructed using OriginPro2015. Statistical analysis was performed using the Mann-Whitney U-test at $p < 0.05$.

3 Results and discussions

Figure 1 shows the dose dependence of the DNA DSB yield in hippocampal cells of mice (8-week-old) formed under the action of Co^{60} γ-rays *in vitro* and *in vivo*. The dependence is linear in both cells irradiated *in vivo* and irradiated *in vitro*. In this case, the number of lesions formed *in vitro* is greater. It was observed that each tissue of a living organism exposed *in vitro* is 2 to 5 times more radiosensitive than that irradiated *in vivo*. These effects are the result of interstitial differences in the repair of DNA strand damage that occur within 3 minutes necessary for irradiation and tissue preparation *in vivo*, these effects disappear when cells are irradiated *in vitro* on ice [13]. In response to *in vivo* irradiation, repair systems work, and there is also a certain contribution of the immune system.

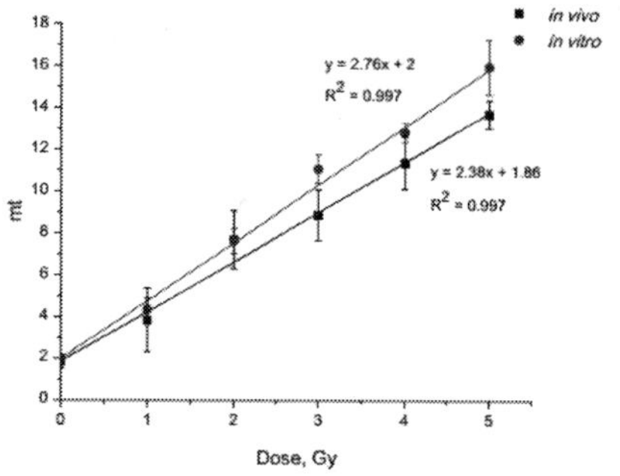

Fig. 1 Induction of DNA DSB in hippocampal cells *in vivo* and *in vitro*.

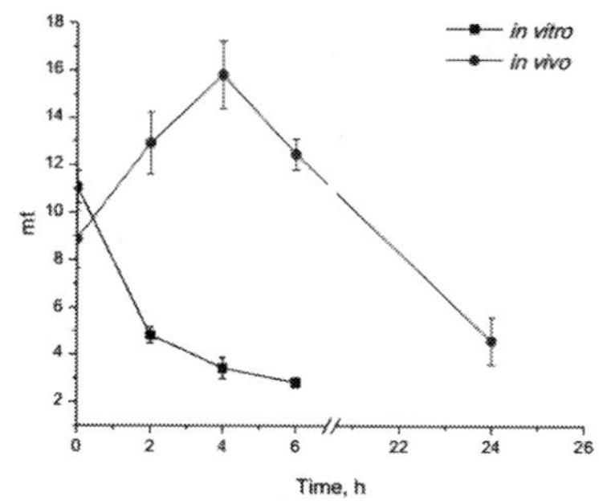

Fig.2 Kinetics of DNA DSB repair in hippocampal cells *in vivo* and *in vitro*.

Figure 2 shows the repair kinetics of DNA DSB in mice (8-week-old) hippocampal cells after exposure to Co^{60} γ-rays at the dose of 3 Gy *in vivo* and *in vitro*. *In vivo* and *in vitro* repair kinetics is significantly different. Repair kinetics *in vivo* has a complex character: after irradiation, a sharp increase of DNA DSB yield is observed up to 4 hours of post-irradiation incubation; then DSB are eliminated, and a majority is repaired; and in 24-hours, their repairs are almost completed. While repair kinetics *in vitro* is characterized by an exponential decrease in the number of lesions in the post-radiation period. These results are consistent with those obtained earlier in other authors *in vivo* [14] and *in vitro* on hippocampal neurons of lines HT22 and HN33 [15]. It can be assumed that in the formation and repair of DNA DSB *in vivo*, an important role is played by the immune and other body systems that affect the features of the course of reconstructive processes in the brain after exposure to ionizing radiation. The repair of most of the induced DNA strand breaks in the culture of mammalian cells is performed within 2 hours after irradiation (X-ray radiation) [13].

The DNA DSB induction in hippocampal cells of different aged mice (8-week-old and 48-week-old) *in vitro* was studied (Fig.3). The amount of DNA DSB, which formed in young adult mice (8-week-old), is dramatically more, that is formed in aged ones (48-week-old). This may be due to the fact that the number of immature and proliferating neurons in the hippocampus of young animals exceeds the number of mature neurons of adult mice approximately tenfold [3]. Casciati et al. [7] showed that the age of mice influences the reaction to irradiation and to neurogenesis in the hippocampus. It has been established that the dynamic interaction of different cell types, such as neural progenitor cells, neurons, astrocytes, endothelial cells, can be associated with the negative consequences of ionizing radiation exposure to the hippocampus. Moreover, changes in proliferative activity lead to memory dysfunction. It was found that a low level of proliferation and differentiation of stem cells correlated with memory impairment [16]. This susceptibility is also explained by the well-known Bergonier and Tribondo law that cells able to proliferate are much more sensitive to radiation than already differentiated cells. This means that the hippocampal cells of older animals are more radioresistant than younger ones.

Repair kinetics of DNA DSB induced by the action of Co^{60} γ-rays at the dose of 4 Gy was researched. A comparative analysis of DNA DSB repair kinetics in hippocampal cells of different aged mice *in vitro* is shown in the Figure 4. Both dependences are characterized by continuous elimination of damage. The greater damage level induced in the young hippocampal cells once again confirms the fact that they are more radiosensitive than aged (48-week-old) animals. The efficiency of repair processes in adult aged animals is reduced in comparison with the efficiency of repair of 8-week-old mice [5, 17].

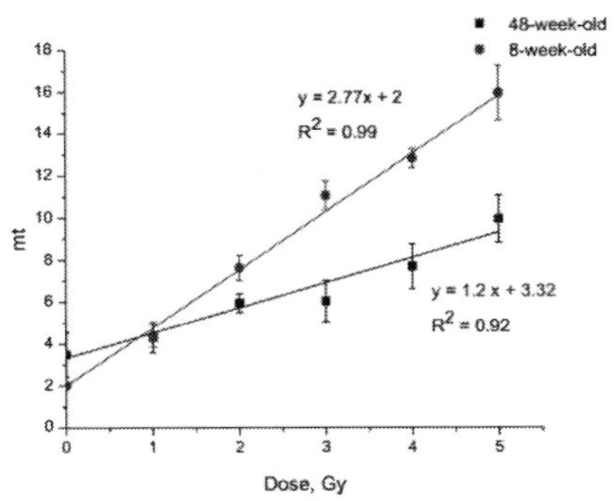

Fig.3 Induction of DNA DSB in hippocampal cells of different aged mice *in vitro*.

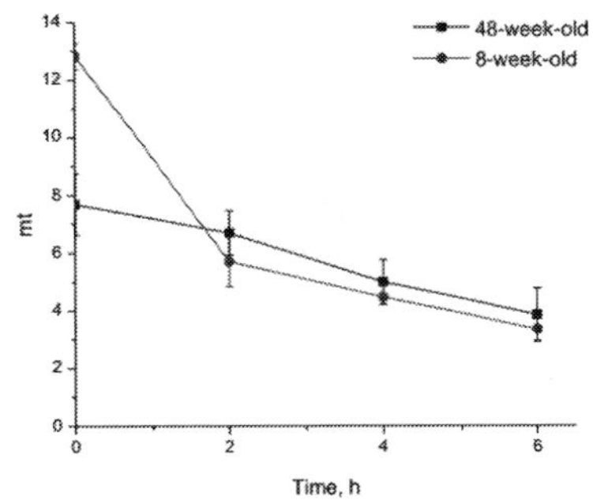

Fig.4 Kinetics of DNA DSB repair in hippocampal cells of different aged mice *in vitro*.

We thank V.N. Gaevsky for carrying out the irradiation and Y.S. Severiukhin for performing cervical dislocation. This work was funded by the RFBR grant 17-29-01007, AYSS JINR grant 17-702-05 and scientific grant of the Moscow region governor.

References

1. A. I. Grigor'ev, E. A. Krasavin, M. A. Ostrovskii, Neuroscience and Behavioral Physiology, **45(1)**, 91-95 (2015).
2. H. A. Cameron, RD. McKay, Nat Neurosci, **2**, 894-897 (1999).
3. A. Fukuda, H. Fukuda, J. Swanpalmer, S. Hertzman, B. Lannering, I. Marky, T. Bjork-Eriksson, K. Blomgren, J Neurochem., **92**, 569-84 (2005a).
4. F.A. Stewart, A.V. Akleyev, M. Hauer-Jensen, J.H. Hendry, N.J. Kleiman, T.J. MacVittie, B.M. Aleman, A.B. Edgar, K. Mabuchi, C.R. Muirhead, R.E. Shore, W.H. Wallace, ICRP, **41(1/2)**, 169 (2012).
5. V. Gorbunova, A. Seluanov, M. Zhiyong, C. Hine, Nucleic Acids Res., **35(22)**, 7466–7474 (2007).

6. M. L. Fishel, M. R. Vasko, M. R. Kelley, Mutat Res; **614 (1-2)**, 24-36 (2006).
7. A. Casciati, K. Dobos, F. Antonelli, A. Benedek, et al., Oncotarget, **7 (19)**, 28040-28058 (2016).
8. Y. Liou, Sun and Andrzej Bartke, Journal of Gerontology: BIOLOGICAL SCIENCES, **62A (2)**, 117–125 (2007).
9. M.L. Seibenhener, M.W. Wooten, J. Vis. Exp., **65**, e3634 (2012).
10. A.V. Boreyko, V.N. Chausov, E.A. Krasavin, I. Ravnachka, S.I. Stukova, Physics of Particles and Nuclei Letters, **8 (4)**, 399-404 (2011).
11. A.V. Boreyko, E.A. Krasavin, A.V. Mozhaeva, I. Ravnachka, S.I. Tiounchik, V.A. Tronov, V.N. Chausov, RADIATS BIOL RADIOECOL, **49 (1)**, 73-77 (2009).
12. K. Konca et al., Mutat. Res. **534 (1–2)**, 15–20 (2003).
13. E. R. Meyn, W. T. Jenkins, D. Murray, Mechanisms of DNA damage and repair, 151-158 (1986).
14. J.T. Lett, A.B. Cox, D.S. Bergtold, Radiat. Environ. Biophys., **25**, 1-12 (1986).
15. S. Eddy, Yang et al., Neuro-Oncology, **13(5)**, 459–470 (2011).
16. R. Coras, F.A. Siebzehnrubl, E. Pauli, et al., Brain, **133(11)**, 3359-72 (2010).
17. V.N. Vyjayanti , K.S. Rao, Neuroscience Letters, **393**, 18–22 (2006).

EPJ Web of Conferences **177**, 06002 (2018) https://doi.org/10.1051/epjconf/201817706002
AYSS-2017

Visualization of complex DNA damage along accelerated ions tracks

Elena Kulikova[1, 2, *], *Alla* Boreyko[1,2], *Tatiana* Bulanova[1,2], *Lucie* Ježková[1,3], *Mariia* Zadneprianetc[1,2] and *Elena* Smirnova[1]

[1]Joint Institute for Nuclear Research, Dubna, Russia
[2]Dubna State University, Dubna, Russia
[3]University of Chemistry and Technology Prague, Prague, Czech Republic

Abstract. The most deleterious DNA lesions induced by ionizing radiation are clustered DNA double-strand breaks (DSB). Clustered or complex DNA damage is a combination of a few simple lesions (single-strand breaks, base damage etc.) within one or two DNA helix turns. It is known that yield of complex DNA lesions increases with increasing linear energy transfer (LET) of radiation. For investigation of the induction and repair of complex DNA lesions, human fibroblasts were irradiated with high-LET ^{15}N ions (LET = 183.3 keV/μm, E = 13MeV/n) and low-LET ^{60}Co γ-rays (LET ≈ 0.3 keV/μm) radiation. DNA DSBs (γH2AX and 53BP1) and base damage (OGG1) markers were visualized by immunofluorecence staining and high-resolution microscopy. The obtained results showed slower repair kinetics of induced DSBs in cells irradiated with accelerated ions compared to ^{60}Co γ-rays, indicating induction of more complex DNA damage. Confirming previous assumptions, detailed 3D analysis of γH2AX/53BP1 foci in ^{15}N ions tracks revealed more complicated structure of the foci in contrast to γ-rays. It was shown that proteins 53BP1 and OGG1 involved in repair of DNA DSBs and modified bases, respectively, were colocalized in tracks of ^{15}N ions and thus represented clustered DNA DSBs.

1 Introduction

The most significant biological effects of ionizing radiation (IR) are caused by the induction of a wide range of DNA damage in cells of living organisms. DNA lesions induced by the action of IR can be divided into individual and clustered DNA damage. Individual DNA damage includes a various types of base damage, single-strand breaks (SSBs), double-strand-breaks (DSBs) etc. Such DNA lesions are usually quickly and successfully recovered, while clustered DNA damage is difficult to repair or is even irreparable. Clustered DNA damage consists of two or more individual lesions within one or two helical turns and is formed especially after exposure to high-LET radiation [1]. Clustered DNA lesions can be either non-DSBs (any combination of two or more oxidized

*Corresponding author: kruglyakovaea@bk.ru

bases, strand breaks, or other DNA lesions that do not form a DSB) or DSBs clusters (DSB(s) in combination with any other DNA lesion). The pathways of individual DNA lesions repair are quite well known, however the exact repair mechanisms of clustered DNA damage give rise to a lot of questions. The most important of them are: 1) how does the presence of one lesion affect the efficiency of recognition and processing of the opposite/ neighboring lesion(s), 2) is there a hierarchy in repair of closely spaced DNA lesions, 3) which repair proteins participate in the processing of these lesions and what mechanisms are used by the cell to eliminate the presence of such lesions? [2]

The goal of this work was to investigate complexity and repair efficiency of clustered DNA double-strand breaks in normal human skin fibroblasts irradiated with accelerated ^{15}N ions. Proteins involved into DSBs (γH2AX and 53BP1) and DNA base damage (OGG1) repair were visualized using immunofluorescent staining procedure and high-resolution fluorescent microscopy. More detailed information about application of this method for studying of the processes of the DNA damage formation and repair can be found in this work [3].

It should be noted that the efficiency of the DNA DSBs repair induced by radiation of different quality based on the γH2AX foci formation and elimination kinetics was also shown in [4]. However in this study only a qualitative analysis of the complexity of the induced lesions was performed. In current study a conclusion about higher complexity of DNA lesions induced by high-LET radiation in comparison with low-LET radiation was made based on a quantitative analysis of the γH2AX/53BP1 foci structure.

2 Material and methods

Cell culture. Primary normal human fibroblasts NHDF 22873 were maintained in Iscove's Modified Dulbecco's Medium (IMDM) supplemented with 10% fetal calf serum (FCS) and 1% gentamicin-glutamine solution (all reagents from Sigma-Aldrich, USA). Fibroblasts were grown at 37°C and 5% CO_2.

Low-LET and high-LET irradiation. Human fibroblasts were irradiated with low-LET ^{60}Co γ-rays and high-LET accelerated ^{15}N ions in two geometries: 90° (perpendicular) or 10° (horizontal) angle between the ion beam and the plane of the cell monolayer (Table 1). Cells were irradiated with accelerated ^{15}N ions using the isochronous cyclotron U-400M (Flerov Laboratory of Nuclear Reaction) and with γ-rays using the Rokus-M facility (Dzhelepov Laboratory of Nuclear Problems) at the Joint Institute for Nuclear Research (JINR, Dubna).

Table 1. Properties of exposure.

Radiation	Geometry	Energy, MeV/n	LET, keV/μm	Dose, Gy	Facility
γ-rays	90°	-	≈ 0.3	1	Rokus-M
^{15}N	90°	13	183.3	2.15	U-400M
^{15}N	10°	13	181.4	1.25	U-400M
^{15}N	10°	35	81.5	0.57	U-400M

Immunostaining procedure. At different time points after irradiation ranging from 5 min to 4 days, fibroblasts were fixed in 4% paraformaldehyde and underwent to standard procedure of immunostaining. For visualization of DNA DSBs (γH2AX and 53BP1) and base damage (OGG1) markers, the specific primary and secondary antibodies were used (primary antibodies: anti-53BP1 rabbit polyclonal, anti- γH2AX (phospho S140) mouse monoclonal (Abcam, USA); anti-53BP1 mouse monoclonal (Millipore, USA) and anti-OGG1 rabbit polyclonal (Novus Biologicals, USA) antibodies; secondary antibodies*:* Texas Red-conjugated goat anti-rabbit, FITC-conjugated goat anti-mouse and Alexa Fluor 488 conjugated goat anti-mouse (Abcam, USA)).

Image acquisition and experimental data analysis. Images of stained human fibroblasts nuclei were obtained using fluorescent microscope AxioImager.M2, equipped with optical light sectioning system ApoTome.2 (Zeiss, Germany). Quantitative analysis of colocalized 53BP1/γH2AX foci was performed manually using the Acquiarium software [5], which enabled the three-dimensional reconstruction of images and inspection of individual foci in 3D space (in x-y, x-z and y-z plane).

3 Results and discussion

We successfully visualized DNA lesions in tracks of accelerated ^{15}N ions (LET = 81.5 keV/μm, E = 35 MeV/n) in human fibroblasts irradiated at sharp-angle (10°) geometry. DNA double-strand breaks were visualized using colocalization of γH2AX and 53BP1 foci (both markers of DSBs; Fig. 1. left panel) and clustered DSBs using colocalization of 53BP1 and OGG1 foci (markers of DNA DSBs and DNA base damage, respectively; Fig. 1. right panel). Analysis of the obtained images revealed that majority of 53BP1 foci were colocalized with OGG1 foci at all time points after irradiation and thus demonstrate that most of DNA DSBs induced by ^{15}N ions were clustered. It was shown that 53BP1 and OGG1 foci were colocalized also after γ-irradiation and therefore we assume that these represent the clustered DNA DSBs (Fig. 2A).

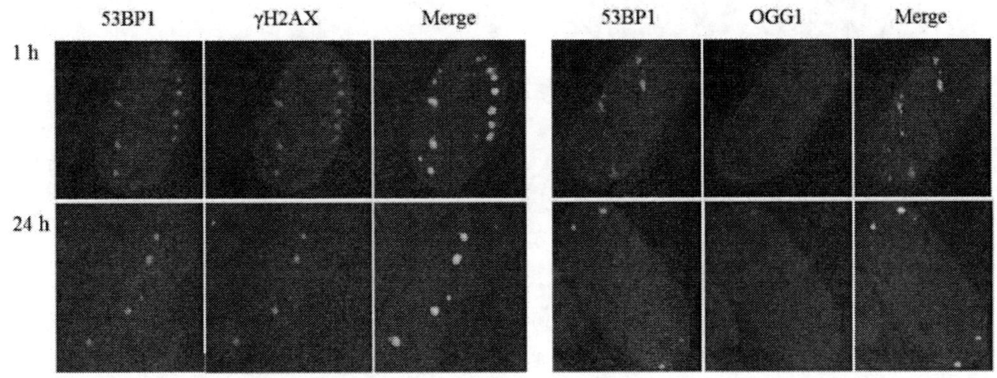

Fig. 1. Visualization of DNA lesions in tracks of accelerated ^{15}N ions (LET = 81.5 keV/μm, E = 35 MeV/n). **Left panel**: Visualization of DNA DSBs using markers 53BP1 (green) and γH2AX (red). **Right panel**: Visualization of DNA DSBs and DNA base damage using markers 53BP1 (green) and OGG1 (red), respectively.

Assuming that the majority of DNA DSBs induced by both high- and low-LET radiation were clustered, we carried out quantitative analysis of γH2AX/53BP1 foci in human fibroblasts irradiated in perpendicular geometry. Quantitative analysis revealed differences in the γH2AX/53BP1 foci formation and elimination after irradiation with accelerated ions and γ-rays (Fig. 2B). Irradiation with the accelerated N ions (LET =183.3 keV/μm, E = 13 MeV/n) led to faster γH2AX/53BP1 foci formation – the maximum number of foci was reached 30 min post irradiation (PI) while after γ-irradiation the maximum was reached within 1 h PI. Four hours post γ-irradiation, about 40% of the maximum numbers of γH2AX/53BP1 foci still persisted in the cells while this fraction was 80% for accelerated ^{15}N ions. After 24 h post γ-irradiation the number of γH2AX/53BP1 foci reduced to 6% and remained at this level up to 96 h PI. Whereas in cells irradiated with ^{15}N ions after 24 h PI the number of foci decreased to 37% and within 96 h reduced to 17%, which is almost in 3 times greater than post γ-irradiation.

The obtained data clearly demonstrate slower repair of nitrogen induced DSBs suggesting much higher complexity of DNA DSBs compared to γ-rays. The slower repair kinetics may also indicate the involvement of specific repair mechanisms in the recovery of clustered DNA DSBs induced by high-LET radiation. There are two main DNA DSBs repair pathways in mammalian cells: non-homologues end joining (NHEJ) and homologous recombination (HR). Evidence indicates that NHEJ is the predominant repair pathway for DSBs induced by low-LET radiation [6], but there is a suggestion that this pathway of DNA repair may be only partially involved in repair of the clustered DSBs induced by high-LET radiation [7]. The colocalization of both markers of DNA DSBs and base damage repair (Fig. 1, 2A) shows the involvement in processing of clustered DSBs not only mechanisms of DSBs repair but also a base excision repair (BER) needed for base damage recovery. Thus, in the next step, involvement of the HR, NHEJ and BER in repair of clustered DNA DSBs induced by accelerated ions will be investigated.

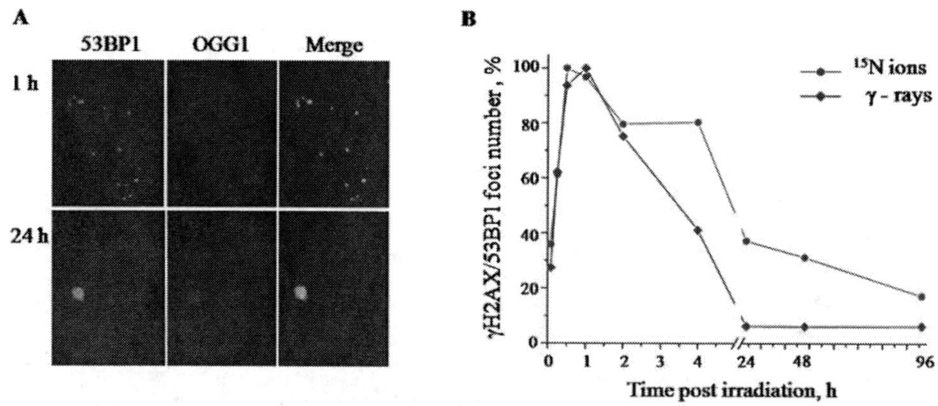

Fig. 2. A: Visualization of clustered DNA DSBs after γ-irradiation: DNA DSBs – 53BP1 (green) and DNA base damage – OGG1 (red); **B:** Formation and elimination of γH2AX/53BP1 foci after ^{60}Co γ-rays and accelerated ^{15}N ions (LET =183.3 keV/μm, E = 13 MeV/n) irradiation in the perpendicular geometry.

Additionally the structure of radiation-induced γH2AX/53BP1 foci in human fibroblasts irradiated by γ-rays and ^{15}N ions (LET = 181.4 keV/μm, E = 13 MeV/n) at the angle of 10° was analyzed. The number of individual γH2AX/53BP1 foci presented in clusters was counted, it was revealed up to 6 individual foci in one high-order γH2AX/53BP1 foci cluster in case of ^{15}N ions irradiation (Fig. 3). The diagrams on Fig. 3 demonstrate the time dependent increasing of the γH2AX/53BP1 foci structure complexity after irradiation by accelerated ions, while the structure of γ-irradiation foci practically didn't change in time. It can be due to the induction of larger number of individual DSBs in one cluster induced by accelerated ions that is primarily caused by physical characteristics of the radiation. Preservation of the complex structure of γH2AX/53BP1 foci after 24 hours after irradiation suggest a difficulty in repairing of clustered DNA DSBs induced by accelerated ^{15}N ions. Difficulty in repairing of clustered DNA DSBs may caused by the competition of several repair mechanisms since in addition to DNA DSB, as it was shown on Fig.1, the DNA base damage include in the clusters too.

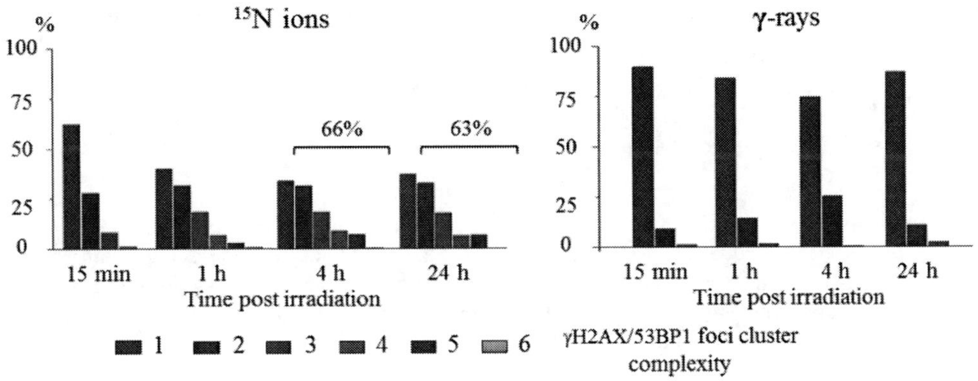

Fig. 3. Comparison of γH2AX/53BP1 foci clusters complexity for ^{15}N ions (LET = 181.4 keV/μm, E = 13 MeV/n) and γ-rays irradiation. The diagrams indicate the percentages of γH2AX/53BP1 foci presented in high-order clusters after 15 min, 1 h, 4 h and 24 h PI.

Thus, the issue of the cluster DNA damage repair is not simple and requires more thorough research and this will be continued in the following study.

References

1. B.M. Sutherland, P.V. Bennet, H. Schenk, O. Sidorkina, J. Laval, J. Trunk, D. Monteleone, J. Sutherland. Phys. Med. **17** (2001) 202.
2. A.G. Georgakilas. Mol. BioSyst. **4** (2008) 30. doi: 10.1039/b713178j
3. D.V. Firsanov, L.V. Solovjeva, M.P. Svetlova, Clin. Epigenet. **2** (2011) 283.
4. F. Antonelli, A. Campa, G. Esposito, P. Giardullo, M. Belli, V. Dini, S. Meschini, G. Simone, E. Sorrentino, S. Gerardi, G. A. P. Cirrone, M. A. Tabocchini, Rad. Res. 183 (2015). DOI: 10.1667/RR13855.1
5. P. Matula, M. Maška, O. Daněk, P. Matula and M. Kozubek, IEEE (2009), 1138
6. A. Gerelchuluun, E. Manabe, T. Ishikawa, L. Sun, K. Itoh, T. Sakae, K. Suzuki, R. Hirayama, A. Asaithamby, D.J. Chen, K. Tsuboi. Radiat. Res. **183** (2015) 345.
7. S. Moore, F.K.T. Stanley, A.A. Goodarzi. DNA Repair **17** (2014) 64

Parallel Algorithm for Solving TOV Equations for Sequence of Cold and Dense Nuclear Matter Models

Alexander Ayriyan[1,⋆], *Ján* Buša Jr.[1,2], *Hovik* Grigorian[1,3], and *Gevorg* Poghosyan[4]

[1] *Laboratory of Information Technologies, JINR*
[2] *Institute of Experimental Physics, SAS, Slovakia*
[3] *Department of Theoretical Physics, YSU, Armenia*
[4] *Karlsruhe Institute of Technology, Germany*

Abstract. We have introduced parallel algorithm simulation of neutron star configurations for set of equation of state models. The performance of the parallel algorithm has been investigated for testing set of EoS models on two computational systems. It scales when using with MPI on modern CPUs and this investigation allowed us also to compare two different types of computational nodes.

1 Introduction

Investigation of cold and dense nuclear matter is one of the key problems in modern particle physics. One of the ways of selecting and testing EoS models is comparison of the mechanical characteristics of compact stars given by models to the observational constraints [1]. To obtain mechanical characteristics of compact stars, one has to calculate the mass and radius relation of static spherical symmetry neutron stars. Such relations are obtained as a solution of the Tolman-Oppenheimer-Volkoff (TOV) equations [2, 3]:

$$\begin{cases} \dfrac{dP(r)}{dr} = -\dfrac{GM(r)\varepsilon(r)}{r^2} \dfrac{\left(1 + \frac{P(r)}{\varepsilon(r)}\right)\left(1 + \frac{4\pi r^3 P(r)}{M(r)}\right)}{\left(1 - \frac{2GM(r)}{r}\right)}, \\ \dfrac{dM(r)}{dr} = 4\pi r^2 \varepsilon(r), \end{cases} \tag{1}$$

where r is the coordinate distance from the center, $M(r)$, $\varepsilon(r)$, and $P(r)$ are mass, energy density, and pressure profiles correspondingly, and G is the gravitational constant. Each of these profiles as a solution of TOV equations depends on central value of energy density $\varepsilon(r = 0)$. The relation between the radius R where the pressure is $P(r = R) = 0$ and the total mass $M = M(r = R)$ are shown in the Fig. 1 for particular cases of realistic models of equation of state HDD [4], DD2 [5], MKVOR and KVORcut02 [6, 7]. These particular models were used as a reference models in investigation of the parallel algorithms.

A parallel algorithm for solving the TOV equations for a set of equations of state (EoS) models has been developed using Message Passing Interface (MPI) [10]. It has been applied to Bayesian

⋆e-mail: ayriyan@jinr.ru

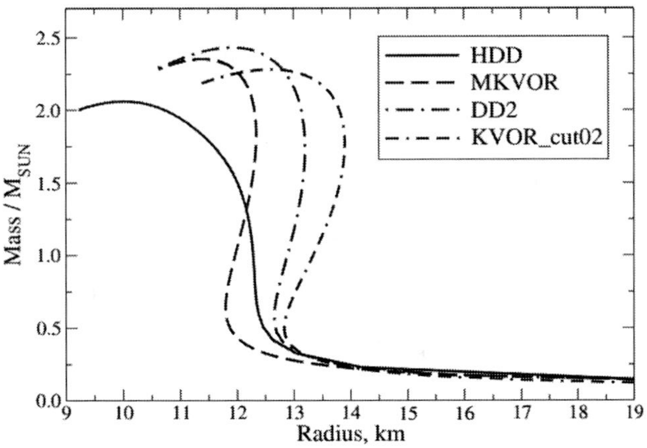

Figure 1. The neutron star mass-radius relations as solutions of TOV equations for different realistic models of equation of state actually used in modern astrophysical applications.

analysis of modern model of stellar nuclear matter [11], where thousands models have been analyzed using the most selective *M–R* constraints.

2 Parallel algorithm

Solving (integrating) TOV equations for one Equation of State model on current CPU is rather fast. With our settings of parameters we needed about 1500 integration steps (calls to Merson integrator) and it takes about 1.5s to solve such set of equations. The complexity comes from the fact, that we need to calculate several thousands of such models. Fortunately this problem is embarrassingly parallel. Each one model can be solved independently on others. This allows us to use MPI approach.

We have decided to use master-slave model. On the input we have one large text file containing data of EoS models. The number of models is not limited. The main process loads data and distributes it among rest of the processes. Master after reading one set of equations and parameters of the model sends these data to the first available slave process and continues to read following model. After all models have been read, master waits until all slave processes finish their calculations and collects results from slaves by combining results slaves have produced into one large output file.

Slave processes, after obtaining data from master, are responsible in analyzing data and checking their integrity (whether input sequences are monotonous as they should be). After this step, the TOV equations are integrated using the Runge-Kutta-Merson method [8, 9]. Obtained results are written into output files with predefined names. In this step we use the fact, that master and slave processes share the same file system and we don't need to communicate the results back to the master which reduces the load and requirements on master.

3 Results

The analysis of parallel calculations has been done for testing set of 3162 EoS models on two computational systems. This set of modern realstic EoS models has been studied using Bayesian Analysis [11]. The cluster of Joint Supercomputer Center of the Russian Academy of Sciences (JSCC) with 4× Intel Xeon Phi 7250 using AVX512 (red line on presented figures) and heterogeneous cluster

 https://doi.org/10.1051/epjconf/201817707001

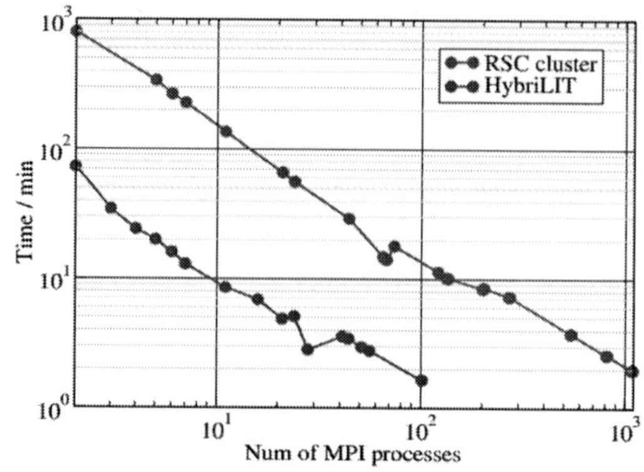

Figure 2. The computational time of parallel computations.

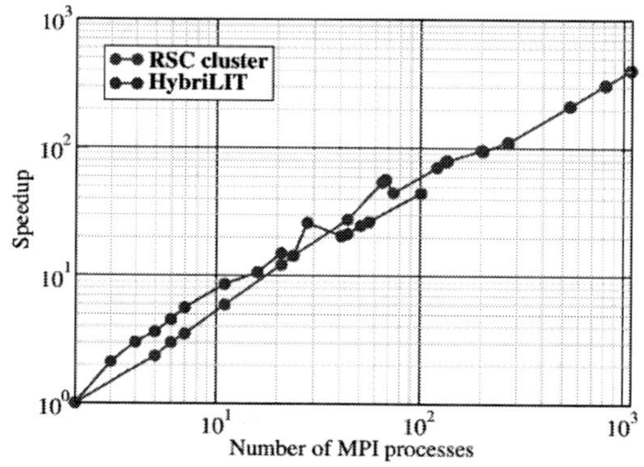

Figure 3. The speedup of parallel computations.

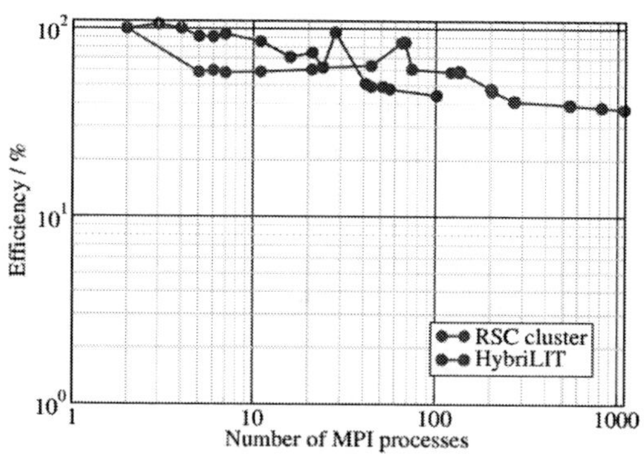

Figure 4. The efficiency of parallel computations.

"HybriLIT" cluster with 2× Intel Xeon E5-2695v3 using AVX2, as it doesn't support AVX512 (blue line). To build our program we have used Intel Compiler with optimization flag -O3.

The computational time, the speedup, and the efficiency of the parallel calculations are shown on Figs. 2, 3, and 4, respectively. On Fig. 2 we can see, that computational time on HybriLIT cluster is by 7-10 times lower than on JSCC cluster even though the base frequency per core of HybriLIT is just 1.4–2.4 times higher than on JSCC cluster. We can see noticeable slow down in speedup (Fig. 3) and efficiency (Fig. 4), which occurs after increasing number of computational processes beyond 28 for HybriLIT and 68 for JSCC cluster. This number of processes coincides with number of individual cores of used CPUs. To use higher number of processes, hyper threading had to be used.

One can see interesting behavior when approaching to the maximal number of physical cores, when the efficiency of calculation reaches 95% for HybriLIT cluster and about 85% for JSCC cluster. This phenomenon can be because of the hardware architecture or a special feature of our data or both of them. Anyway it will need further investigation to identify.

4 Conclusions and Discussion

We have introduced the algorithm solving TOV equations for set of EoS models, which is parallelized on the level of different models. Such algorithm is important for investigation of models of cold and dense nuclear matter. On testing set of modern realistic EoS models we have shown how it scales when using with MPI on modern CPUs. This tests allowed us also to compare two different types of computational nodes. Ones based on Intel Xeon Phi 7250 multi-core processors and ones based on Xeon E5-2695v3 processors. We could see, that efficiency of Intel Xeon Phi is higher when using more than 28 cores, which is number of physical processors provided by 2× Xeon E5-2695v3 which we had in our testing system. Nevertheless the computational times were much shorter when using Xeon E5. The real advantage of using Xeon Phi would be when running much larger number of MPI processes in parallel (up to 1088 for our setting of 4× Xeon Phi).

Note, that the difference in running times for both systems for the same number of threads cannot be explained only by difference in frequencies. The performance of the Intel Xeon Phi 7250 is 10~11 times smaller than of Xeon E5-2695v3 when using the same amount of threads while Xeon E5-2695v3 has only about 1.64 times higher frequency. Such dramatic difference of the performances can be explained by the features of the architecture, and first of all, in memory access organization. For more details see the extensive performance comparison of the modern Intel architectures [14].

For problems with several thousands of models or with larger number of free parameters, also calculation of single model as well as transfer of data will start to play role. In this case we will consider to use different approach to load files (move from text to binary files) and to transfer data – probably send just pointers to where the data is stored and how to obtain it. This will reduce communication costs, which are for small problems negligible. Also we will consider CUDA as even calculating solution to TOV equations of single EoS model is embarrassingly parallel. We have not used MPI to integrate TOV equations for single EoS model in parallel as the communication costs would be much higher compared to the calculation itself, but using CUDA we are not be bound by waiting for MPI signals and therefore we can get some performance boost.

Acknowledgements

The research was carried out under financial support of the Russian Science Foundation (project No. 17-12-01427).

References

[1] Fridolin Weber. *Pulsars as Astrophysical Laboratories for Nuclear and Particle Physics* (CRC Press, Boca Raton, 1999), 696 pp.

[2] R. C. Tolman, Phys. Rev. **55**(4), pp. 364–373 (1939)

[3] J. R. Oppenheimer and G. M. Volkoff, Phys. Rev. **55**(4), pp. 374–381 (1939)

[4] D. Blaschke, H. Grigorian, and D. N. Voskresensky, Phys. Rev. C **88**, 065805 (2013)

[5] S. Typel, G. Röpke, T. Klähn, D. Blaschke, and H. H. Wolter, Phys. Rev. C 81, 015803 (2010)

[6] K. A. Maslov, E. E. Kolomeitsev and D. N. Voskresensky, Phys. Rev. C **92**(5), 052801 (2015)

[7] K. A. Maslov, E. E. Kolomeitsev and D. N. Voskresensky, Nucl. Phys. A **950**(64) (2016)

[8] R. H. Merson, Proc. Symp. Data Processing, Weapons Res. Establ. Salisbury, Salisbury (1957) pp. 110–125

[9] J. Christiansen, Numer. Math. **14**, pp. 317–324 (1970)

[10] *MPI: A Message-Passing Interface Standard Version 3.1.* (2015), http://mpi-forum.org/docs/mpi-3.1/mpi31-report.pdf

[11] D. Alvarez-Castillo, A. Ayriyan, S. Benic, D. Blaschke, H. Grigorian, and S. Typel. European Physical Journal A **52**:69, (2016), doi:10.1140/epja/i2016-16069-2

[12] Joint Supercomputer Center of the Russian Academy of Sciences, http://www.jscc.ru/eng/index.shtml (14.11.2017)

[13] Heterogeneous cluster "HybriLIT", http://hybrilit.jinr.ru/en/ (14.11.2017)

[14] I. Jabbie, G. Owen, and B. Whiteley. SIAM Undergraduate Research Online (SIURO) **10**, (2017), doi:10.1137/17S015896

EPJ Web of Conferences **177**, 07002 (2018) https://doi.org/10.1051/epjconf/201817707002
AYSS-2017

Ef: Software for Nonrelativistic Beam Simulation by Particle-in-Cell Algorithm

A. Yu. Boytsov[1,⋆] and *A.A.* Bulychev[2]

[1] *Veksler And Baldin Laboratory of High Energy Physics, Joint Institute for Nuclear Research, 141980 Dubna, Moscow region, Russia*
[2] *Independent Researcher, Moscow, Russia*

Abstract. Understanding of particle dynamics is crucial in construction of electron guns, ion sources and other types of nonrelativistic beam devices. Apart from external guiding and focusing systems, a prominent role in evolution of such low-energy beams is played by particle-particle interaction. Numerical simulations taking into account these effects are typically accomplished by a well-known particle-in-cell method. In practice, for convenient work a simulation program should not only implement this method, but also support parallelization, provide integration with CAD systems and allow access to details of the simulation algorithm. To address the formulated requirements, development of a new open source code – *Ef* – has been started. It's current features and main functionality are presented. Comparison with several analytical models demonstrates good agreement between the numerical results and the theory. Further development plans are discussed.

1 Introduction

Nonrelativistic electron and ion beams are widely used in nuclear and particle physics, nanotechnology, microelectronics, medicine and other areas of science and technology. Numerical simulations of charged particles dynamics is essential step in order to achieve required device parameters. Nonrelativistic beams have a range of specific properties that simulation software should take into account.

First, at low energies particle-particle interaction (space charge) plays prominent role and has to be considered alongside with external fields. Particle-in-cell algorithm is typically used for this purpose (see the next section for more details). Besides, nonrelativistic kinematics can be utilized to simplify calculations without too much loss in accuracy, whereas relativistic regime can be provided as an option. Due to large number of particles (tens of millions and more) and long time durations to simulate (milliseconds, while characteristic time scale is nanoseconds) the problems demand extreme amounts of computational resources. Parallelization is the only means to provide them, and, therefore, it is required to accomplish the modelling in reasonable amount of time.

Another important aspect is device geometry, which plays major role at low beam energies. Integration with CAD systems is highly desirable to simplify geometry specification. A useful feature would be an option to automate search for optimal device configuration and it's various parameters.

⋆e-mail: boytsov@jinr.ru

For practical means, such properties as intensity, emittance, current fluctuations, particle energy spectrum, phase profiles and particle trajectories are of most interest. An ideal program should provide tools to display and analyze them easily.

To comply with principle of reproducible research [1], the source code of the program should be open and available for examination.

Apart from these general requirements, there are application-specific concerns. In particular, authors' motivation stems from modelling of electron beam ion sources (EBIS). Such sources are commonly used to produce high-density beams of highly charged ions. Their principle of operation employs an electrostatic trap to hold the ions while ionizing them to desired charge state within an electron beam. In EBIS different spatial and time scales are involved: first, slow ions and fast electrons have incomparable time evolution scales, second, transversal size of the beam (less than 1 mm radius) is much less than longitudinal trap length (50-100 cm). Besides, various instabilities may form that can destroy steady state of the electron beam or heat confided ions, leading to ion losses and lack of ionization efficiency. Such properties require fine control over details of the simulation algorithm so that it is possible to modify and experiment with it.

A representative list of existing software implementing particle-in-cell method can be found at [2, 3]. Warp, PIConGPU and Smilei [4–6] can be mentioned specially: all of them are professionally developed and overall of very high quality. Unfortunately, while source code of these programs is open, it seems that massive changes in the architecture would be required to incorporate certain necessary features. Apart from that, there are several freely distributed (SNOP [7]) and commercial packages (CST-Studio, Opera3d, SIMION, EGUN-IGUN, etc). However, they do not provide access to the source code which makes it impossible to customize them.

To address the aforementioned requirements, development of a new program – named *Ef* – specialized for low-energy beams modelling has been started [8]. While writing new code is a laborious task, it is an optimal solution in the long-term perspective. Currently the program focuses on simulation of electron beam ion sources, but it also should be useful for modelling of other low-energy charged particle beam and (possibly) plasma devices.

2 Implementation Details

The basis of the created program is a so-called "particle-in-cell" method. Initially suggested in the 1950s for hydrodynamics simulations and popularized in the 1960s for plasma research, it is well described in the literature [9, 10]. The method aims to model dynamics of a large number of charged particles in external electromagnetic fields taking into account particle-particle interaction. First, the equations of motion of individual particles are discretized, so that time evolution can be viewed as a sequence of time steps. At each time step, forces acting on particles consist of two components: the external fields and particle self-interaction. The external fields are considered preset. As for the particle-particle interaction, the idea is to approximately compute electromagnetic fields generated by the particles instead of calculating binary Coulomb forces between each pair of them. Such collective fields are obtained by numerically solving Maxwell's equations, which requires to perform spatial discretization of the simulation volume, approximate contribution of each particle to charge density and current at each node, and solve for potentials on the mesh using finite difference or finite element method. After the combined action of the external and particle-generated fields on each particle is computed, integration of the dynamics equations is performed.

The current version of the program [8] implements nonrelativistic dynamics with Boris time integration scheme [11], which is notable for it's long term energy conservation [12].

For interparticle interaction electrostatic particle-in-cell method is used. It takes into account only electric fields generated by particles, so that on each time step the Poisson equation $\Delta\phi = -\rho/\epsilon_0$ is solved. The Laplace operator Δ is discretized using finite difference scheme. To compute charge density ρ, macroparticle's charge is distributed between the nodes of enclosing cell in reverse proportion to distance to the node (for each particle, only enclosing cell nodes are used). Such algorithm is surprisingly efficient for large array of applications, and more advanced versions can be implemented in the future.

To run the simulation, it is necessary to prepare a configuration file describing the simulation domain. Currently the program supports specification of several sources of particles with various properties, such as charge, mass, mean momentum, etc. It is possible to choose between particle-in-cell method and noninteracting particles for computations of space charge. External electric and magnetic fields can be specified, either uniform or by simple analytical expression. It is possible to define inner regions, that absorb particles on collisions and act as conductors under fixed potential. During the computation, the program outputs files with simulation results, containing all the information inside the simulation domain (HDF5 library and file format are used [13]). Such files can be used as a starting point for new computation, which is convenient to resume interrupted execution.

The program is implemented in C++. It's source code is free and open and available for modification and redistribution under MIT license.

3 Examples

To test correctness of the code, we have performed comparisons of the numerical results with several analytical models.

First tests have been devoted to evolution of a single particle in free space and uniform electric and magnetic fields. The particle-in-cell algorithm is not used in these computations, and only time integration scheme is tested. Numerical results demonstrate good agreement with the theory [8].

The simplest example with particle-particle interaction is propagation of an electron beam in free space. For a beam with rectangular cross section (ribbon beam) moving along the Z-axis, with "height" (dimension along the Y-axis) much larger than "width" (dimension along the X-axis), under a number of simplifying assumptions it is possible to derive a trajectory of an electron located on the edge of the beam [14] (which can be interpreted as it's envelope):

$$x(t) = x_0 + \frac{qI_0}{4\epsilon_0 m v} t^2, \qquad z(t) = vt.$$

In these relations, I_0 is linear current density per unit of "height" and v is particles speed. To check them numerically it is possible to perform a simulation of a beam propagating in free space and plot projections of particles positions on the X-Z plane. The result are shown on the left panel of figure 1.

In practice, it is common to use magnetic field to prevent beam divergence. In the case of ribbon beam propagating in uniform magnetic field, trajectory of the electron on the edge of the beam is given by the following equation [14]:

$$x(t) = x_0 + \frac{I_0 m}{2\epsilon_0 q v B^2}\left(1 - \cos(\omega_g t)\right), \qquad z(t) = vt, \qquad \omega_g = \frac{qB}{m}.$$

Numerical results can be compared with analytical predictions using the same procedure as in the field-free case. The results are shown on the right panel of figure 1. In both cases numerical results are in good agreement with the theory.

Another example we would like to consider is propagation of a cylindrical electron beam inside a conducting tube. By solving Poisson equation

$$\Delta\phi_1 = -\rho/\epsilon_0, \quad 0 \le r \le r_{beam}$$
$$\Delta\phi_2 = 0, \quad r_{beam} \le r \le r_{wall}$$

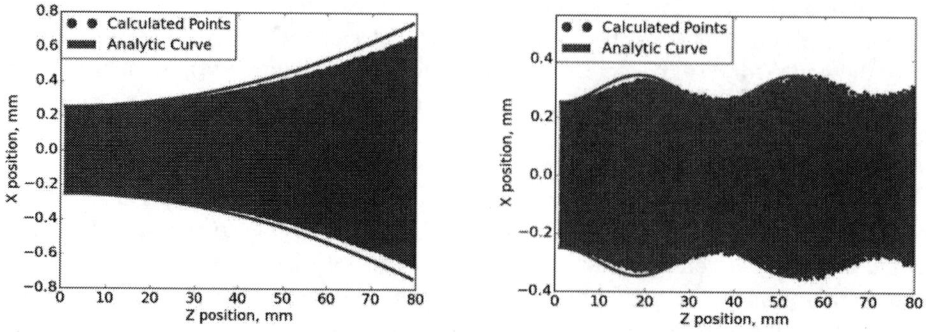

Figure 1. Numerical simulations and analytical expressions for envelope of an electron beam propagating in free space (left) and uniform magnetic field (right). Total simulation time is 5 [ns], time step is $5 \cdot 10^{-12}$ [s]. Beam current is 0.008 [A], particles' speed corresponds to 1 [keV] energy. 500 new macroparticles are added each time step. For the right figure, magnetic field is $B = 0.02$ [T].

with boundary conditions $\phi_1(r_{beam}) = \phi_2(r_{beam})$, $\phi_1'(r_{beam}) = \phi_2'(r_{beam})$, $\phi_2(r_{wall}) = U_0$ it is possible to obtain potential distribution inside the tube. Numerical and analytical results are compared on the figure 2. It can be seen, that agreement is decent.

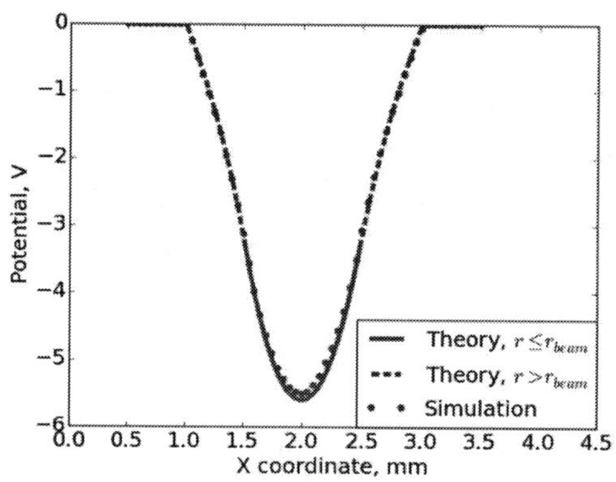

Figure 2. Potential well inside a conducting tube with an electron beam. Beam current $I = 0.01$ [A], beam radius $r_{beam} = 0.5$ [mm], inner radius of the tube $r_{wall} = 1$ [mm], and potential on the tube $U_0 = 0$ [V].

4 Conclusion

Simulation of nonrelativistic low-energy beams has a number of features that distinguish them from other areas of charged particles modelling. To take them into account, development of new specialized code has been started. The current version of the program has functionality sufficient to perform basic simulation tasks and numerical results are in good agreement with predictions of several simple analytical models.

Speaking of development plans, our current long-term goal is simulation of an electron beam ion source in reflective operation mode [15]. This requires various improvements to physical and mathematical models, such as support for particle-particle magnetic interaction, collisions and ionization, multigrid Poisson solver, advanced macroparticle shapes, etc. From a technical side, top priority is GPU acceleration. Besides, current experimental features: graphical user interface, import of 3d models and precomputed external fields - need to be improved.

Authors would like to thank Dr. E.E. Donets for help and discussion. One of the authors (A.Yu.) would like to acknowledge support from JINR AYSS Grant. Computations were held on the basis of the heterogeneous computing cluster HybriLIT (LIT, JINR).

References

[1] R.D. Peng, Science **334**, 1226 (2011)

[2] *Accelerator physics codes*, Wikipedia, `https://en.wikipedia.org/wiki/Accelerator_physics_codes`

[3] *Electromagnetic particle-in-cell computational applications*, Wikipedia, `https://en.wikipedia.org/wiki/Particle-in-cell#Electromagnetic_particle-in-cell_computational_applications`

[4] A. Friedman, R.H. Cohen, D.P. Grote, S.M. Lund, W.M. Sharp, J.L. Vay, I. Haber, R.A. Kishek, IEEE Transactions on Plasma Science **42**, 1321 (2014)

[5] M. Bussmann, H. Burau, T.E. Cowan, A. Debus, A. Huebl, G. Juckeland, T. Kluge, W.E. Nagel, R. Pausch, F. Schmitt et al., *Radiative Signatures of the Relativistic Kelvin-Helmholtz Instability*, in *Proceedings of the International Conference on High Performance Computing, Networking, Storage and Analysis* (ACM, New York, NY, USA, 2013), pp. 5:1–5:12

[6] J. Derouillat, A. Beck, F. Pérez, T. Vinci, M. Chiaramello, A. Grassi, M. Flé, G. Bouchard, I. Plotnikov, N. Aunai et al., Computer Physics Communications (2017)

[7] V.L. Smirnov, S.B. Vorozhtsov, *SNOP-Beam dynamics analysis code for compact cyclotrons*, in *Proceedings of the XXI Russian Accelerator Conference* (2012), pp. 325–327

[8] *Ef: Software for nonrelativistic beam simulation by particle-in-cell algorithm*, `https://github.com/epicf/ef/wiki`

[9] R.W. Hockney, J.W. Eastwood, *Computer Simulation Using Particles* (CRC Press, 1988)

[10] Y.N. Grigoryev, V.A. Vshivkov, M.P. Fedoruk, *Numerical "Particle-in-Cell" Methods: Theory and Applications* (VSP, 2002)

[11] J. Boris, *Relativistic Plasma Simulation—Optimization of a Hybrid Code*, in *Proceedings of the Fourth Conference on Numerical Simulation of Plasmas* (Naval Research Laboratory, Washington DC, 1970), pp. 3–67

[12] H. Qin, S. Zhang, J. Xiao, J. Liu, Y. Sun, W.M. Tang, Physics of Plasmas **20**, 084503 (2013)

[13] The HDF Group, *Hierarchical Data Format, version 5* (1997-2017), http://www.hdfgroup.org/HDF5/

[14] I.V. Alyamovsky, *Electron Beams and Electron Guns* (Soviet Radio, 1966)

[15] E.D. Donets, D.E. Donets, E.E. Donets, V.V. Salnikov, V.B. Shutov, Y.A. Tumanova, V.P. Vadeev, M. Bjorkhage, A. Lundquist, L. Liljeby et al., Review of Scientific Instruments **73**, 679 (2002)

EPJ Web of Conferences **177**, 07003 (2018) https://doi.org/10.1051/epjconf/201817707003
AYSS-2017

U-238 fission and Pu-239 production in subcritical assembly

Magdalena Grab[1,*] and *Andrzej* Wojciechowski[2,3,**]

[1] *Institute of Physics, Jagiellonian University, ul. prof. Stanislawa Lojasiewicza 11 30-348 Krakow, Poland*
[2] *National Center for Nuclear Research, Otwock-Swierk, Poland*
[3] *Joint Institute for Nuclear Research, 141980 Dubna, Russia*

Abstract. The project touches upon an issue of U-238 fission reactions and Pu-239 production reactions in subcritical assembly. The experiment took place in November 2014 at the Dzhelepov Laboratory of Nuclear Problems (JINR, Dubna) using PHASOTRON.Data of this experiment were analyzed in Laboratory of Information Technologies (LIT). Four MCNPX models were considered for simulation: Bertini/Dresnen, Bertini/Abla, INCL4/Drensnen, INCL4/Abla. The main goal of the project was to compare the experimental data and simulation results. We obtain a good agreement of experimental data and computation results especially for detectors placed besides the assembly axis. In addition, the U-238 fission reactions are more probable to be observed in the region of a higher particle energy spectrum, located closer to the assembly axis and the particle beam as well and vice versa Pu-239 production reactions were dominant in the peripheral region of geometry.

1 Introduction

The electronuclear method of neutron generation using a spallation reaction in Accelerator Driven System (ADS) has been studied by many scientist worldwide [1-9]. The efficiency of the neutron production based on spallation reaction depends on many parameters such as: beam energy, kind of particles and kind and size of target [3, 4, 6]. These neutrons can be employed for transmutation of the long-lived isotopes existing in nuclear waste [5, 7, 9] and for nuclear energy production from thorium or spent uranium fuel [1, 2, 8, 9]. The main aim of this work is to analyze the U-238(n,f) and U-238(n,g) reactions in QUINTA assembly [7, 9]. This work presents a comparison of experimental data and calculated results predicted by MCNPX code [10].

2 The Experiment

The experiment was employed QUINTA assembly run with a PHASOTRON in Dzhelepov Laboratory of Nuclear Problems in JINR, Dubna (see [8] for more details). Experimental data come from experiment run by Professor Voronko and his group of scientists from Ukraine.

*e-mail: magdalena.grab@doctoral.uj.edu.pl
**e-mail: Andrzej.Wojciechowski@ncbj.gov.pl

The research described in this paper was a part of experimental collaboration "Energy and Transmutation of Radioactive Waste" as well as the experiment which is the source of data. Research connected with particle beam profile was run by Lukas Zavorka from Czech Republic. Experimental data were obtained for activation detector of natural uranium. Activation detectors were located on six aluminum holders at the distance R from assembly axis equal to 0, 4, 8, and 12 cm. These holders were placed between all sections and before the first and after the last section (Z =0, 12.3, 25.4, 38.5, 51.6, 64.7 cm). In this experiment were measured the fission products: Zr-97, I-131, I-133 and Ce-143 by yield of gamma-lines 743.36 keV, 364.49 keV, 529.9 keV, and 293.3 keV respectively. The number of the fission reactions was determined by measured value of fission product and following neutron fission yield: Zr-97- 5.7%, I-131- 3.6% , I-133- 6.3%, Ce143-4.3%[8]. It means that, experimental data of the number of the fission reactions include neutron induced fission reaction only. The number of Pu-239 production was determined by measuring number of decay reactions Np-239(β-). It was assumed that the axis of the assembly and the beam are parallel. This experiment employed a of 0.66 GeV proton beam.

3 Calculation results and experimental data

This section gives a graphical presentation of computation results vs. experimental data. On every plot black, bold lines and full symbols depict experimental data. On Fig.1 (left) shows the spallation neutron yield vs. proton energy for natural uranium target compared to calculations. The Bertini/Dresnen model is in the best accordance to experimental data (Fig.1 left). The number of neutrons produced by spallation reaction depends on the initial particle energy and this magnitude rises with the energy.[3, 4] The results of Bertini/Dresnen model shows that the number of neutrons occurring due to spallation reaction induced by proton with energy E=0.66 GeV is approximately 20 neutrons/proton but when the energy of a particle is rised to E=1.5 GeV the amount of neutrons also rises to approximately 60. For energy levels smaller or equal to 660 MeV models give results less than experimental data about 17%. The radial distribution of plutonium production reaction U-238(n,g) is a decreasing function of distance R from the assembly axis (Fig.1, right), compare [1, 2]. Calculation for Z=123 turned out to be more accurate comparing to experimental data. The axial distributions of U-238(n,g) and U-238(n,f) reactions have local maximum for Z=254 for different distance from assembly axis R=40, 80, 120 (Fig.2).The number of reactions decreases when the radial location increases. This effect was expected because of the geometry in the middle of section there is the smallest escape probability of the particles inducing the both reactions Please note, that agreement of calculation results and experimental data for fission reactions are significantly better than for capture reaction (Fig.2). However the radial distribution of fission reactions show a big disagreement for detectors placed on the assembly axis i.e. for R=0 (see Fig.3 left). This figure presents comparison number of fission reactions induced by neutron only for both experimental data and calculation results. If we take into account proton and neutron induced fission reaction in calculation results we obtain better agreement (Fig.3 right). Important magnitude useful for this research is a spectral index defined by the following formula:

$$SI = \frac{N[U - 238(n, g)]}{N[U - 238(n, f)]} \tag{1}$$

Where:

SI- Spectal Index
N[U-238(n,g)] – number of U-238(n,g) capture reactions

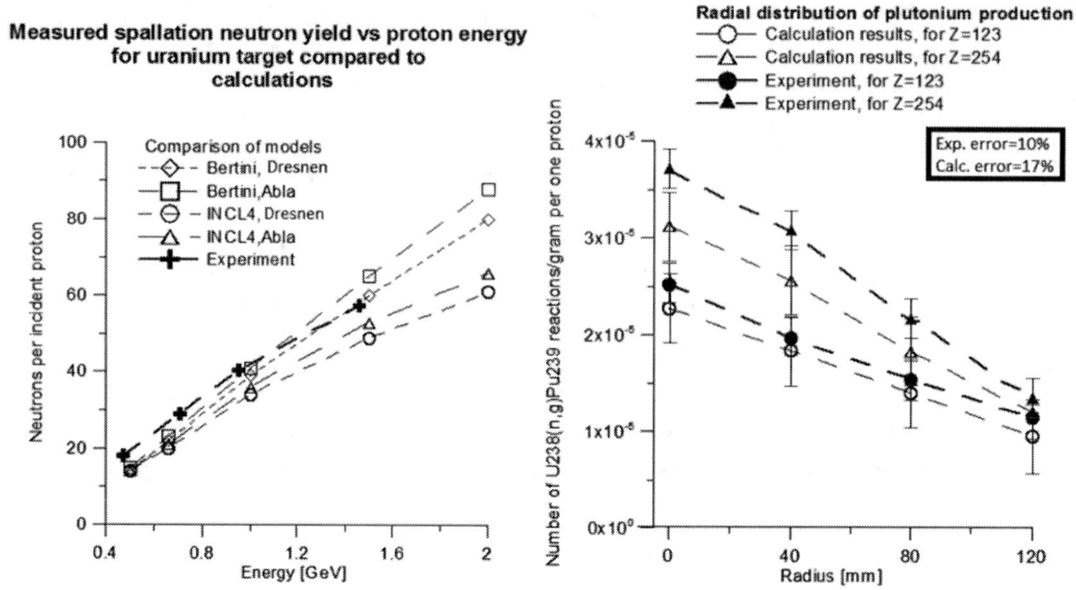

Figure 1. Spallation neutron yield for uranium target (left; experimental data from [4, 6]) and radial distribution of U-238(n,g) reaction (right).

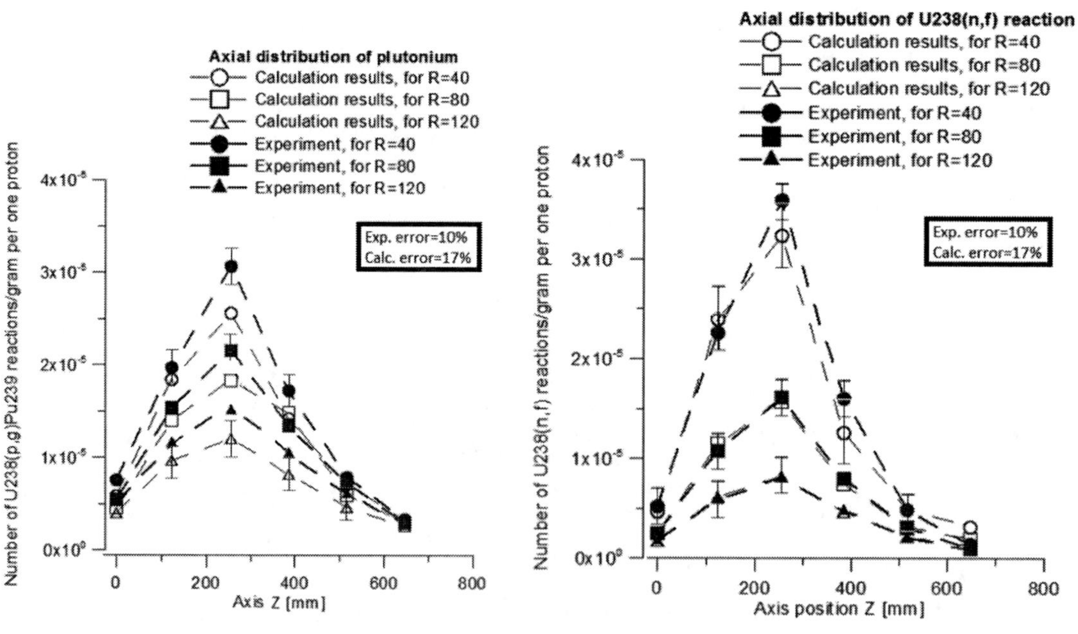

Figure 2. Axial distribution of plutonium production (left) uranium fission (right).

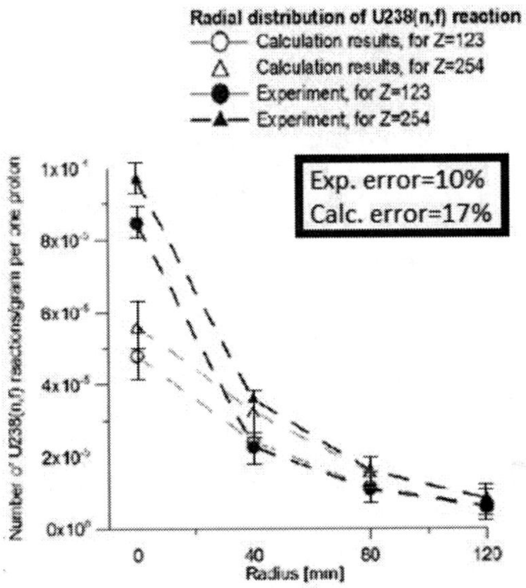

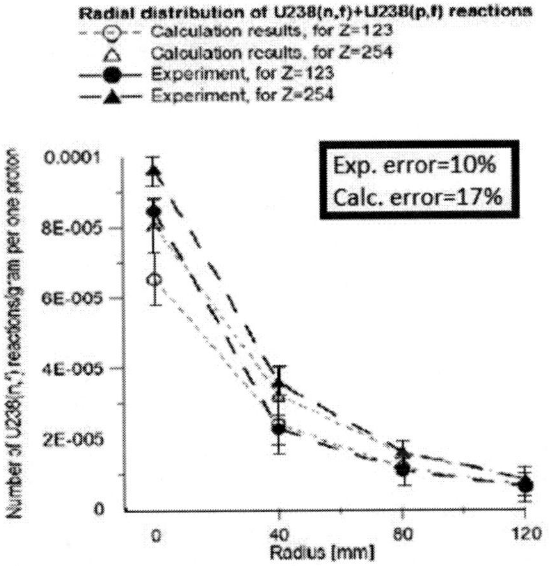

Figure 3. Radial distributions of U-238 fission reactions before (left) and after (right) considering fission reactions induced by protons.

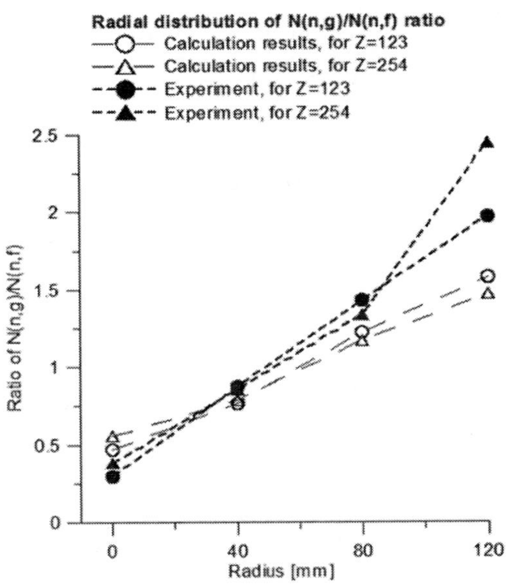

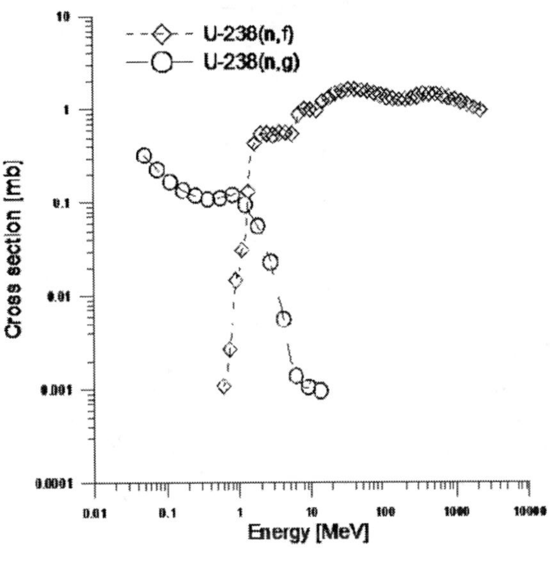

Figure 4. Radial distribution of Spectral Index (left) and cross section of U-238(n,f) and U-238(n,g) reactions - the evaluated data come from JEDL library. (right).

N[U-238(n,f)] – number of U-238(n,f) fission reactions

Spectral index is often used because it does not depend on normalizing method of neutron flux. In other words, it is not sensitive on normalizing errors. The spectral index value is an increasing function of radius (Fig.4 left). The reason is simple-the average neutron energy is decreasing function of radius because neutron loses its energy during scattering reactions when it moves away from the neutron beam radially. Additionally radially decreasing neutron energy cause decreasing U-238(n,f) fission cross section and increasing U- 238(n,g) capture cross section (see Fig.4 right). That is why the denominator decreases and the nominator increases of Eq.(1) simultaneously and as the whole fraction increases.(compare [1, 2]).

4 Conclusions

The calculation results for U-238 fissions reactions are in good accordance to experimental data besides detectors placed on the assembly axis (compare Fig.3 left and right). The reason for this discrepancy was that there was not taken into account an experimental number of fission reaction induced by high energy particles of beam and not enough precisely determined the beam parameters. Our recent studies confirm that taking into account proton and neutron induced fission reaction in experimental data and correcting the angle between the assembly and the beam axis, one can a obtain good accordance for all detectors. For energy levels smaller or equal to 660 MeV the models give results less than experimental data about 17% (Fig.1). Maximal difference between the presented models is equal to about 20% (Fig.1). The difference between experimental data and calculation results for U-238(n,g) reactions (Fig.2) is greater than experimental and calculation errors. This result suggests that the low energy fraction of the neutron flux in the experiment was greater than in calculation simulation. The reasons this situation can be in that simulation model not take into account all materials used in experiment for instance: pieces of paper or scotch tape or plastic box around detectors. As we know these materials have a very low mass but they are very good moderators and are placed directly on the detectors.

5 Acknowledgment

One of the authors (M. Grab) thanks the Polish Ministry of Science and Higher Education (through JINR Dubna) for the support.

References

[1] A. Wojciechowski *et al.*, JINR Reports **12-13**, 102-106 (2012).

[2] A. Wojciechowski *et al.*, JINR Reports **12-13**, 98-101 (2012).

[3] L. Pienkowski *et al.*, PHYSICAL REVIEW C **Vol.56, No. 4**, 1909-1917 (1997).

[4] J. Carpenter, www.neutron.anl.gov, **NeutronProduction**, 1-23 (2004).

[5] A. Baldin *et al.*, et.jinr.ru/preprints **E1-2011-24**, 1-15 (2011).

[6] J. Frazer *et al.*, Measured Spallation Neutron Yield vs. Proton Energy for Various Targets,(1965).

[7] L. Zavorka *et al.*, Validation of Monte Carlo simulation of neutron production in a spallation experiment, Annals of Nuclear Energy **80**, 178-187 (2015).

[8] I. Adam, A. Baldin *et al.*, The study of Spatial Distribution of Neutron Capture and fission Reactions in massive Uranium Target Irradiated by deuterons with Energies of 1-8 GeV (Quinta setup), Preprint of JINR, **P12012-147**, (2012).

[9] I. Adam *et al.*, Study of Deep Subcritical Electronuclear Systems and Feasibility of their application for Energy Production and Radioactive Waste Transmutation, Preprint JINR, **E1-2010- 61**, (2010).

[10] D.B. Pelowitz *et al.*, *MCNPX 2.7.0 Extensions*, (Los Alamos National Laboratory, Los Alamos, 2011) 80.

EPJ Web of Conferences **177**, 07004 (2018) https://doi.org/10.1051/epjconf/201817707004
AYSS-2017

Effective Methods for Solving Band SLEs after Parabolic Nonlinear PDEs

Milena Veneva[1,⋆] and *Alexander* Ayriyan[1,⋆⋆]

[1] *Joint Institute for Nuclear Research, Laboratory of Information Technologies, Joliot-Curie 6, 141980 Dubna, Moscow region, Russia*

Abstract. A class of models of heat transfer processes in a multilayer domain is considered. The governing equation is a nonlinear heat-transfer equation with different temperature-dependent densities and thermal coefficients in each layer. Homogeneous Neumann boundary conditions and ideal contact ones are applied. A finite difference scheme on a special uneven mesh with a second-order approximation in the case of a piecewise constant spatial step is built. This discretization leads to a pentadiagonal system of linear equations (SLEs) with a matrix which is neither diagonally dominant, nor positive definite. Two different methods for solving such a SLE are developed – diagonal dominantization and symbolic algorithms.

1 Introduction and Mathematical Model

In this note, we focus on solving pentadiagonal (PD) and tridiagonal (TD) systems of linear algebraic equations (SLEs) which are obtained after the discretization of parabolic nonlinear partial differential equations (PDEs), using finite difference methods (FDM) of second-order approximation. Such a problem was solved in [1]. There, a finite difference scheme of first-order approximation was built that leads to a TD SLE with a diagonally dominant coefficient matrix. The system was solved using the Thomas method ([2]). The following nonlinear model of a cylindrical multilayer structure is considered:

$$\rho^m(u)\, c_v^m(u)\, \frac{\partial u}{\partial t} = \frac{1}{r}\frac{\partial}{\partial r}\left(r\,\lambda^m(u)\,\frac{\partial u}{\partial r}\right) + \sum_{\alpha=1}^{n} \frac{\partial}{\partial z_\alpha}\left(\lambda^m(u)\,\frac{\partial u}{\partial z_\alpha}\right) + \varphi^m(u); \tag{1}$$

$$\frac{\partial u}{\partial r} = 0 \qquad \forall\, r \in \{r_{\min}, r_{\max}\}; \tag{2}$$

$$-\lambda^m(u)\,\frac{\partial u}{\partial r}\bigg|_{r=r^m_{i^*}-0} = -\lambda^{m+1}(u)\,\frac{\partial u}{\partial r}\bigg|_{r=r^m_{i^*}+0} \quad \text{and} \quad u|_{r=r^m_{i^*}-0} = u|_{r=r^m_{i^*}+0}, \tag{3}$$

where $(r, \vec{z}) \in \Omega \cup \partial\Omega$, $t \geq 0$; m – index of the subdomain. Equation (1) represents the conservation of heat inside a multilayer structure. It is an energy equation with conduction heat transfer, where the densities, the specific heat capacities, and the thermal conductivities depend on the temperature. For

⋆e-mail: milena.p.veneva@gmail.com
⋆⋆e-mail: ayriyan@jinr.ru

instance, in the two-dimensional case, Eq. (1) could be defined in a domain similar to the one shown in Fig. 1. Homogeneous Neumann boundary conditions (2) are applied on the outer boundaries in the radial direction. The ideal contact conditions (3) model the heat flux on the inner boundaries in the radial direction, where $r^m_{i^*}$ denotes the point of discontinuity. The numerical algorithms for solving the multidimensional governing equation (GE), using FDM (e.g. ADI algorithms ([3])), ask for a repeated SLE solution.

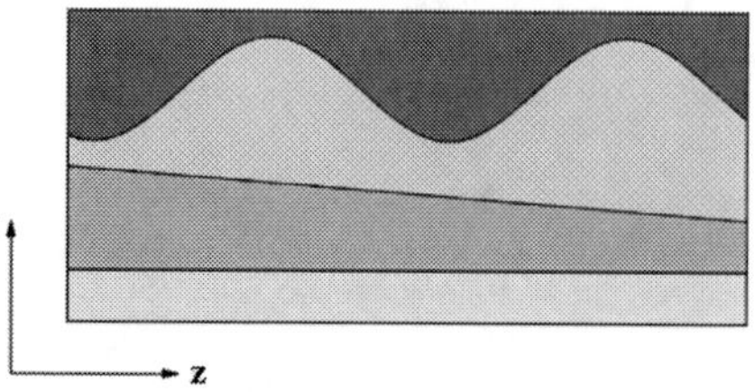

Figure 1. Example of a rectangular domain in cylindrical coordinates (a longitudinal section of a multilayer cylinder), where the thermal coefficients are different in all the subdomains and they have a discontinuity of the first kind at the borders in-between the subdomains

2 Discretization of the Problem

Focusing on the radial term of the GE with the assumption that the other terms will be moved to the right-hand side (RHS), we consider the following special mesh with grid points on the inner boundaries: $\overline{\omega}_r = \{(t,r) \,|\, t_k = k\,h_t,\, k \in \mathbb{N}_0;\, r_0 = r_{\min},\, r_{i+1} = r_i + h_{i+1},\, i = 0,\dots,N-2\}$. A finite difference scheme with a second-order approximation has the following form (three-point stencils are taken for the GE and the outer boundary conditions (BC), and five-point stencil for the inner BC):

$$\hat{\rho}_i\,\hat{c}_{v_i}\frac{\hat{u}_i - u_i}{\tau} = \frac{1}{r_i}\frac{1}{\hbar_i}\left[r_{i+1/2}\,\hat{\lambda}_{i+1/2}\frac{\hat{u}_{i+1}-\hat{u}_i}{h_{i+1}} - r_{i-1/2}\,\hat{\lambda}_{i-1/2}\frac{\hat{u}_i-\hat{u}_{i-1}}{h_i}\right] + \hat{\varphi}_i; \tag{4}$$

$$-\frac{h_2\,(2h_1+h_2)\hat{u}_0 - (h_1+h_2)^2\hat{u}_1 + h_1^2\hat{u}_2}{h_1\,h_2\,(h_1+h_2)} = 0; \tag{5}$$

$$\hat{\lambda}^m_{i^*}\,\frac{h_{i^*-1}\,(2h_{i^*}+h_{i^*-1})\hat{u}_{i^*} - (h_{i^*}+h_{i^*-1})^2\hat{u}_{i^*-1} + h_{i^*}^2\hat{u}_{i^*-2}}{h_{i^*}\,h_{i^*-1}\,(h_{i^*}+h_{i^*-1})} = \tag{6}$$

$$= -\hat{\lambda}^{m+1}_{i^*}\,\frac{h_{i^*+2}\,(2h_{i^*+1}+h_{i^*+2})\hat{u}_{i^*} - (h_{i^*+1}+h_{i^*+2})^2\hat{u}_{i^*+1} + h_{i^*+1}^2\hat{u}_{i^*+2}}{h_{i^*+1}\,h_{i^*+2}\,(h_{i^*+1}+h_{i^*+2})};$$

$$\frac{h_{N-2}\,(2h_{N-1}+h_{N-2})\hat{u}_{N-1} - (h_{N-1}+h_{N-2})^2\hat{u}_{N-2} + h_{N-1}^2\hat{u}_{N-3}}{h_{N-1}\,h_{N-2}\,(h_{N-1}+h_{N-2})} = 0, \tag{7}$$

where

$$\hat{\lambda}_{i\pm1/2} = \lambda\left(\frac{\hat{u}_i+\hat{u}_{i\pm1}}{2}\right), \qquad \hbar_i = \frac{h_{i+1}+h_i}{2}, \qquad r_{i\pm1/2} = \frac{r_i+r_{i\pm1}}{2}.$$

The matrix form of the considered system is: $A\vec{\hat{u}} = \vec{\varphi}(\hat{u})$, where A is a PD sparse matrix which does not have any special properties, e.g. diagonally dominance or positive definiteness. In order to preserve the band structure of the matrix, we cannot use the Gaussian elimination with pivoting. Within this, we could obtain division by zero at some point of the procedure, which is going to make our algorithm unstable. For that reason, we alter the initial PD matrix by adding the minimum values to the diagonal elements so as to transform the matrix into a weakly diagonally dominant one:

$$A_{DD}\,\vec{\hat{u}} = \vec{\varphi}(\hat{u}) + P\,\vec{\hat{u}}, \qquad \text{where} \qquad A_{DD} = A + P;$$

$$P = \operatorname{diag}\left(2h_1^2,\, \delta_{i^*,j}\,p_{i^*,j},\, 2h_{N-1}^2\right), \qquad \text{where} \qquad p_{i^*,i^*} = \frac{2\lambda^m_{i^*}\,h_{i^*}}{h_{i^*-1}\,(h_{i^*}+h_{i^*-1})} + \frac{2\lambda^{m+1}_{i^*}\,h_{i^*+1}}{h_{i^*+2}\,(h_{i^*+1}+h_{i^*+2})}.$$

The Gaussian elimination with pivoting (the procedure we use to transform the PD matrix into a TD one) does not preserve the diagonal dominance of the matrix. The use of the Gaussian elimination to the initial matrix A (not A_{DD}) yields a transformed matrix $\tilde{A}$. Then, the diagonal dominantization method is used in order to transform the obtained TD matrix $\tilde{A}$ into a diagonally dominant one. To that purpose, the nondiagonal elements are added to the diagonal ones:

$$\tilde{A}_{DD}\,\vec{\hat{u}} = \vec{\varphi}(\hat{u}) + \tilde{P}\,\vec{\hat{u}}, \qquad \text{where} \qquad \tilde{A}_{DD} = \tilde{A} + \tilde{P};$$

$$\tilde{P} = \operatorname{diag}\left(|\tilde{A}_{0,1}|, \delta_{i^*,j}\,\tilde{p}_{i^*,j}, |\tilde{A}_{N-2,N-1}|\right), \qquad \text{where} \qquad \tilde{p}_{i^*,i^*} = \sum_{\beta\in\{-1,1\}} |\tilde{A}_{i^*,i^*+\beta}|.$$

3 Numerical and Symbolic Algorithms

Two different approaches for solving the SLE are considered – numerical and symbolic. The complexity of all the suggested numerical algorithms is $O(N)$. Since it is unknown what stands behind the symbolic library, evaluating the complexity of the symbolic algorithms is a very complicated task.

Numerical Algorithms. Two different numerical algorithms are applied to the system with a PD matrix. Both of them are based on LU decomposition. The first one ([4]) is intended for a dense PD matrix. In the case of the considered problem, the matrix is sparse. For that reason, a modified algorithm is built. The main idea is that after the mesh was defined, the indexes of the discontinuity points are known. Since these indexes coincide with those of the matrix rows which are not sparse, we can reference them to the algorithm and conduct the full calculation only for them. For the rest of the rows, the algorithm is reduced to a problem similar to solving a system with TD matrix. This way, the complexity of the algorithm is decreased but at the cost of additional $N + 2$ check-ups for the non-sparse rows. In the case of the TD matrix, the system is solved using the Thomas method.

Symbolic Algorithms. The symbolic algorithm in the case of a PD matrix is also based on LU decomposition ([4]). For the TD matrix, a symbolic version of Thomas method is created. As it is known, Thomas method is not suitable for nondiagonally dominant matrices ([2]). In order to cope with this problem, in the case of a zero quotient of two subsequent leading principal minors, a symbolic zero is assigned instead and the calculations are continued. At the end of the algorithm, this symbolic zero is substituted with zero. The same approach is suggested in [5].

Table 1. Results from solving SLE

	Wall-clock time [s]				
N	NPDM	MNPDM	SPDM	NTDM	STDM
10^3	0.000036	0.000034	0.088669	0.000021	0.043690
10^4	0.000403	0.000373	8.467241	0.000245	2.971745
10^5	0.004709	0.003916	3547.020851	0.002416	799.533587
10^8	3.159357	2.682258	–	1.652945	–
$\max_N \|\|y - \bar{y}\|\|_\infty$	2.22×10^{-16}	2.22×10^{-16}	0	2.22×10^{-16}	0
Complexity:	$19N - 29$	$13N + 7K - 14$	–	$9N + 2$	–
OS: Fedora 25; Processor: Intel Core i7-6700 (3.40 GHz); Compiler: GCC 6.3.1 (-O0).					

4 Implementation and Results

All the numerical algorithms are implemented using C++. The matrix needs to be nonsingular and diagonally dominant so as the methods to be stable. Two symbolic algorithms are implemented, using the GiNaC library (version 1.7.2) ([6]) of C++. The symbolic algorithms require the matrix to be nonsingular only. In Table 1 one can find the wall-clock time results from the conducted experiments. Since the largest supported precision in the GiNaC library is **double**, during all the experiments double data type is used. The notation is as follows: **NPDM** stands for numerical PD method, **MNPDM** – modified numerical PD method, **SPDM** – symbolic PD method, **NTDM** – numerical TD method, **STDM** – symbolic TD method. The achieved accuracy is summarized, using infinity norm. On the penultimate row of the table, one can find the complexity of all the considered methods. On the last row, the characteristics of the computer which is used are described. The number of considered discontinuity points is $K = 11$.

5 Discussion and Conclusions

A nonlinear heat transfer equation in a multilayer domain was considered. The suggested discretization scheme always has a first-order approximation. In the case of piecewise constant thermal conductivities or when $\|\hat{\lambda}_{i+1/2} - \hat{\lambda}_{i-1/2}\|_\infty = O(\|h_i\|_\infty)$, the order of approximation is going to be second. Focusing on the radial term, a SLE with a PD matrix was obtained. Then, applying Gaussian elimination, a TD matrix was derived. For both these matrices, a diagonal dominantization procedure was suggested. This approach ensures the stability of the suggested methods. A modified version of the numerical method for solving a SLE with a PD matrix was built. Since the complexity of this method is lower than the complexity of the general algorithm (usually $K \ll N$), better computational time was achieved. The fastest numerical algorithm was found to come from the Thomas method. All the experiments gave an accuracy of an order of magnitude of 10^{-16}. As a next step symbolic algorithms were used. They do not require the matrices to be of a special form and are exact. However, they are not comparable with the numerical algorithms with respect to the required time in the case of a numerical solving of the heat equation when one needs to solve the SLE many times. On the other hand, these symbolic methods are not as restrictive as the numerical ones when it comes to the matrix properties. Another upside of the symbolic algorithms is that in the case of a piecewise linear equation, they do not add nonlinearity to the RHS of the system and hence, there is no need of iterations for the time step to be executed. In future, the approach suggested in this note will be investigated in detail.

Acknowledgements

The authors want to express their gratitude to the Summer Student Program at JINR, and Ján Buša Jr. (JINR). A. Ayriyan thanks the JINR grant No. 17-602-01.

References

[1] A. Ayriyan, J. Buša Jr., E. Donets, H. Grigorian, and J. Pribiš, Applied Thermal Engineering **94**, 151–158 (2016)

[2] Nicholas J. Higham, *Accuracy and Stability of Numerical Algorithms* (SIAM, 2002) 680

[3] A. A. Samarskii, *The Theory of Difference Schemes* (Taylor & Francis Inc, New York, 2001) 788

[4] S. S. Askar and A. A. Karawia, Mathematical Problems in Engineering **2015**, 9 (2015)

[5] M. El-Mikkawy, Applied Mathematics **3**, 342–345 (2012)

[6] C. Bauer, A. Frink, and R. Kreckel, J. of Symb. Comp. **33**, 1–12 (2002)

EPJ Web of Conferences **177**, 07005 (2018) https://doi.org/10.1051/epjconf/201817707005
AYSS-2017

Application of Turchin's method of statistical regularization

Mikhail Zelenyi[1,2,*], *Mariia* Poliakova[1,2], *Alexander* Nozik[1,2,**], and *Alexey* Khudyakov[1,***]

[1] *Institute for Nuclear Research RAS*
[2] *Moscow Institute of Physics and Technology*

Abstract. During analysis of experimental data, one usually needs to restore a signal after it has been convoluted with some kind of apparatus function. According to Hadamard's definition this problem is ill-posed and requires regularization to provide sensible results. In this article we describe an implementation of the Turchin's method of statistical regularization based on the Bayesian approach to the regularization strategy.

1 Introduction

1.1 Problem

Consider the usual situation: detector measures a signal ($\varphi(x)$) and produces observed data ($f(y)$) via convolution with its own apparatus function ($K(x, y)$). The resulting observed data $f(y)$ contains a random noise factor both from initial statistical uncertainty of $\varphi(x)$ and additional noise from measurement procedure. In order to reconstruct initial signal, one needs to solve Fredholm equation of 1-st kind:

$$f(y) = \int dx\, K(x, y)\varphi(x), \tag{1}$$

but this equation is ill-possed: a small error in the measurement of $f(y)$ leads to big instability of $\varphi(x)$.

1.2 Solution

Solving such ill-possed problems requires additional operation called regularization. It means we need to introduce additional information to problem well-posed one. The idea of statistical regularization [1, 2] is to look on the problem from the point of view of Bayessian statistics approach: unknown statistical value $\varphi(x)$ could be reconstructed using measured value $f(y)$, the model ($K(x, y)$) and some prior information about $\varphi(x)$ behavior. Main features of statistical regularization:

- Based on Bayesian approach and decision theory (choice theory).
- It's able to accommodate various forms of prior information, such as smoothness, constraints on boundary conditions, non-negativity, etc.
- It allows estimation of errors for reconstructed function φ

*e-mail: mihail.zelenyy@phystech.edu
**e-mail: altavir@gmail.com
***e-mail: alexey.skladnoy@gmail.com

- Gives clear data-driven procedure for choice of regularization strength.

While the idea of the regularization was proposed a long time ago, it was not published in the journal with sufficient visibility and therefore left unknown to the broad scientific auditory. The aim of this work is both to reconstruct the initial work (create a modern implementation for described algorithms), and study its performance under different conditions.

Since original work is very hard to find and was published only in Russian language, we will present the short description of the method.

2 Description of statistical regularization's method

2.1 Strategy

Consider eq. 1 in operator form:

$$\hat{K}\varphi = f. \tag{2}$$

According to statistical decision theory ([3]), one can define a reconstruction strategy $\hat{\varphi} = \hat{S}[f]$, which uses a prior information. Good strategy minimize an impact of prior information (which could be wrong). That could be achieved by introducing the loss-function:

$$L(\varphi, \hat{S}[f]) = ||\varphi - \hat{S}[f])||_{L_2},$$

For this loss-function:

$$\hat{S}[f] = E[\varphi|f] = \int \varphi P(\varphi|f)d\varphi \tag{3}$$

Strategy depend on a prior information $P(\varphi)$:

$$P(\varphi|f) = \frac{P(\varphi)P(f|\varphi)}{\int d\varphi P(\varphi)P(f|\varphi)}$$

Errors of solution:

$$D(x_1, x_2) = E[\varphi(x_1) - \hat{S}[f](x_1)][\varphi(x_2) - \hat{S}[f](x_2)]$$

Let us consider what a prior information can we use?

2.2 A prior information

We expect that φ is relatively smooth and doesn't contain wild oscillations which, for example, we usually produce in use naive least squares. Common way of measuring function's smoothness is $\langle\varphi, \hat{\Omega}\varphi\rangle$, where $\hat{\Omega} = |\frac{d^2}{dx^2}\rangle\langle\frac{d^2}{dx^2}|$. At the same time we want to minimize information content of prior.

Reasonable choice is ensemble of functions with average smoothness ω which maximizes Shannon entropy (or is equivalent to minimize that value of Shannon's information, which we add to solution). Which means we need to maximize:

$$H[P(\varphi)] = -\int \ln P(\varphi)P(\varphi)d\varphi$$

Subject to constraints:

$$\int P(\varphi)d\varphi = 1 \quad \text{and} \quad \int \langle\varphi, \hat{\Omega}\varphi\rangle P(\varphi)d\varphi = \omega, \tag{4}$$

The resulting prior probability density for the Gaussian random process:

$$P(\vec{\varphi}|\alpha) = \frac{\det(\alpha\Omega)^{1/2}}{(2\pi)^{N/2}} \exp\left(-\frac{1}{2}\langle\vec{\varphi}, \alpha\Omega\vec{\varphi}\rangle\right), \tag{5}$$

where $\alpha = \alpha(\omega)$ — parameter of smoothness. In result, we get α-depend solution $\hat{S}_\alpha[f]$ and now we have problem of picking α. In rare cases we can select α manually using known empirical smoothness, but we proposal another way. Information about its value almost never available so we could consider it random variable as well. This yields us hierarchical Bayes model $P(f, \varphi, \alpha) = P(f|\varphi)P(\varphi, \alpha)P(\alpha)$. Since we don't have any information about α we chose hyperprior $P(\alpha)$ to be flat. Now we have several options to pursue. In general case, we can use fully Bayesian approach and compute solution to marginalize by α and to use eq. 3:

$$P(\varphi) = \int P(\varphi|\alpha)P(\alpha)\, d\alpha \tag{6}$$

or, to marginalize by φ and to integrate by $P(\alpha|f)$:

$$P(\alpha|f) \sim \int d\varphi\, P(f|\varphi)P(\varphi|\alpha)P(\alpha) \qquad \text{and} \qquad \hat{S}[f] = \frac{1}{Norm}\int d\alpha \hat{S}_\alpha[f]P(\alpha|f) \tag{7}$$

using MCMC for sampling φ and α (for Gaussian errors it could be reduced to numerical integral). In particular case we can simplify calculations by using following approximations. Also if distribution $P(\alpha|f)$ is narrow, we can pick α corresponding to most probable α_0 and use $P(\varphi|\alpha_0)$ as prior. This is called empirical Bayes. For our problem it produces almost identical results.

Derivation above assumed that Ω if positive definite. It possible to accommodate singular Ω. Then det in (5) should be replaced with pseudodeterminant. Positive definiteness is desirable since regularization is guaranteed to work only for non-singular Ω and singular or even near-singular Ω may and likely will cause numerical problems.

Also note that Ω we use is singular since $\varphi = kx + b$ will have smoothness equal to zero. One way fix it is to fix value of φ to 0 and boundaries or to put priors on values on boundaries[4]. In that case $\alpha\Omega$ in (5) should be replaced with $\sum_i \alpha_i\Omega_i$ where different Ω_i corresponds to smoothness, boundary values etc.

Also it's possible to add any other prior information about function. For example it possible to add non-negativity:

$$P(\varphi) \sim P_\alpha(\varphi)P(\varphi(x) > 0)$$

2.3 Discretization

In order to find numerical solution, one need to go from continuous functions in eq. 2 to discrete matrix form:

$$\varphi(x) = \sum_n \varphi_n T_n(x), \qquad \text{and} \qquad K_{mn} = (\hat{K}T_n(x))(y_m),$$

$$\Omega_{ij} = \int_a^b \left(\frac{d^pT_i(x)}{dx}\right)\left(\frac{d^pT_j(x)}{dx}\right) dx, \tag{8}$$

$$D[\varphi(x)] = D[\sum_n \varphi_n T_n(x)] = \sum_{i,j} \varphi_i\varphi_j cov(T_i(x), T_j(x)),$$

$$\vec{f} = K\vec{\varphi},$$

where $T_n(x)$ - some basis in function space. For example, cubic spline, Fourier series and Legendre polynomials.

2.4 Solution for Gaussian distributed noise

The most common experimental case of noise distribution is normal (or Gaussian) distribution:

$$P\left(\vec{f}|\vec{\varphi}\right) = \frac{1}{(2\pi)^{M/2}|\Sigma|^{1/2}} \exp\left(-\frac{1}{2}(\vec{f} - K\vec{\varphi})^T \Sigma^{-1} (\vec{f} - K\vec{\varphi})\right) \tag{9}$$

Using the most probable α, one can get the best solution in a simple form.

$$\hat{S}_{\alpha}[\vec{f}] = (K^T \Sigma^{-1} K + \alpha\Omega)^{-1} K^T \Sigma^{-1T} \vec{f}, \tag{10}$$

$$\mathrm{cov}_{\alpha}(\varphi_m, \varphi_n) = ||(K^T \Sigma^{-1} K + \alpha^* \Omega)^{-1}||_{mn} \tag{11}$$

For definition α, we use a posterior probability, which is calculated using Bayes' rule:

$$P(\alpha|\vec{f}) = C\alpha^{\frac{Rg(\Omega)}{2}} \sqrt{|(B + \alpha\Omega)^{-1}|} \exp(-\frac{1}{2} b^T B^{-1} b) \exp(\frac{1}{2} b^T (B + \alpha\Omega)^{-1} b), \tag{12}$$

where norming factor C doesn't depend on α. If we want to get more accurate solution, we should average solution over this posterior probability as it is describe in integrals 7.

3 Application

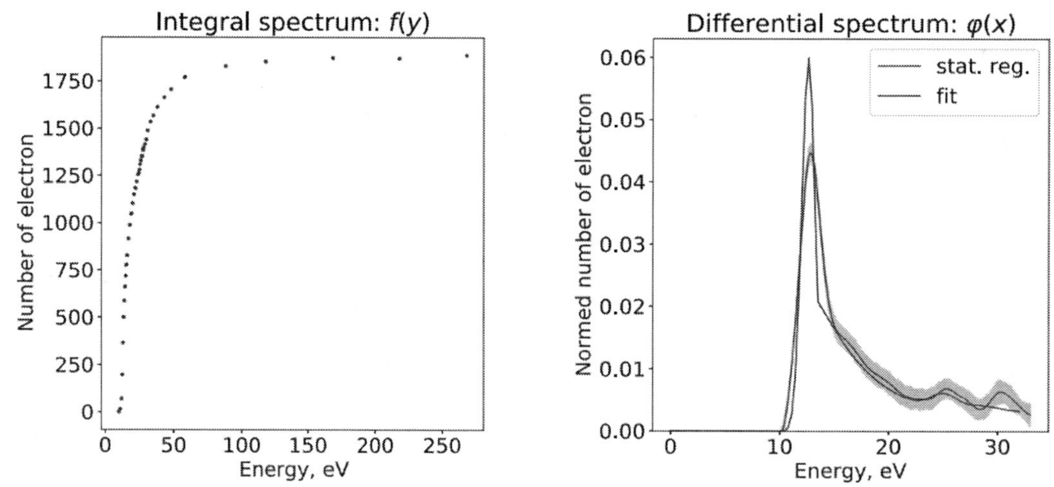

(a) (b)

Figure 1: The spectrum of electron scattering (Troitsk ν-mass data)

The basic concepts of the described method (alongside some improvements) were implemented in a new prototype program written in the Python language. Then the program was used to reconstruct the differential cross-sections of electron scattering on hydrogen isotopes measured in Troitsk nu-mass experiment. In this experiment gaseous hydrogen (as well as deuterium and tritium) was irradiated by electrons with energies of about 20 keV from electron gun. Troitsk nu-mass spectrometer on the other side of gas volume registers the integral electron spectrum (like at fig 1a) with relative resolution of about 10^{-5}. In previous works ([5] and [6]) the differential cross-section of scattering electrons on $H2$ molecules was reconstructed using fit procedure against 5 or 6 parameters shape. That analysis had three major flaws:

- The parametric shape for fit does not have sufficient physics basis and therefore result is subjective in respect to the parametrization.
- The chosen parametrization has too many parameters and produces very tight correlations between these parameters, making the fit results instable.
- There is not clean way to determine uncertainties for any given point of reconstructed function.

Turchin's regularization allows to avoid all these problems and produces clean model-independent result with uncertainties. The result of regularization and previously published fit are presented at Fig. 1b. The regularization reconstruction is in good agreement with fitted curve. Also there are two very small peaks picture shows two additional peaks at 25 and 30 eV.The peak on 25 eV is known to be produced by double scattering events. Peak on 30 eV is previously unknown. It could be either statistical anomaly (errors are quite large in this point), or possibly point to existence of additional dissociative scattering.

4 Conclusion

Turchin's statistical regularization is a very powerful tool to solve ill-posed inversed problems. It provides a flexible way to introduce almost any kind of prior knowledge into reconstruction problem. Also it opens a new way to solve some problems which previously were solved only by direct approach. In this work, we created an implementation of Turchin's regularization using modern language and successfully applied it to the Troitsk nu-mass data for electron scattering on hydrogen isotopes. The work was preceded by a lot of testing of the algorithm with different artificial functions and discretization options. The results of those tests are beyond the scope of this article.

This work is supported by the Ministry of Education and Science of the Russian Federation under the contract No. 3.3008.2017/PP.

References

[1] V.F. Turchin, V.Z. Nozik, in *Physics of atmosphere and ocean* (Nauka, 1969), v. 5, `https://books.google.ru/books?id=N3RUAAAAMAAJ`
[2] L.S. Turovceva, Ph.D. thesis (1973)
[3] F. James, *Statistical Methods in Experimental Physics* (World Scientific, 2006), ISBN 9789812705273, `https://books.google.ru/books?id=S8N\_QgAACAAJ`
[4] D. Calvetti, J.P. Kaipio, E. Somersalo, Int. J. Math **1**, 63 (2006)
[5] V. Aseev, A. Belesev, A. Berlev, E. Geraskin, O. Kazachenko, Y. Kuznetsov, V. Lobashev, R. Ostroumov, N. Titov, S. Zadorozhny et al., The European Physical Journal D - Atomic, Molecular, Optical and Plasma Physics **10**, 39 (2000)
[6] D.N. Abdurashitov et al., Physics of Particles and Nuclei Letters **14**, 892 (2017), `1603.04243`

Magnetic field errors tolerances of Nuclotron booster

Andrey Butenko, *Olha* Kazinova*, *Sergey* Kostromin, *Vladimir* Mikhaylov, *Alexey* Tuzikov, *Hamlet* Khodzhibagiyan

Joint Institute for Nuclear Research, Dubna, Russia

Abstract. Generation of magnetic field in units of booster synchrotron for the NICA project is one of the most important conditions for getting the required parameters and qualitative accelerator operation. Research of linear and nonlinear dynamics of ion beam $_{197}Au^{31+}$ in the booster have carried out with MADX program. Analytical estimation of magnetic field errors tolerance and numerical computation of dynamic aperture of booster DFO-magnetic lattice are presented. Closed orbit distortion with random errors of magnetic fields and errors in layout of booster units was evaluated.

1 Magnetic lattice of booster

The main task of this article is to evaluate maximum orbit distortion and dynamic aperture according to requirements specification for booster magnetic measurements and to find out magnetic field errors tolerances. For start let's have a look on the booster lattice. Magnetic lattice of the booster consists of 4 quadrants. Each quadrant includes 6 DFO-periods, 5 periods contain 2 dipole sector magnets and 1 period contains rectilinear gap. The booster will accelerate $_{197}Au^{31+}$ ions to 578 MeV/n. The perimeter of the accelerating ring is 210.96 m.

Simulation of beam motion was carried out with MAD-X program [1]. Dynamic functions in one quadrant of booster are shown in Fig.1. There are two types of errors of magnetic field:

1. nonlinearity of third and higher orders - systematic errors, existing in every unit and determining magnet design;
2. random errors, appearing because of magnet production inaccuracy and inaccuracy of placement along equilibrium closed orbit of the accelerator.

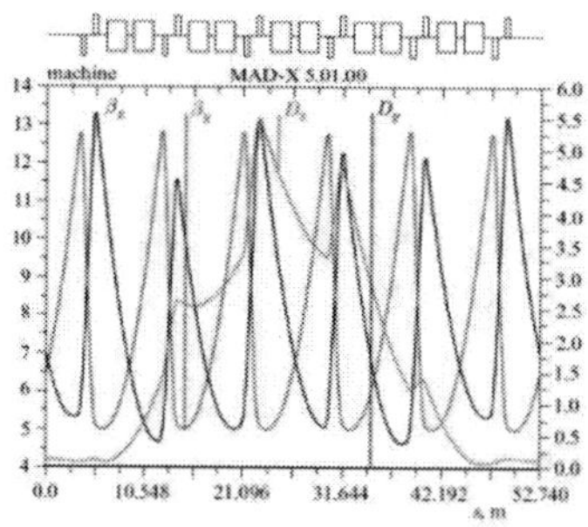

Fig. 1. $\beta_{x,y}$ and dispersion in one superperiod of magnetic lattice.

2 Analytical estimation of magnetic field errors tolerance

The main orbit distorters are:

1. effective length spread L_{eff} and inductions $<\Delta B_z/B>_{o,m}$ in dipole magnets;
2. magnets rotation of median plane by $<\alpha_s>_m$;
3. transverse displacement of longitudinal magnetic axis for quadrupole lenses by $\langle\Delta x\rangle_l$ and $\langle\Delta y\rangle_l$.

Maximum deviation of closed orbit was estimated with harmonic method [2]. Four harmonics with numbers $m = 3$, 4,5, 6 are the closest harmonics to betatron oscillation tunes. Mean square amplitudes of m-harmonic for Fourier orbit expansion determine as:

$$\langle Y_m \rangle = \frac{Q_y \beta^{1/2}{}_{y,max}}{\left|Q_y{}^2 - m^2\right| \pi\rho} \left[\sum_j m_j \langle \Delta B / B \rangle_{O,j}^2 L_J \int_0^{S_J} \beta_y ds \right]^{1/2} \tag{1}$$

where j – type of unit (magnet or lens), L_J - effective length, m_j - number of units of the same type, ρ – radius of curvature for dipole magnets. Maximum orbit deviation in accordance with procedure [2] with 98% probability doesn't exceed the value:

$$Y_{max} = 2\left(\sum_{m=3}^{6} \langle Y_m \rangle^2 \right)^{1/2} \left(1 + |\delta| - |\delta|^2\right) \tag{2}$$

---

* O.Kazinova: kazinova.olga@gmail.com

where $|\delta|=|Q_y-m|$, m - harmonic number of the closest integer resonance disturbance.

For almost symmetrical magnetic lattice of booster on gradients the coefficients in (1) and (2) for x- and y-planes are approximately equal, therefore the below estimation covers horizontal and vertical orbit projection. For $|\delta|=0.1$ mean square amplitudes of 3-6 harmonics connect with field and adjustment errors by next ratios:

$$\begin{aligned}\langle Y_3\rangle^2 &= 1.4\langle \Delta B/B\rangle^2_{O,M} + 12.6\langle \Delta Y\rangle^2_L\\ \langle Y_4\rangle^2 &= 5.2\langle \Delta B/B\rangle^2_{O,M} + 43.7\langle \Delta Y\rangle^2_L\\ \langle Y_5\rangle^2 &= 134.8\langle \Delta B/B\rangle^2_{O,M} + 1156.2\langle \Delta Y\rangle^2_L\\ \langle Y_6\rangle^2 &= 1.9\langle \Delta B/B\rangle^2_{O,M} + 14.8\langle \Delta Y\rangle^2_L\end{aligned} \qquad (3)$$

where linear displacement $\langle \Delta Y\rangle_L$ and amplitudes $\langle Y_m\rangle$ are given in meters. Tolerances for magnetic field spread induction in dipole magnets, adjustment errors and possible displacements of units using eq.(1) are determined from inequality:

$$632.1\langle \Delta B/B\rangle^2_{O,M} + 5338.5\langle \Delta Y\rangle^2_L < Y^2_{max} \qquad (4)$$

If magnetic induction spread $\langle \Delta B/B\rangle_{O,M} = 5\cdot 10^{-4}$, rotation angle of longitudinal axis for dipole magnets $\langle \alpha_s\rangle_m = 5\cdot 10^{-4}$ and transverse displacement of quadrupole magnets $\langle \Delta x\rangle_l = \langle \Delta y\rangle_l = 0.1$ mm, maximum deviation of horizontal and vertical orbit projection is approximately 10 mm. As can be seen from these inequalities, main contribution to the orbit distortion adds by $\langle Y_5\rangle$. The percentage of its impact is 80%. According to the technical project of the NICA accelerator complex [3], maximum distortion for the uncorrected orbit is 15 mm. So, there is no exceeding the limits.

The effect of systematic and random errors on betatron tune characteristics were analyzed by standard method [4,5] with help of shortened equations:

$$\begin{aligned}I'_x &= n_x|H_r| I_x^{\frac{|n_x|}{2}} I_y^{\frac{|n_y|}{2}} \sin\varphi\\ I'_y &= n_y|H_r| I_x^{\frac{|n_x|}{2}} I_y^{\frac{|n_y|}{2}} \sin\varphi\\ \Phi'_x &= \sum_{n=1}^{10}\sum_{j=1}^{n_o}\sum_{k=0}^{j-1}(j-k)H_{n,j,k} I_x^{j-k-1} I_y^{k} + |n_x||H_r| I_x^{\frac{|n_x|-2}{2}} I_y^{\frac{|n_y|}{2}}\cos\varphi\\ \Phi'_y &= \sum_{n=1}^{10}\sum_{j=1}^{n_o}\sum_{k=1}^{j-1} k H_{n,j,k} I_x^{j-k} I_y^{k-1} + |n_x||H_r| I_x^{\frac{|n_x|}{2}} I_y^{\frac{|n_y|-2}{2}}\cos\varphi\end{aligned} \qquad (5)$$

where " ' " – azimuth θ differentiation, $I_{x,y}(\theta)$ and $\Phi_{x,y}(\theta)$ - slowly varying function: square amplitudes and horisontal and vertical betatron tune phases,

$$H_{n,j,k} = \frac{(-1)^k n! R}{k!(j-k)!(n+1-2j)!\pi\rho r^n 2^{j+1}}\left(\frac{\Delta p}{p}\right)^{n+1-2j}\int_0^{2\pi}\beta_x^{j-k}\beta_y^{k}\psi^{n+1-2j}\left(\frac{\Delta B_y}{B}\right)_n d\theta \qquad (6)$$

- coefficients which determine frequency shift and spread relatively to the selected position of operation point, n_0 - integral part of $(n+1)/2$. Parameters which define impact on the beam from one of resonances $n_xQ_x + n_yQ_y = m$ are described in (6) by the following expression:

$$\begin{aligned}|H_r| &= Rh_r\int_0^{2\pi}\beta_x^{\frac{|n_x|}{2}}\beta_y^{\frac{|n_y|}{2}} e^{i(n_x\gamma_x + n_y\gamma_y - m\theta)}\left(\frac{\Delta B}{B}\right)_{n_r-1} d\theta\\ \varphi &= n_x\Phi_x + n_y\Phi_y + \Phi_r + \delta\theta\end{aligned} \qquad (7)$$

where $h_r = \dfrac{(n_r-1)!}{2^{n_r/2}\pi\rho r^{n_r-1}|n_x|!|n_y|!}$, $\delta = n_xQ_x + n_yQ_y - m$ - resonance detuning, $n_r = |n_x| + |n_y|$ - resonance order, $\gamma_{x,y} = \mu_{x,y} - Q_{x,y}\theta$ - argument of Floquet function, Φ_r - phase perturbation.

Tolerance estimation for systematic errors of magnetic field was carried out in no-resonance effect on the beam assumption. Then the values of $\Phi_{x,y}$ are frequency shifts and in general case depend on amplitudes and particle momentum (with constancy of betatron tune amplitude). The estimated fractional increase of emittance $\Delta\varepsilon_{x,y}/\varepsilon_{x,y}$ on the basis of detuning modulus $|\delta|$ is much bigger than the resonance band half-width. Therefore $\Phi_{x,y}(\theta)$ dependence can be neglected and the equations for square of amplitude will be:

$$\Delta I_{x,y} = n_{x,y}|H_r| I_x^{\frac{|n_x|}{2}} I_y^{\frac{|n_y|}{2}} \sin\varphi\Delta\theta \qquad (8)$$

Whereas mean value $\sin\varphi$ equals $2/\pi$ in the range $0<\varphi<\pi/2$ and assuming $\Delta Q=\pi/(2|\delta|)$, we will get

$$\frac{\Delta\varepsilon_{x,y}}{\varepsilon_{x,y}} = \frac{|\Delta I_{x,y}|}{I_{x,y}} = \frac{|n_{x,y}|}{|\delta|}|H_r| I_{x,y}^{\frac{|n_{x,y}|-2}{2}} I_{y,x}^{\frac{|n_{y,x}|}{2}} \qquad (9)$$

where $\langle H_r\rangle$ for resonances exciting by random errors of magnetic field doesn't exceed with 98% probability

$$\langle H_r \rangle = h_r \left[\sum_j m_j \left\langle \frac{\Delta B}{B} \right\rangle^2 S_j \int_0^{S_j} \beta_x^{|n_x|} \beta_y^{|n_y|} ds \right]^{1/2} \quad (10)$$

Magnetic field errors tolerances for booster calculated by the equations (9) and (10) are shown in Table 1. According to the results of dynamic calculations such increase of beam emittance is totally acceptable.

Table 1. Magnetic field random errors tolerances.

Resonance	|δ|	Tolerance
$3Q_x = 14$	0.4	$32\langle \Delta B_z / B \rangle^2_{1,M} + \langle \Delta G / G \rangle^2_{\lambda} < 4.2 \cdot 10^{-5}$
$Q_x - Q_z = 0$	0.05	$40\langle \Delta B_x / B \rangle^2_{1,M} + \langle \alpha_s \rangle^2_{\lambda} < 8.9 \cdot 10^{-7}$
$Q_x - 2Q_z = -5$	0.1	$\langle \Delta B_x / B \rangle^2_{2,M} + 0,09\langle \Delta B_x / B \rangle^2_{2,\lambda} < 1.1 \cdot 10^{-5}$
$2Q_x = 10$	0.4	$\langle \Delta B_z / B \rangle^2_{3,M} + 0,01\langle \Delta B_z / B \rangle^2_{3,M} < 1.2 \cdot 10^{-5}$
$2Q_z = 10$	0.3	$\langle \Delta B_x / B \rangle^2_{3,M} + 0,05\langle \Delta B_x / B \rangle^2_{3,M} < 5.6 \cdot 10^{-5}$

For example, for the first resonance $3Q_x$=14 $\langle\Delta B_z/B\rangle=\langle\Delta G/G\rangle=5*10^{-4}$ (as given in requirements specification). The inequality is $8*10^{-6}<4.2*10^{-5}$. So, these tolerances are satisfied the requirements and can be applicable to the assessment of magnetic field quality.

3 Numerical simulation of magnetic field random errors

Simulation of beam motion was carried out in MAD-X program over the period of 10^3 turns. Dynamic aperture (DA) is defined as stable motion region after particle passing of given number of turns.

3.1 Effective length spread for dipole magnets

Effective length spread for dipole magnets bring about closed orbit deviation from the central position. Fig.2 presents a) the example of closed orbit deviation when $\sigma(BdL)=1*10^{-3}$; b) maximum orbit deviation to effective length spread diagram. Triple maximum value *(3σ)* was used for tolerance feasibility assessment on effective length spread. As it can be seen from Fig.2b the orbit deviation in horizontal direction doesn't exceed 9 mm in the case of $5\cdot10^{-4}$ relative spread.

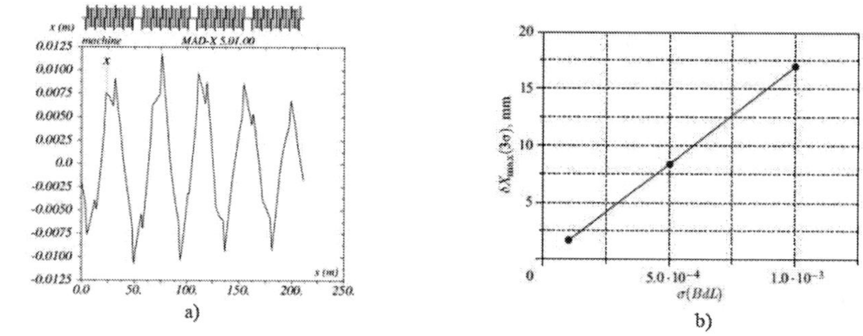

Fig. 2. a) Closed orbit deviation along booster ring with $\sigma(BdL)=1*10^{-3}$ versus longitudinal coordinate; b) maximum orbit deviation (3σ) by 100 statistical samplings of effective length spread *σ(BdL)*.

3.2 Effective length spread for quadrupole magnets

Effective length spread for quadrupole magnets can cause excitation of parametric resonance and appropriate particle loss of accelerating beam. This effect can be estimated by DA attenuation. Figure 3 below provides the stable motion region on transverse phase planes with different values of effective length spread. 8 sextupoles were used for chromaticity correction (2 sextupoles for one quadrant). Gradients of magnetic field are S_f: k_2=4.2·25=105 T/m², S_d: k_2=-3.1·25=-77.5 T/m². As can be seen from Fig.3, stable motion region is fully preserved if the relative spread is less than $5\cdot10^{-4}$.

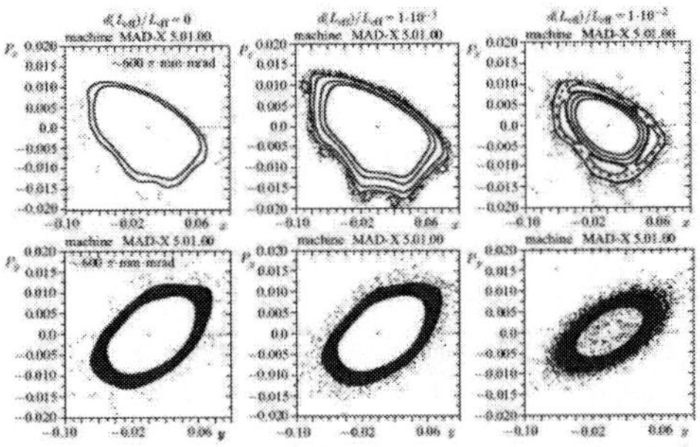

Fig. 3. Stable motion regions with different values of effective length spread in quadrupole magnets.

3.3 Transverse displacements of quadrupole magnets around longitudinal axis

Such displacements affect on generation of vertical and horizontal dipole components on ideal closed orbit line and its adequate distortion. This effect can be estimated by decrease of accelerator DA. Figure 4 a) presents the results of maximum allowed displacements simulation. Calculations were made similarly to 3.1. As can be seen from this figure maximum disturbance of the closed orbit doesn't exceed 7 mm.

3.4 Rotation around longitudinal axis for dipole magnets

Rotation around longitudinal axis results in horizontal component generation of magnetic field in accelerator median plane that in turn causes vertical orbital perturbation. The results of closed orbital perturbation versus the mean-square spread of rotation angle for all dipole magnets in the ring are shown on the Fig.4 b). Maximum orbit deviation doesn't exceed 11 mm in our case.

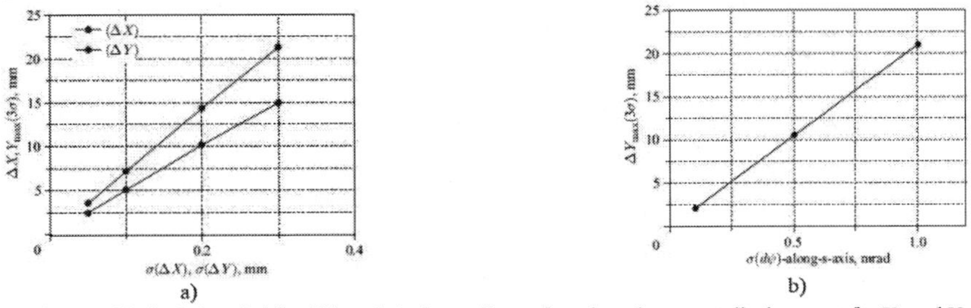

Fig. 4. a) Maximum orbit deviation (3σ) by 100 statistical samplings of quadrupole magnet displacement for X- and Y- directions; b) Maximum orbit deviation (3σ) by 100 statistical samplings of rotation angle for dipole magnets.

3.5 Rotation around longitudinal axis for quadrupole magnets

This effect results in connection between movements in X- and Y-directions. Such situation can reinforce the action on the coupled resonance bunch. Booster operating point (4.8, 4.85) is placed sufficiently far from coupled resonances. As can be seen from Fig.5 stable motion region on phase planes with 5 mrad rotation angle doesn't change in comparison with the ideal layout under sextupoles for chromaticity correction.

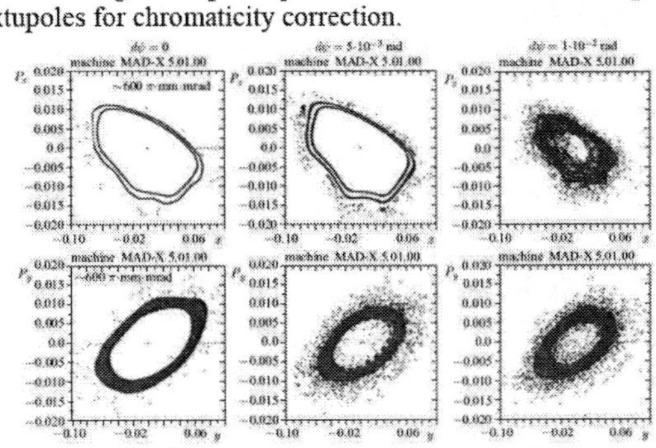

Fig. 5. Stable motion regions with different values of mean-square rotation angle in quadrupole magnets.

On the ground of the received results, tolerances on the effective length spread and adjustment errors for booster unit are formulated in Table 2.

Table 2. Tolerances on the effective length spread and adjustment errors.

Error	Dipole magnet	Quadrupole magnet
Relative effective length spread	$5\cdot10^{-4}$	$5\cdot10^{-4}$
Transverse displacement around longitudinal axis, mm	–	±0.1
Rotation around longitudinal axis, rad	$5\cdot10^{-4}$	10^{-3}

3.6 Booster closed orbit and dynamic aperture under main influencing factors

The results of booster closed orbit simulation and dynamic aperture are shown below (Fig.6 a) and b) respectively). They include the following errors of magnetic field and adjustment:

1. Effective length spread for dipole magnets $\sigma(BdL)=5\cdot10^{-4}$;
2. Rotation around longitudinal axis for dipole magnets $\sigma(d\varphi)=5\cdot10^{-4}$ rad ;
3. Transverse displacement of quadrupole magnets $\sigma(\Delta X)=\sigma(\Delta Y)=0.1$mm ;
4. Effective length spread for quadrupole magnets $\sigma(GdL)=5\cdot10^{-4}$;
5. Rotation around longitudinal axis for quadrupole magnets $\sigma(d\varphi)=10^{-3}$ rad.

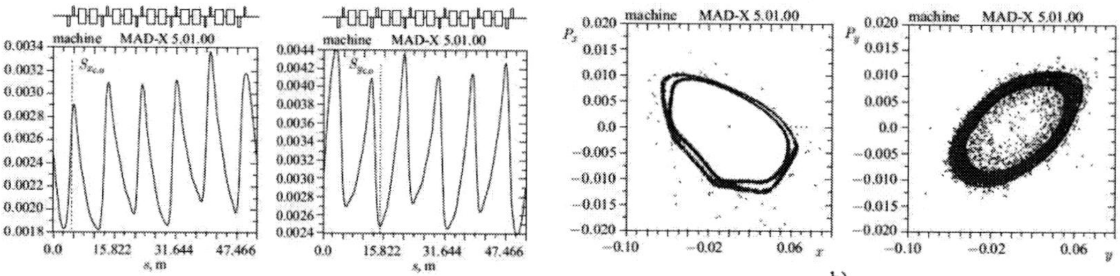

a) b)

Fig. 6. a) Mean-square value of closed orbit deviation (1σ); b) Stable motion regions on the transverse phase.

For more details, please, see [6].

Conclusions

As the result of the research in linear and nonlinear dynamics of ion beam 197Au31+ in booster we have investigated the influence of random errors and adjustment errors on the closed orbit distortion and the betatron tune amplitude. Based on the results, tolerances about the quality of magnetic field and about alignment accuracy for booster units have been formulated. Maximum orbit deviation doesn't exceed 10 mm by both transverse directions. Analytical estimation and numerical simulation are conformed with each other.

References

1. MAD – Methodical Accelerator Design // CERN – BE/ABP Accelerator Beam Physics Group http://mad.web.cern.ch/mad/
2. P.L.Gluckstern, *Distribution of the maximum orbit distortion for random distributed misalignment*. PA, **8**, p.203-209 (1978).
3. I.N.Meshkov *et al.*, *Technical project of the NICA accelerator complex*. Vol.2 (2015).
4. G.A.Guignard, *General Treatment of Resonances in Accelerators*. CERN-78-11, p.72 (1978).
5. V.I.Balbekov, P.N.Chirkov, *The Octupole Correction System*. IHEP 84-167 (1984).
6. A.V.Butenko et al., *Magnetic field errors tolerances of Nuclotron booster*. JINR P9-2017-18 (2017).

EPJ Web of Conferences **177**, 08002 (2018) https://doi.org/10.1051/epjconf/201817708002
AYSS-2017

ESIS ions injection, holding and extraction control system

E.D. Donets, E.E. Donets, D.E. Donets, D.A. Lyuosev, D.O. Ponkin, A.Yu. Ramsdorf, A.Yu. Boytsov, V.V. Salnikov, I.V. Shirikov[1]

[1]Laboratory of High Energy Physics, Joint Institute for Nuclear Research, 6 Joliot-Curie, Dubna, Moscow region, 141980, Russia

Electron string ion source (ESIS) KRION-6T is one of the main parts of the NICA injection complex [1]. During the work on creation of a new ion source for the NICA/MPD project the new ion motion control system was developed, produced and successfully put into operation. Modules development process and operation results are described.

1. Introduction

The Electron String type of Ion Sources (ESIS), was developed, constructed and tested firstly in the Joint Institute for Nuclear Research [2]. ESIS KRION-6T is currently designed at LHEP JINR in order to produce heavy ions [3]. KRION-6T operation is based on step-by-step ionization of the ions by hitting with electrons of an electron string and the main task is to produce the Au^{32+} ions for the NICA/MPD project. During the preparation of physics experiments with the ion source KRION-6T, it is necessary to apply special potential distribution for particle ionization and complex signal for beam extraction (Figure 1).

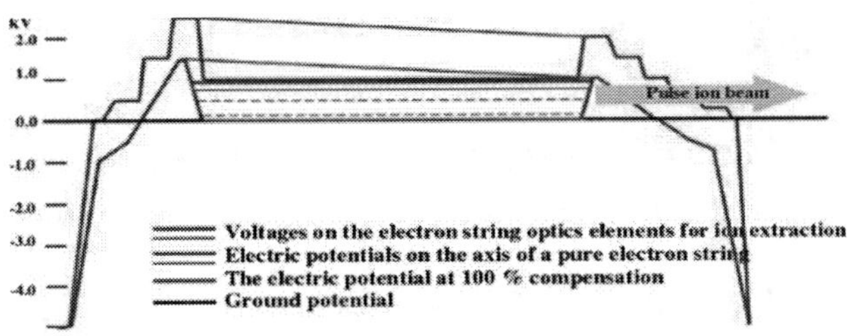

Figure 1. KRION-6T drift structure potentials

2. General description

Ions injection, holding and extraction control system contains several types of modules connected via RS-485 Modbus interface. The system is controlled from PC using Ethernet/optics and optics/RS-485 converters (Figure 2). Each module has its own Modbus ID.

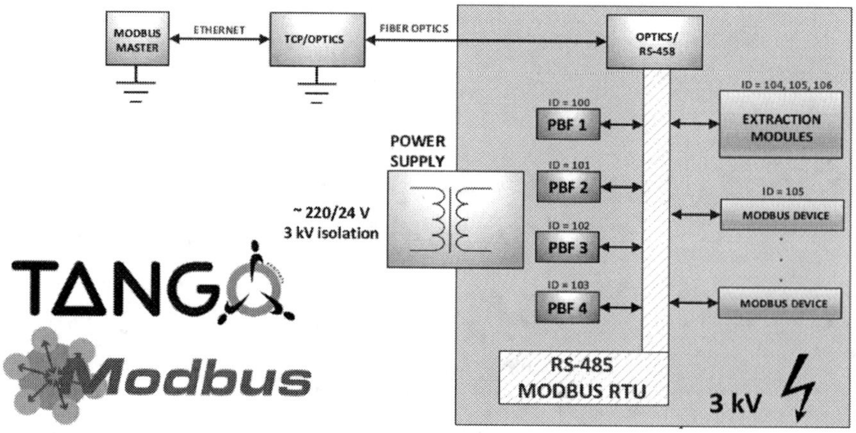

Figure 2. Functional diagram of the system

To complete ions holding task, potential barrier formation modules series (PBF) and DC high-voltage (HV) (HVFM 10.30) modules have been designed, produced and tested. The systems power supply module was also designed. To complete ions extraction task, extraction modules series were designed, produced and tested. In addition, several secondary devices were designed, produced and tested, as well.

Figure 3. Power supply module

The special 24V 3A DC supply module with 4 kV electrical insulation supplies the system (Figure 3). Electrical insulation is necessary to provide the floating ground of the whole system, which is required for initial ion beam energy. The specifications:

- Supply: AC 220 V;
- Output: DC 5 - 24V, up to 3 A;
- Electrical insulation: 3 kV;
- Output protection: short circuit, overvoltage, overcurrent.

The module is based on a PWM switching controller TL494. All the injection, holding and extraction control systems modules are connected to a single network using flat cable with common +24V line from this power supply unit. Total current consumption is about 1.8 A.

The potential barriers formation modules (PBF) are pulsed HV modules with fiber optics trigger input (figure 4). The main idea of the module is based on a PWM switching controller, which supplies HV power transformer and HV IGBT transistors. Pulsed HV signal appear on the output at the time of the input trigger signal is applied. Thus, input trigger signal opens transistor, which is connected into the output circuit in a key mode. 32-bit ARM MCU performs output voltage and load current measurements, implements Modbus interface and PID-regulator logic. It allows controlling output voltage magnitude from PC using 12-bit DAC in the same way as using front panel variable resistor. To improve output signal linearity the MCU DAC based PID-regulator was implemented. The PBF modules specifications:

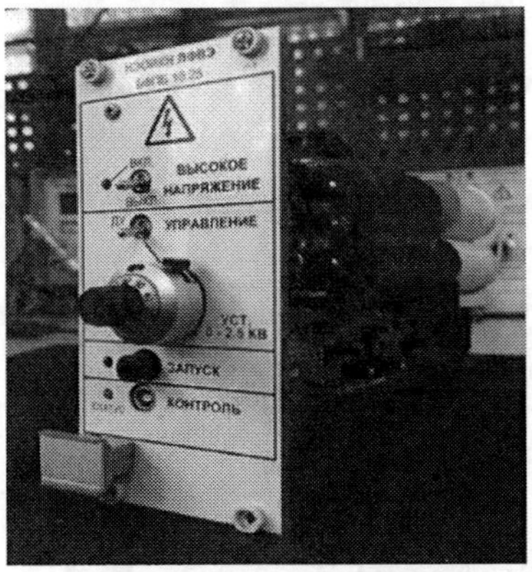

Figure 4. PBF 10.25 module

- Output: 0 … +3 kV;
- Rise time: ~ 10 us;
- Pulses width: 50 us – 10 s;
- Max load current: 8 mA;
- Supply: + 24 V, 300 mA;
- Overcurrent, short circuit protection;
- Modbus interface;
- Manual & Remote control.

Figure 5 shows +3 kV 1 s pulse output signals difference in cases of manual mode control using front panel variable resistor and MCU-based control with PID-regulator.

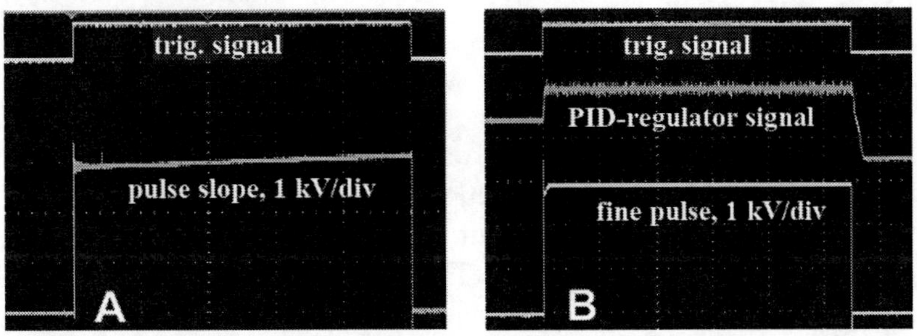

Figure 5. PBF 10.25 module output signal for two cases: A) manual mode control, B) remote control

The DC HV (HVFM 10.30) modules are another ions injection, holding and extraction control systems elements, which creates static potential barriers, positive potential biases and produce HV for modulator of the beam extraction system. The HVFM 10.30 modules working principle is also based on a PWM switching controller, which supplies HV power transformer. MCU logic is the same as in PBF 10.25 module described above with the only difference lying in input trigger signal absence necessity. The HVFM 10.30 modules specifications:

- Output: 0 … +3 kV;
- Max load current: 10 mA;
- Supply: + 24 V, 300 mA;
- Overcurrent, short circuit protection;
- Modbus interface;
- Manual & Remote control.

The HVFM 10.30 modules have the same as the PBF 10.25 PC program GUI functionality. User can remotely control the output state, magnitude level and analyze status and measurements.

However, from the point of view of the team of authors, the most interesting devices are beam extraction modules, which consist of the DC HV source (HVFM 10.30), HV modulator (Figure 6) and capacitive storage device. To extract the ion beam it is necessary to create a special pulse of complex shape in case of slow extraction (100 – 300 us width) and fast rising edge pulse for quick extraction case (1 – 10 us width). This pulsed potential is applied to one of the beam drift structure parts simultaneously with of extraction process.

Figure 6. ESIS extraction system HV modulator: A) front panel, B) PCB view

Thus, the HV modulator is a special device, which allows user to create external triggered complex shape or fast rising edge HV pulse. External trigger is based on the fiber optics to TTL converter circuit. Fiber optics is used due to the systems floating ground. The main idea of the HV modulator is PWM-based HV IGBT transistors control. In this case, the transistor output is connected to the capacitive load and the HV modulator output is connected to the point of their connection.

To achieve a special slope of the pulse rising edge the different duty cycle PWM sections are applied. As well as duty cycle, user can also specify PWM section width. Thus, user can implement up to four time slots with different PWM duty cycle. Obviously, the higher PWM signal duty cycle is the faster is pulse rising edge. This is how pulse complex shape is achieved. Figure 6 B shows the HV modulators PCB a where couple of HV-connectors is shown. The top connector is DC HV input, the bottom one is modulated signal output. Figure 7 shows two pulses for different beam extraction cases which have been created using the HVFM 10.30, HV modulator and capacitive storage modules (beam extraction system).

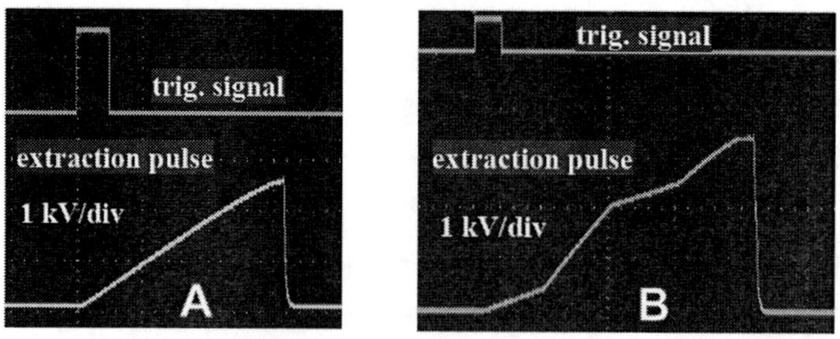

Figure 7. Beam slow extraction pulse for two cases: A) common, B) complex shape

3. Conclusion

The new ions injection, holding and extraction control was successfully put into operation (Figure 8). The basic stand-alone PC software was also designed to control and monitor beam motion processes. The software was modified according to requests from users after first test run and currently works without any errors and bugs. It is based on the Visual C# WPF technology where Modbus RTU over TCP transport class is designed by the authors. The choice in favor of Visual C# WPF instead of TANGO controls [4] was made based on the real-time HV monitoring requirements. The next step is to connect the system with ESIS TANGO controls database on the monitoring level. The group of authors is also going to design similar HV devices with the maximum output voltage up to ±12 kV.

Figure 8. ESIS ions motion control system: A) during KRION-6T work, B) at the test bench

References

1. NICA (Nuclotron-based Ion Collider fAcility). http://nica.jinr.ru/

2. E.D. Donets, D.E. Donets, E.E. Donets, V.V. Salnikov, V.B. Shutov, S.V. Gudkov, Yu.A. Tumanova and V.P. Vadeev, Rev. Sci. Instrum., Vol. 75, 1543-1545 (2004).

3. D.E. Donets, E.E. Donets, T. Honma, K. Noda, A.Yu. Ramsdorf, V.V. Salnikov, V.B. Shutov and E.D. Donets, «Physics research and technology development of electron string ion sources», Rev. Sci. Instrum. 83, 02A512 (2012).

4. V. A. Andreev, V. I. Volkov, E. V. Gorbachev, V. A. Isadov, A. E. Kirichenko, S. V. Romanov, and G. S. Sedykh, TANGO Standard Software to Control the Nuclotron Beam Slow Extraction, Physics of particles and nuclei letters vol. 13 no. 5 pp. 951–956, (2016).

EPJ Web of Conferences **177**, 08003 (2018) https://doi.org/10.1051/epjconf/201817708003
AYSS-2017

Compact 2.45 GHz ECR Ion Source for generation of singly-charged ions

Riyaz Fatkullin[1,2,*], *Sergey* Bogomolov[2], *Konstantin* Kuzmenkov[2] and *Andrey* Efremov[2]

[1]Institute for Theoretical and Experimental Physics (ITEP), 117218, Moscow, Russia
[2]Flerov Laboratory of Nuclear Reactions (FLNR JINR), 141980, Dubna, Moscow region, Russia

Abstract. 2.45 GHz ECR ion sources are widely used for production of protons, single charged heavy ions and secondary radioactive ion beams. This paper describes the development of a compact ECR ion source based on 2.45 GHz coaxial resonator. The first results of extracted current measurements at different resonator configuration as a function of UHF frequency, power and gas flow are presented.

1 Introduction

Several ECR ion sources for production of radioactive ion beams operating at 2.45 GHz were developed at the Flerov Laboratory of Nuclear Reactions (FLNR JINR) [1]. Such ion sources are used for production of $^{6}He^{+}$ ions in the DRIBs project [2] and at the MASHA mass-spectrometer [3]. The magnetic system of such sources is composed of NdFeB permanent magnet rings, and it provides the creation of pseudo-closed resonant surfaces (875 Gs). The plasma chamber of such sources is based on a single-mode cylindrical resonator with an internal diameter of 90 mm and a length of 100 mm. The measured gas efficiency of those sources is about 90% for noble gases (Ar, Kr). The transformation time of atoms into ions of this ion source has not been investigated yet, however, according to the results of paper [4], this time decreases with the decrease of the plasma chamber volume.

The volume of plasma chamber can be reduced if the chamber will be based on a coaxial resonator loaded with a capacitor. Pseudo-closed surfaces should be located in the gap of the capacitor to achieve optimum conditions for plasma confinement.

1.1 Development of compact ECR ion source

The schematic structure of the source is shown in Fig.1. The magnetic system of the source is composed of a radially magnetized ring with an external diameter of 52 mm, an inner diameter of 22 mm and a thickness of 10 mm. The distribution of the magnetic field on the axis of the ring is shown in Fig. 2, the lines of an equal field are shown in Fig. 3. It can be seen that this magnetic system provides creation of pseudo-closed surfaces with a field level from 875 Gs up to 1750 Gs.

* Corresponding author: rdfatkullin@gmail.org

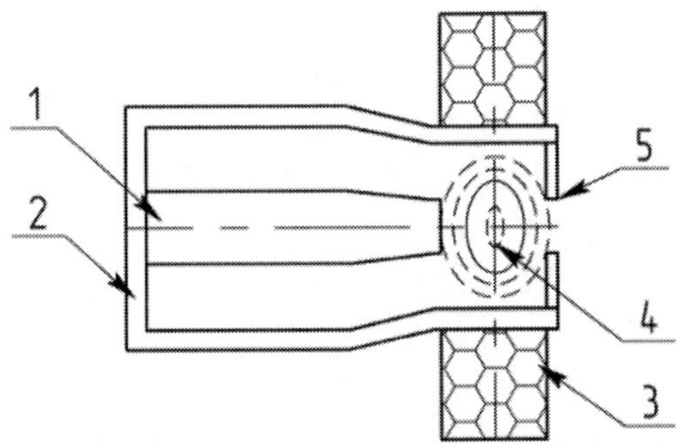

Fig. 1. Schematic view of the ECR ion source based on a coaxial resonator. 1 - central conductor of a coaxial resonator; 2 - resonator casing; 3 - magnetic ring; 4 - closed surfaces of the magnetic field; 5 - extraction hole.

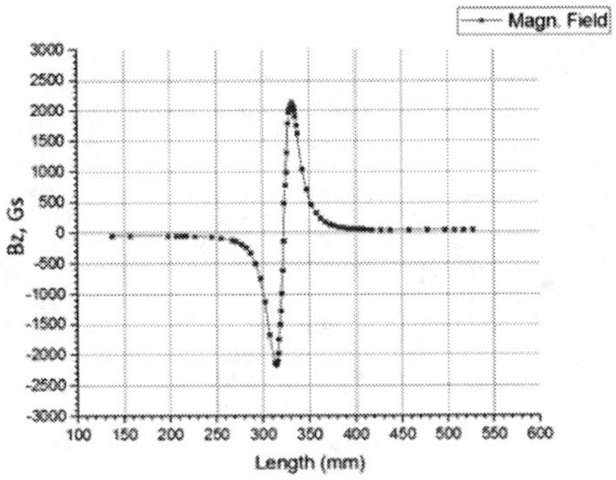

Fig. 2. Axial magnetic field distribution of a ring magnet.

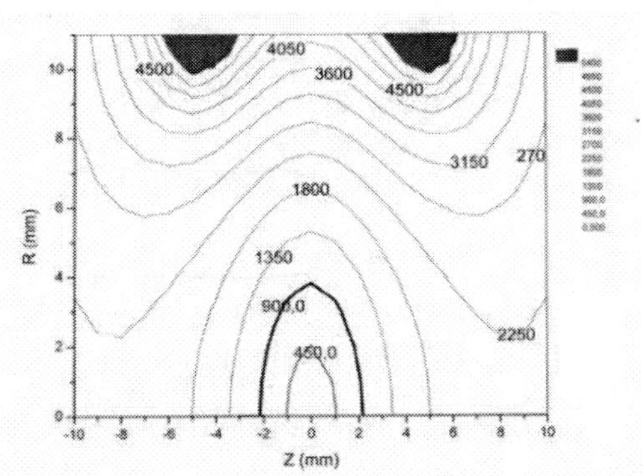

Fig. 3. The lines of equal magnetic field

The plasma chamber of the source represents the quarter wave resonator. The results of the resonator modeling are described in paper [5]. Based on the modelling results, an ECR ion source was designed and manufactured. Assembly drawing of the source is presented in Fig. 4. The view of the source is presented in Fig.5.

For changing the resonator gap three sets of the replaceable internal parts of resonator and replaceable plasma electrodes were used, providing the gap sizes of 5, 8 and 10 mm. The diameter of the extraction hole in plasma electrodes is 2 mm.

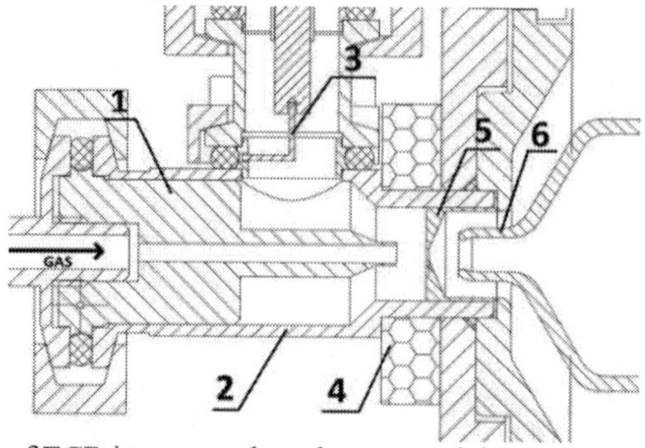

Fig. 4. Assembly drawing of ECR ion source based on a coaxial resonator: 1 - replaceable internal part of the resonator, 2 - external part of the resonator, 3 – coupling loop, 4 – permanent magnet ring, 5 - replaceable plasma electrode, 6 - pulling electrode.

Fig. 5. The view of ECR ion source: 1 – coaxial input, 2 – magnet ring, 3 – plasma electrode, 4 - resonator

2 Experimental stand and tests

Test of the source included: measuring of the extracted ion current at different gap sizes, gas flow and experimental definition of minimum discharge ignition power at different generator frequencies. Ar, Ne and He were used as working gases.

2.1 Experimental stand

The source was mounted on the DN160 flange and connected to the vacuum volume pumped by the 150 l/s turbopump providing the base vacuum about of $1.5 \div 3.0\times10^{-6}$ torr. During the source operation the pressure in vacuum chamber was about of $1.2\text{x}10^{-5}$ torr. Solid state UHF generator with the output power up to 100 W in the frequency range of $2.4 \div 2.5$ GHz was used to feed the source. The UHF power was fed to the coupling loop through a coaxial cable and tuner. Control of the gas flow was performed by the piezoelectric leak valve controlled by the high voltage power supply (up to 1500 V). A negative potential about of 300 V was applied to the pulling electrode. During tests the current of ions incident on the pulling electrode was measured. The distance between the plasma electrode and pulling electrode equals to 5 mm in all experiments.

2.2 Results

The dependence of the ion current on the power of the generator was measured for different gases (Fig. 6) at 2500 MHz with the resonator gap 8 mm. The highest current was achieved with helium. All other measurements were performed with argon.

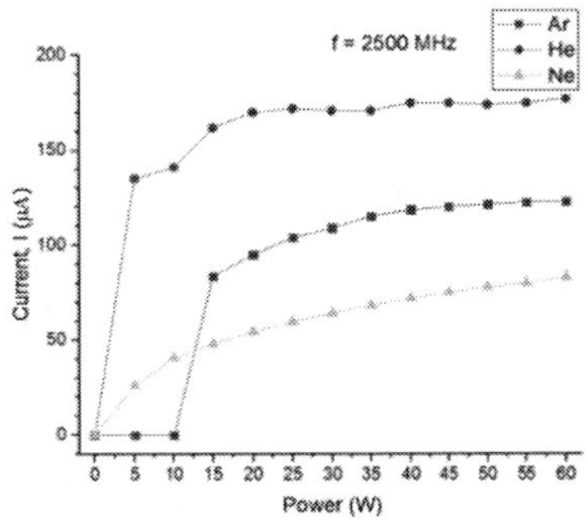

Fig. 6. Dependence of the ion current on the injected power for different gases.

Fig. 7 presents the dependence of the ion current on the gas flow at 2450 MHz and 20 W of injected power. The resonator gap equals to 8 mm. By varying the gas flow, one can tune the source for the maximal value of ion current.

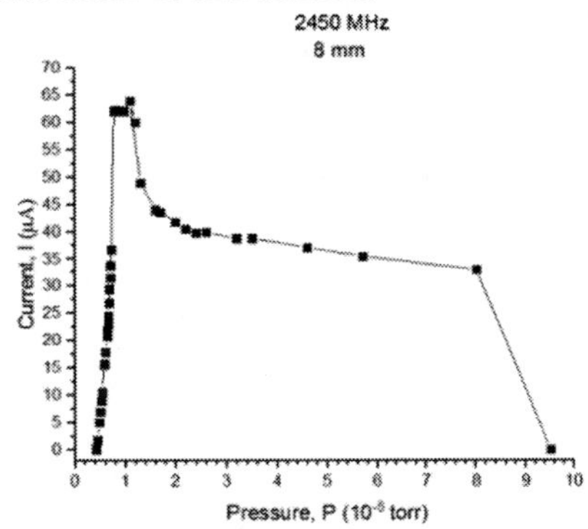

Fig. 7. Dependence of the Ar ion current on gas flow.

The dependence of the ion current on the frequency of the generator was measured for various resonator gaps (Fig. 8). For all values of resonator gap the current increases with the frequency of generator. The highest current was produced with resonator gap of 8 mm.

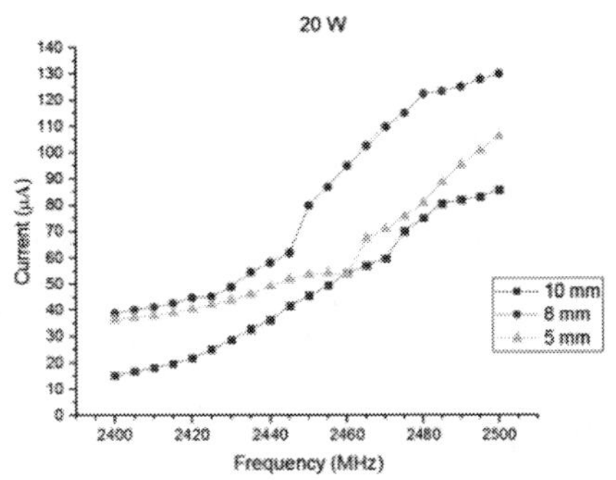

Fig. 8. Dependence of the ion current on the generator frequency.

The dependence of the minimal ignition power on the generator frequency was investigated for various gaps at fixed gas flow (Fig. 9). It was possible to establish that in a certain frequency range 2450 ÷ 2465 MHz for 10 mm gap, 2430 ÷ 2460 MHz for 8 mm gap and 2425 ÷ 2435 MHz for 5 mm gap, the lowest power (about of 0.5 W) required for ignition is observed. The average required power over the whole range is minimal for 5 mm, maximal for 10 mm.

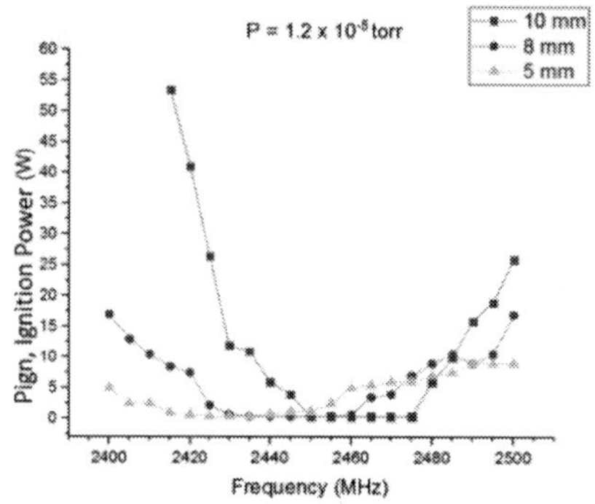

Fig. 9. Dependence of ignition power on the generator frequency.

3 Discussion

From the experimental data it can be concluded that the optimal value of the coaxial resonator gap is 8 mm. With this configuration of the source the highest extracted current was measured. Moreover, in this configuration of the source the minimal value of power for discharge ignition (about of 0.5 W) is maintained in a wider frequency range (2430 ÷ 2460 MHz) than for others gaps value. Extracted current increases with the power and shows slight saturation for He and Ar for the power above 25 ÷ 30 W.

The extracted current increases with the frequency for all values of resonator gap without noticeable saturation. It can be explained by the increase of resonance surface with the increase of frequency and shortening the distance between the resonance surface and plasma electrode.

The setting of the source for the maximal extracted current requires the fine tuning of the gas flow. The minimal gas flow, which allows maintain discharge corresponds to vacuum chamber pressure about of 4.2 x 10^{-6} torr.

The highest current produced in these experiments constitutes about of 180 μA for He. The further increase of the extracted current can be achieved by the increase of the extraction hole diameter and the frequency.

Conclusions

Compact and low power ECR ion source was designed and tested. The source is suitable for production of low and medium currents of singly charged ions. The further study will include the measurement of ions spectrum generated by the source, measurements of the gas efficiency and time response of the ion source. The beam quality should be also studied because it is an important feature of the ion source.

References

1. A. Efremov et al. Nucl. Instr. Meth. Phys. Res. B **204,** 368–371 (2003)
2. Yu. Ts. Oganessian, Proc. of the Int. Conf. *Nuclear shells – 50 years*, Dubna, Russia, 1999, World Scientific, 2000, p61.
3. Yu. Ts. Oganessian et al. Nucl. Instr. Meth. in Phys. Res. B **204**, 606–613 (2003)
4. P. Jardin et al. Nucl. Instr. Meth. Phys. Res. B **204,** 377–381 (2003)
5. R. Fatkullin, S. Bogomolov, *Design of the coaxial resonator for ion source*, (Dubna, Russia, 2016) http://students.jinr.ru/en/jinr2016-participants

EPJ Web of Conferences **177**, 08004 (2018) https://doi.org/10.1051/epjconf/201817708004
AYSS-2017

The effect of simplifications of a numerical mesh on the results of electromagnetic analysis of the Nuclotron-type cable

Łukasz Tomków[1]*

[1] Przedsiębiorstwo Badawczo-Wdrożeniowe Frako-Term sp. z o.o., Toruń, ul. Morwowa

Abstract. The model of a single Nuclotron-type cable is presented. The goal of this model is to assess the behaviour of the cable under different loads. Two meshes with different simplifications are applied. In the first case, the superconductor in the cable is modelled as single region. Second mesh considers individual strands of the cable. The significant differences between the distributions of the electric current density obtained with both models are observed. The magnetic field remains roughly similar.

1 Introduction

The Nuclotron-type cable consists of 23 superconducting strands wound around a copper tube, which carries liquid helium to cool down the superconductor below its critical temperature. The assembly is surrounded by several layers of mechanical support including NiCr wire and Kapton. The structure of the cable is shown in figure 1. The cable of this type was first applied in the superconducting particle accelerator Nuclotron in JINR to construct dipole and quadrupole magnets, hence the name.

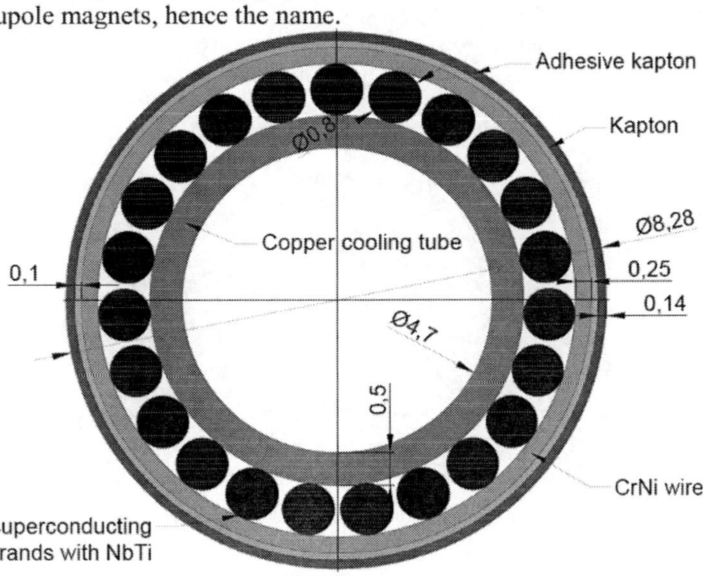

Fig. 1. Structure of a Nuclotron-type cable

Currently the cables of this type are applied in newly constructed accelerator complexes NICA and FAIR. At FAIR they will be used not only to build magnets magnets, but also at cryogenic transfer lines [1]. The transfer lines will carry electric current and liquid helium between the sections of the accelerator. The superconducting material applied in the cable is NbTi. In this work, the model of the Nuclotron-type cable for the application at the line is presented. The results obtained using two different meshes are considered.

2 Methods

The model used in the analysis calculates magnetic field and the electric current distribution in the cable. To find these values the H-formulation is used. The form of the formulation used for the described model was described by Brambilla et al. [2]. The model is two-dimensional with two components of the magnetic field and a single component of electric current density and electric field. To account for high non-linearity the formula for electric field E takes the form (1).

$$\boldsymbol{E} = \left(\frac{J-J_c}{J_c} \cdot \mathcal{H}(J - J_c)\right)^n \tag{1}$$

$\boldsymbol{E}$ in this formula is electric field, J is the local density of electric current found as the curl of the magnetic field. J_c is critical current density found with Bottura fit [3]. Density of superconducting strands is added as an additional regulating parameter in the calculations of critical current density. In this model the thermal dependence of the critical current density on temperature is not considered. $\mathcal{H}$ is the Heaviside function, which attributes the value of 0 if the value under the function is lower than 0 and 1 if it is higher. n is the exponent of power law.

The second basic formula for the model is Faraday's law giving the interaction between magnetic field and electric field (2).

$$\frac{\partial \boldsymbol{B}}{\partial t} + \nabla \times \boldsymbol{E} = 0 \tag{2}$$

In this formula $\boldsymbol{B}$ is the magnetic induction.

The two meshes used in the analysis are shown in figure 2. The one to the left leads to more complex and longer analysis with each strand modelled individually. The right mesh models the superconductor as a single region. The calculations performed using the latter one are significantly faster.

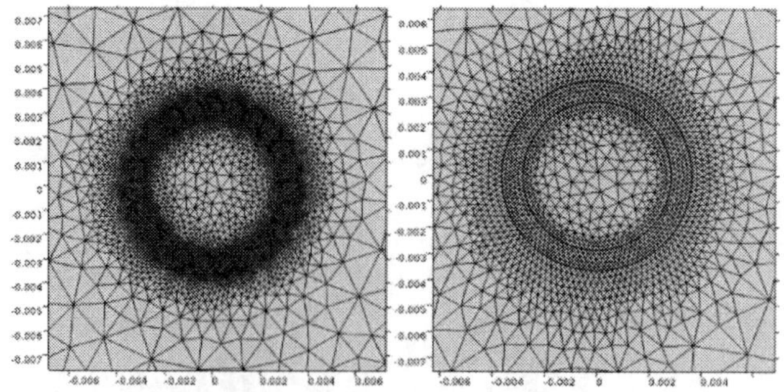

Fig. 2. Meshes applied for the electromagnetic analysis

The boundary condition for the model is the value of the magnetic field on the external wall of the model. It is calculated based on the magnetic field generated by the point current. In the sufficient distance from the centre of the cable it is a good estimation.
The magnetic field varies with time to reflect the operation of the accelerator in the most demanding testing regime. In this regime the triangular current is applied, which rises from 0 kA to 13 kA with the frequency of 1 Hz. That gives the rate of increase of the current of 26 kA/s.
The additional constraint of the total carried current (expressed as an integral of the current density over the area) is placed on superconducting region. This further ensures the proper value of the carried current.

3 Results and discussion

The model yielded the results consistent with initial predictions and literature data. Figure 3 shows the results of the calculations for the magnetic induction with more complex mesh. Figure 4 shows the results on magnetic induction obtained with the simpler mesh.

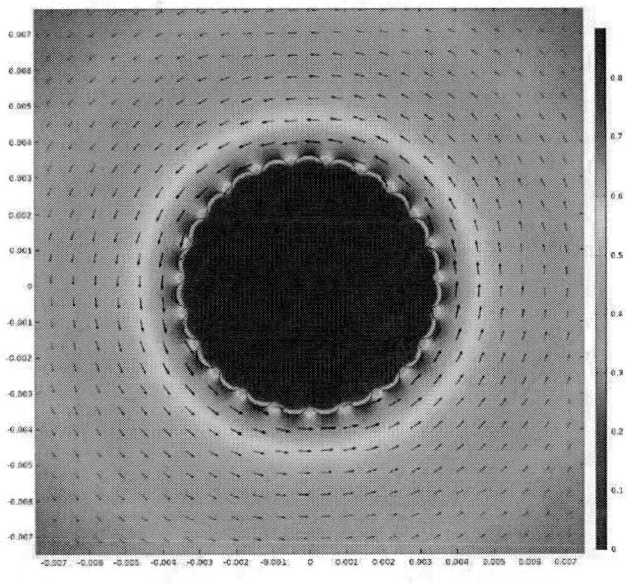

Fig. 3. Magnetic induction in T calculated with the complex mesh at t=0.3 s

The magnetic induction can be seen to decrease inside the cable as expected. In the case of the more complex mesh some irregularities in the vicinity of the strands can be observed. Further from the ring of the strands the field becomes regular again, similarly to the case of simpler mesh. The differences disappear less than 0.1 mm from the strands. Therefore, they do not affect any objects located outside of the cable. It shows that the simplified model is sufficient to calculate the distribution of magnetic field and perform the electromagnetic analysis. The results are very clear and free from the numerical artefacts.

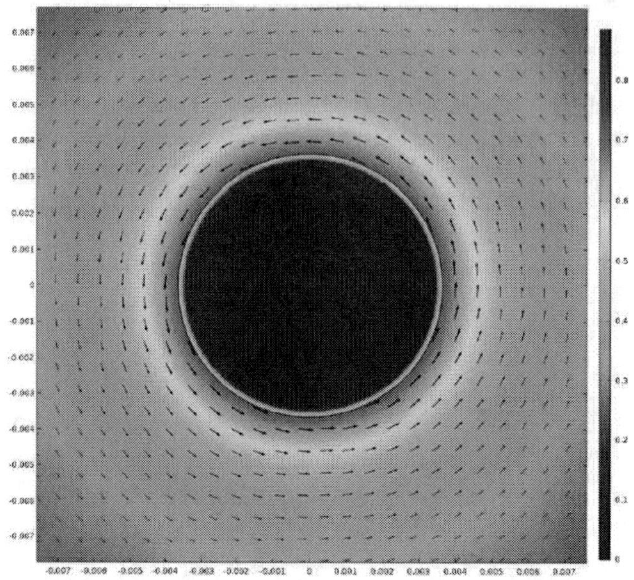

Fig. 4. Magnetic induction in T calculated with the simpler mesh at t=0.3 s

Figures 5 and 6 present the analogical results of the calculations of electric current density.

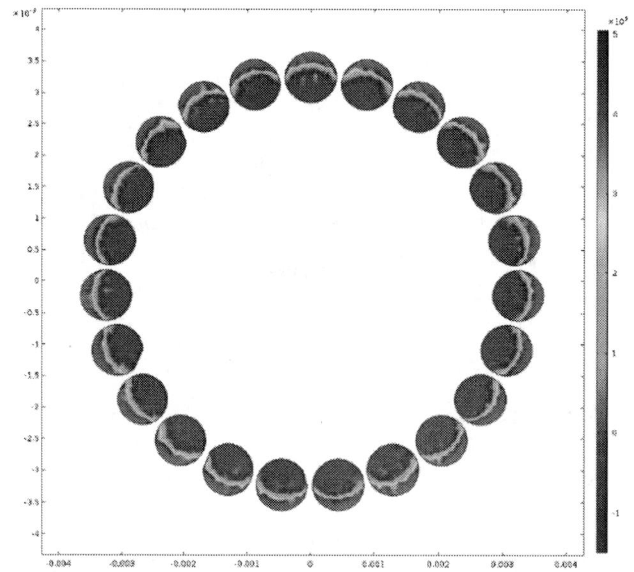

Fig. 5. Electric current density in A/m^2 calculated with the complex mesh at t=0.3 s

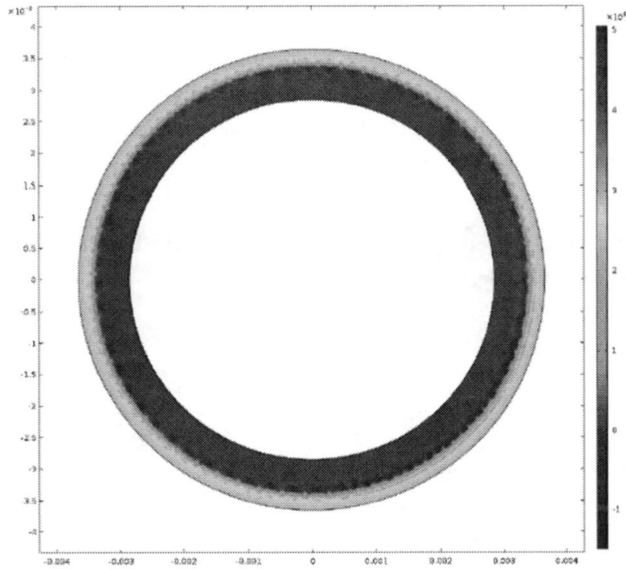

Fig. 6. Electric current density in A/m^2 calculated with the simpler mesh at t=0.3 s

Significant differences between the results from two analyses can be observed. In the case with more complex mesh the maximum value of the current is significantly higher. It can be expected, as the transport area in the second model is significantly larger. However, that means that the calculated heat generation will be significantly different. Therefore the simpler mesh cannot be applied for thermal analysis of the cable or the calculation of losses occurring during the operation.

Some numerical artefacts can be observed on the interface between the current-carrying region and the region with zero magnetic field. This is especially pronounced in the model obtained with more complex mesh. Their area is relatively small, therefore they do not affect strongly the integral of the current density.

4 Conclusions

The simpler mesh cannot be applied for the thermal analysis of the cable due to the significant differences occurring in this model. However, it is well fit for the electromagnetic analysis. The irregularities disappear well within the volume of the cable. The calculations of interactions between the cables in the line will be valid. It is especially important in the assessment of the quality of power delivered to the magnets. It strongly affects the particle beam as well as the safety of operation of the accelerator.

References

1. A. Bleile, E. Fischer, W. Freisleben, A. Mierau, P. Schnizer, and P. Szwangruber Physics Procedia **67** (2015)
2. R. Brambilla, F. Grilli, and L. Martini Supercond.Sci. Technol **20** (2007)
3. L. Bottura IEEE Transactions on Applied Superconductivity **10** (2000)

EPJ Web of Conferences **177**, 09001 (2018) https://doi.org/10.1051/epjconf/201817709001
AYSS-2017

False vacuum decay in quantum mechanics and four dimensional scalar field theory

Maxim Bezuglov[1,2,⋆]

[1] *Bogoliubov Laboratory of Theoretical Physics, Joint Institute for Nuclear Research, Joliot-Curie 6, 141980 Dubna, Moscow region, Russia*
[2] *Moscow Institute of Physics and Technology (State University), 9 Institutskiy per., 141701 Dolgoprudny, Moscow Region, Russian Federation*

Abstract. When the Higgs boson was discovered in 2012 it was realized that electroweak vacuum may suffer a possible metastability on the Planck scale and can eventually decay. To understand this problem it is important to have reliable predictions for the vacuum decay rate within the framework of quantum field theory. For now, it can only be done at one loop level, which is apparently is not enough. The aim of this work is to develop a technique for the calculation of two and higher order radiative corrections to the false vacuum decay rate in the framework of four dimensional scalar quantum field theory and then apply it to the case of the Standard Model. To achieve this goal, we first start from the case of d=1 dimensional QFT i.e. quantum mechanics. We show that for some potentials two and three loop corrections can be very important and must be taken into account. Next, we use quantum mechanical example as a template for the general d=4 dimensional theory. In it we are concentrating on the calculations of bounce solution and corresponding Green function in so called thin wall approximation. The obtained Green function is then used as a main ingredient for the calculation of two loop radiative corrections to the false vacuum decay rate.

1 Introduction

As is well known, within the Standard model all particles acquire their masses through the Higgs mechanism. If we talk about it in a simplified way - the expectation value of Higgs field slides down to a minimum of some potential. At tree level this potential has one minimum. However, if we account for loop corrections, then at high energies it can form an additional deeper minimum [1–8]. Whether this second minimum arises or not depends on the masses of top quark and Higgs boson. With the known values of Higgs and top quark masses we get into the region in which this minimum is indeed formed. This means that the Higgs field can quantum tunnel from its current vacuum state into this second true vacuum, what will lead to an end of the known universe. The latter is not happening however. There are two possible solutions (not excluding each other) to this problem. The first is that there is some new physics connected with the Higgs sector that changes Higgs potential in such a way that all bad minimums there disappear. The second possibility is that the tunneling time to this minimum is much longer than the lifetime of the universe. Since we do not know anything about

⋆e-mail: bezuglov.ma@phystech.edu

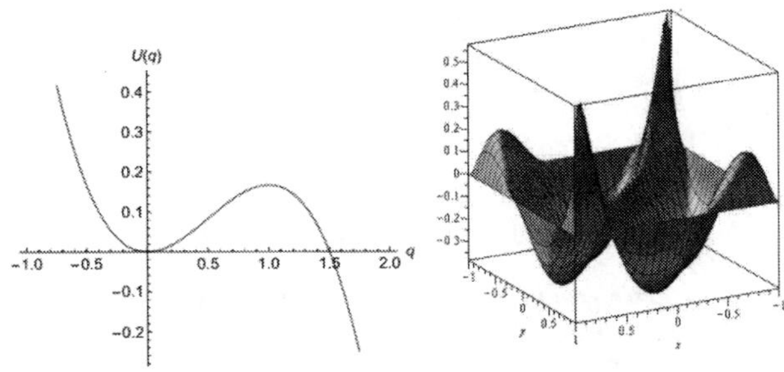

Figure 1. $U(q) = q^2/2 - \lambda q^3/3$ on the left and corresponding instanton based Green function on the right

new physics, it makes sense to discuss the second option. Calculations show that the lifetime of the present false vacuum is large, but these calculations are carried out only at one loop[1], which is apparently not enough. The aim of this work is to develop a technique for the calculation of two and higher order radiative corrections to the false vacuum decay rate in the framework of four dimensional scalar quantum field theory. To achieve this goal, we first start from the case of quantum mechanics and then we will use this example as a template for the general four dimensional theory.

2 Quantum mechanics

From the quantum mechanics we know that for a particle in a potential well there is a nonzero probability to jump out of it even if particle does not have enough energy for it in a classical sense. In a mathematical language it means that particles's energy acquires some imaginary part characterizing the tunneling probability. The mentioned imaginary part can be further extracted from the thermal partition function. However, it could not be obtained by means of perturbation theory at false vacuum. The main contribution to the imaginary part of the particles's energy is given by the factor $\exp(-S_c)$, where $-S_c$ is the euclidean action for the extremal trajectory minimizing euclidean action. Such a trajectory is called bounce or a instanton solution and to find it - it is enough to solve the usual equations of motion for the inverted potential. The latter corresponds to going to euclidean action by means of Wick time rotation.

Let's now discuss the corrections for the complex part of energy and consider the potential $U(q) = q^2/2 - \lambda q^3/3$, see left part of Figure 1. The imaginary part of the ground state energy could be written as:

$$\Im E_0(\lambda) = \sqrt{\frac{15 S_c}{2\pi}} e^{-S_c} \left(1 + \frac{I_2}{S_c} + \frac{I_3}{S_c^2} + O\left(\frac{1}{S_c^3}\right)\right) \tag{1}$$

where $S_c = 6/5\lambda^2$ is the bounce action and I_2, I_3 are the two and three loop radiative corrections. To calculate two and three order corrections, we are following the method developed in [18–21] which is based on the Feynman diagram technique in coordinate representation. The main ingredient of this technique is the instanton based Green function. For the potential with cubic interaction we managed to find corresponding Green function analytically [22], the plot of which could be found on the right-hand side of Figure 1.

[1]For tunneling rates calculations in Standard Model and its extensions see [9–17] and references therein.

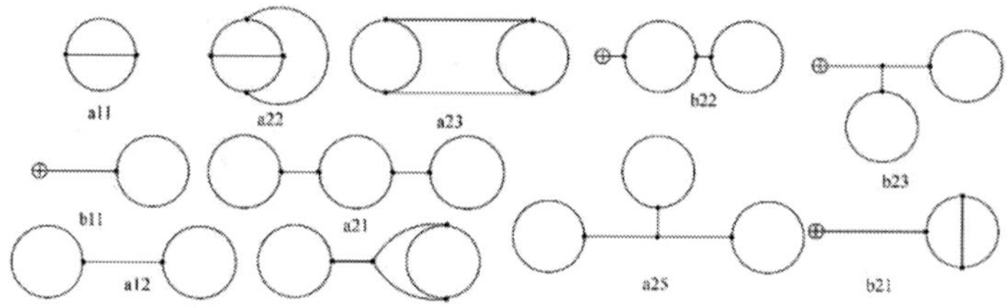

Figure 2. Two and tree loop Feynman diagrams for quantum mechanical case of the potential with the cubic term

The two-loop correction includes contributions from three Feynman diagrams, while there are already eight such diagrams at three-loops, see Figure 2. All these contributions were calculated numerically with the help of CUBA library [23] which uses the Monte Carlo technique. The results are $I_2 = -1.4083307 \pm 7.85 * 10^{-6}$ and $I_3 = -1.543 \pm 0.0045$. The numerical error in both cases is less than 1%, which means that the calculations are sufficiently accurate, at least for qualitative conclusions. Both these contributions turn out to be of the order of unity and negative. Substituting these values into (1), we see that these corrections can significantly reduce the probability of decay in the case when $S_c \sim O(1)^2$. If $S_c \gg 1$, then these corrections do not have a qualitative effect on the decay.

More details and discussions about the quantum mechanical case can be found in our work [22] and references therein.

3 False vacuum decay in d=4 dimensional scalar quantum field theory

Now, with the lessons learned in the case of quantum mechanics we may proceed with four-dimensional scalar field theory. For simplicity we will assume O(4) spherical symmetry of the bounce solution, so that the latter will depend only on one space coordinate. This reduces the problem of finding the bounce solution to that in one-dimensional case [3], the only difference is the presence of the so-called dumping term which can be interpreted as some external force. On the the right part of the Figure 3 we plotted bounce solutions for the scalar potential of the Higgs type ($U(\varphi) = -m^2\varphi^2/2! + g\varphi^3/3! + \lambda\varphi^4/4!$, see left part of the Figure 3). Solutions were obtained numerically as this problem can not be solved analytically in the general case.

Nevertheless, some analytic approximation can still be done in the case when the true and false vacuums are practically degenerate, so that we can neglect the cubic term in the potential. This is the so-called thin-wall approximation. In this case the true vacuum is inside the bubble separated from the false vacuum by a wall, which thickness is much smaller than the size of the bubble.

To calculate higher order radiative corrections we again need to know an expression for Green function in instanton background. As usual, in order to find it, it is necessary to solve a second-order partial differential equation. The easiest way to solve it is to make use of spherical symmetry and expand the Green function in a series of spherical harmonics. For the potential discussed above, the coefficients of these series were calculated by Garbrecht and Millington within thing-wall approximation in [24]. In the case of zero harmonics their expression should be however corrected and is given

[2]Instead of the bounce action(S_c), we can speak about the height of the barrier between the true and false vacuum, because these quantities are directly proportional to each other.

[3]Note, that the false vacuum decay in d=4 dimensional case does not reduce entirely to one-dimensional, the similarity consists only in finding the bounce solution and the general principles on which the perturbation theory is constructed.

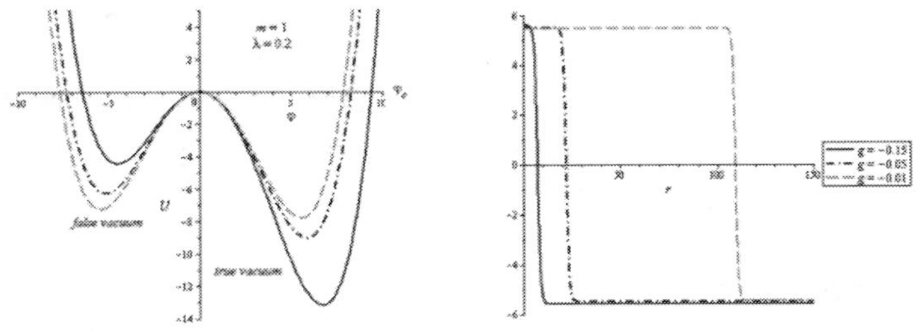

Figure 3. $U(\varphi) = -m^2\varphi^2/2! + g\varphi^3/3! + \lambda\varphi^4/4!$ on the left and the corresponding bounce solution on the right

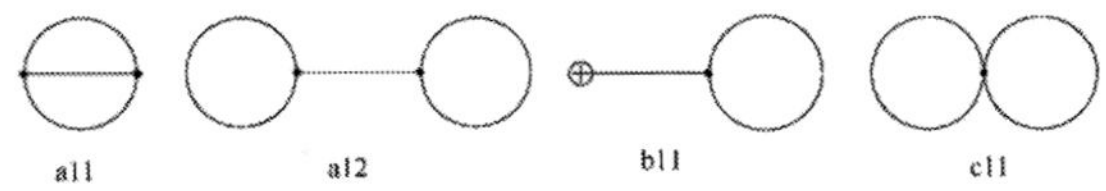

Figure 4. Two loop Feynman diagrams for the d=4 dimensional scalar quantum field theory

by

$$G(x, y, 2) = \frac{g(x, y)}{4\gamma}\left\{2 - xy + \frac{|x - y|}{4}(11 - 3xy) + (x - y)^2\right\} + \\ + \frac{3}{32\gamma}(1 - x^2)(1 - y^2)\left(\log g(x, y) - \frac{11}{3}\right) \quad (2)$$

where: $g(x, y) = \frac{1-|x-y|-xy}{1+|x-y|-xy}$, $\gamma = \frac{\mu}{\sqrt{2}}$, $x = \tanh(\gamma(r - R))$, $y = \tanh(\gamma(r' - R))$. Here r and r' is a radius in O(4) spherical coordinate system, $\mu^2 = -m^2$ and $R = \frac{\sqrt{12\lambda}}{g}$ is the bubble radius. Next, both one-loop functional determinant and higher order corrections could be calculated knowing this Green function. There is however one additional difficulty related to renormalization compared to quantum mechanical case. The ultraviolet divergences may be further regularized using momentum cutoff and so-called continuum approximation. Within the latter the summation over harmonics is rewritten as an integral over an almost continuous quantity. This technique and the method for finding the one-loop correction has been analyzed in detail in [24].

Now we can finally begin to study the higher order radiative corrections. The perturbation theory is constructed by the same principles as in quantum mechanics. The integrations are now however four-dimensional. Full two-loop correction acquire contributions from four diagrams, Figure 4. Here is an example of a contribution from "double bubble" diagram which is labelled "c11" on Figure 4:

$$I_{c11} = -\frac{\lambda\gamma^3 R^3}{4\pi^2}\left[\frac{1}{240}\left\{-135 + 10\sqrt{3}\pi - \frac{9}{7}\pi^2 + \left(4\sqrt{3}\pi + 45\left(2 + \log\frac{\gamma^2}{\Lambda^2}\right)\right)\log\frac{\gamma^2}{\Lambda^2}\right\} - \\ - \frac{\Lambda^2}{48\gamma^2}\left(\sqrt{3}\pi + 9\log\frac{\gamma^2}{\Lambda^2}\right)\right] \quad (3)$$

where Λ is ultraviolet cut-off regularization parameter.

4 Conclusions

In this work we used the method developed in [18–21] to calculate higher order radiative corrections to false vacuum decay rate in quantum mechanics and scalar quantum field theory. We show that in some cases these corrections may be significant and should be taken into account. At a moment we do not yet have a full two loop result in the case of scalar field theory. In this case we do have some technical difficulties related to the regularization of the sums over spherical harmonics in instanton Green functions. Still, we believe that this difficulties could be resolved and will be the subject of one of our future publications.

5 Acknowledgements

I would like to thank A.I. Onishchenko for interesting and stimulating discussions as well as for general assistance in writing this work. This work was supported by RFBR grants # 17-02-00872-a.

References

[1] J. Elias-Miro, J. R. Espinosa, G. F. Giudice, G. Isidori, A. Riotto and A. Strumia, Phys. Lett. B **709**, 222 (2012).

[2] G. Degrassi, S. Di Vita, J. Elias-Miro, J. R. Espinosa, G. F. Giudice, G. Isidori and A. Strumia, JHEP **1208**, 098 (2012).

[3] F. Bezrukov, M. Y. Kalmykov, B. A. Kniehl and M. Shaposhnikov, JHEP **1210**, 140 (2012).

[4] S. Alekhin, A. Djouadi and S. Moch, Phys. Lett. B **716**, 214 (2012).

[5] I. Masina, Phys. Rev. D **87**, no. 5, 053001 (2013).

[6] D. Buttazzo, G. Degrassi, P. P. Giardino, G. F. Giudice, F. Sala, A. Salvio and A. Strumia, JHEP **1312**, 089 (2013).

[7] J. R. Espinosa, G. F. Giudice, E. Morgante, A. Riotto, L. Senatore, A. Strumia and N. Tetradis, JHEP **1509**, 174 (2015).

[8] A. V. Bednyakov, B. A. Kniehl, A. F. Pikelner and O. L. Veretin, Phys. Rev. Lett. **115**, no. 20, 201802 (2015).

[9] G. Isidori, G. Ridolfi and A. Strumia, Nucl. Phys. B **609**, 387 (2001).

[10] Z. Lalak, M. Lewicki and P. Olszewski, JHEP **1405**, 119 (2014).

[11] A. D. Plascencia and C. Tamarit, JHEP **1610**, 099 (2016).

[12] M. Endo, T. Moroi, M. M. Nojiri and Y. Shoji, JHEP **1601**, 031 (2016).

[13] Z. Lalak, M. Lewicki and P. Olszewski, Phys. Rev. D **94**, no. 8, 085028 (2016).

[14] O. Czerwińska, Z. Lalak, M. Lewicki and P. Olszewski, JHEP **1610**, 004 (2016).

[15] J. R. Espinosa, M. Garny, T. Konstandin and A. Riotto, Phys. Rev. D **95**, no. 5, 056004 (2017).

[16] M. Endo, T. Moroi, M. M. Nojiri and Y. Shoji, Phys. Lett. B **771**, 281 (2017).

[17] M. Endo, T. Moroi, M. M. Nojiri and Y. Shoji, arXiv:1704.03492 [hep-ph].

[18] A. A. Aleinikov and E. V. Shuryak, Yad. Fiz. **46**, 122 (1987).

[19] S. Olejnik, Phys. Lett. B **221**, 372 (1989).

[20] M. A. Escobar-Ruiz, E. Shuryak and A. V. Turbiner, Phys. Rev. D **92**, no. 2, 025046 (2015) Erratum: [Phys. Rev. D **92**, no. 8, 089902 (2015)].

[21] M. A. Escobar-Ruiz, E. Shuryak and A. V. Turbiner, Phys. Rev. D **92**, no. 2, 025047 (2015).

[22] M. A. Bezuglov and A. I. Onishchenko, Phys. Rev. D **96**, no. 3, 036001 (2017).

[23] T. Hahn, *CUBA: A Library for multidimensional numerical integration*, Comput. Phys. Commun. **168**, 78 (2005) [hep-ph/0404043].

[24] B.Garbrecht and P.Millington, Phys. Rev. D **91**, 105021 (2015).

EPJ Web of Conferences **177**, 09002 (2018) https://doi.org/10.1051/epjconf/201817709002
AYSS-2017

Some Applications of Holography to Study Strongly Correlated Systems

Neha Bhatnagar

Department of Physics, Banaras Hindu University,
Varanasi-221005, India,
bhtngr.neha@gmail.com,

Abstract. In this work, we study the transport coefficients of strongly coupled condensed matter systems using gauge/gravity duality (holography). We consider examples from the real world and evaluate the conductivities from their gravity duals. Adopting the bottom-up approach of holography, we obtain the frequency response of the conductivity for (1+1)-dimensional systems. We also evaluate the DC conductivities for non-relativistic condensed matter systems with hyperscaling violating geometry.

Keywords: Holography, bottom-up approach, strongly coupled systems

1 Introduction

Recently, the discovery of high T_c superconductors and strange metals challenged the old concept of Fermi-liquid theory and the BCS (Bardeen Cooper Schrieffer) theory. These newly found materials are considered inherently strongly coupled with no well-defined quasi-particles. Thus, perturbative methods and conventional approaches fail to probe them completely. The remarkable progress in this area occurs with the advent of AdS/CFT duality[1]. The duality relates d-dimensional strongly coupled gauge theory with weakly coupled gravity theory in (d+1)-dimensions. The extra dimension in bulk corresponds to the energy scale of the boundary theory. Holography relates high T_c superconductivity with the spontaneously broken U(1) gauge symmetry modeled by introducing complex scalar field in the gravity background [2, 3]. The next significant development in this area occurs with the introduction of momentum dissipation term in the gravity action and obtain finite DC conductivity. In this work, we discuss the transport coefficients for two different strongly correlated systems using the holographic approach in a simplified manner.

2 (1+1)-dimensional systems

We first explore (1+1)-dimensional systems using holography. Lower dimensional systems such as nanowires, Luttinger liquids, and spin chains are useful for technology development but are difficult to analyse theoretically because of strong correlated nature. Applying the holographic approach, we capture the frequency response of the conductivity for the system. The gravity action in the soft-wall model is given by,

$$S = \int d^3x\sqrt{-g}e^{-2\phi}\left(\frac{1}{2\kappa^2}(R-2\Lambda)+\frac{1}{4g^2}F^2\right) \quad (1)$$

where, F^2 is the gauge field strength, $\kappa^2 = 8\pi G_3 = 1$ and $\Lambda = -2/L^2$. Here the dilaton field ϕ, softly truncates the IR boundary. The equations of motion are given in the form,

$$R_{\mu\nu} - \frac{1}{2}g_{\mu\nu}R - \Lambda g_{\mu\nu} = T_{\mu\nu}, \quad \text{and} \quad \nabla_\mu(e^{-2\phi}F^{\mu\nu}) = 0 \tag{2}$$

where $T_{\mu\nu}$ is the energy-momentum tensor. We consider the metric ansatz as [4, 5],

$$ds^2 = -f(r)dt^2 + r^2dx^2 + \frac{dr^2}{f(r)} \tag{3}$$

$$\text{where} \quad f(r) = r^2 - r_h^2 + \frac{\mu^2}{2r^2}\ln\left(\frac{r_h}{r}\right), and \quad A_t' = \mu e^{2\phi}\ln(\frac{r_h}{r}). \tag{4}$$

In this model, μ corresponds to the chemical potential and we set, $r_h = 1$ (radius of black hole horizon). We introduce perturbations in the gauge field and metric tensors in the Fourier space as, $A_x(r,t) = a_x(r)e^{-i\omega t}$ and $\mathbf{h}_{xt}(r,t) = h_{xt}(r)e^{-i\omega t}$. The linearised equations of motion are given by,

$$h_t'^x - A_t' z^2 a_x = 0, \tag{5}$$

$$a_x'' + \left(\frac{f'}{f} + \frac{1}{r} - 2\phi'\right)a_x' - \frac{A_t' h_t'^x}{f} + \frac{\omega^2}{f^2}a_x = 0. \tag{6}$$

The boundary asymptotics of the gauge field takes the form, $a_x = -a_x^{(0)}\ln r + a_x^{(1)}$. Now applying the in-going boundary condition to maintain the casual behavior the optical conductivity is obtained using [4, 5] , $\sigma(\omega) = \frac{i a_x^{(1)}}{\omega a_x^{(0)}} = \frac{i(a_x + r a_x' r \ln[r])}{\omega r a_x'}\Big|_{r\to\epsilon}$. We consider two different dilaton profiles, $\phi(r) = \frac{1}{r}$(Model-I) and $\phi(r) = \frac{1}{r^2}$ (Model-II) to investigate the frequency response of the system with varying chemical potential shown in Fig. 1. and Fig. 2 respectively.

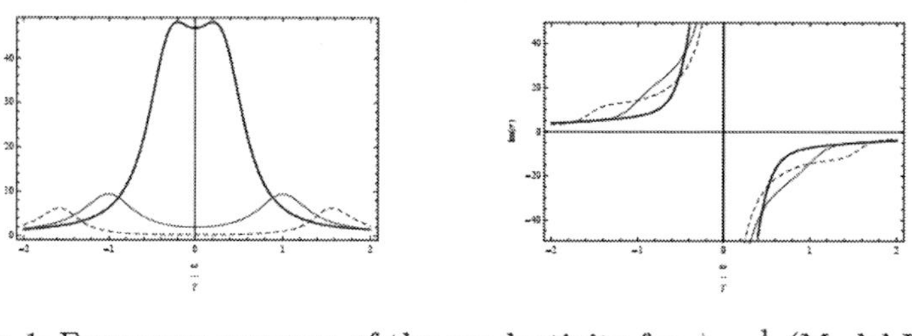

Fig. 1: Frequency response of the conductivity for $\phi = \frac{1}{r}$ (Model-I)

We attempt to capture the response of (1+1)-dimensional boundary theory with finite chemical potential. At lower frequency, we obtained Drude peak indicating the normal phase of the holographic system. With the increase in μ, we observe flattening of the peak in $Re[\sigma]$. Different dilaton profiles correspond to different condensate in the holographic superconductor[6]. Applying the soft-wall model for our work gave us the freedom to explore the specific features of strongly coupled condensed matter systems in a phenomenological simple way. We have captured the Fermi-Luttinger liquid behavior of the boundary theory using the soft-wall model for two different dilaton profiles.

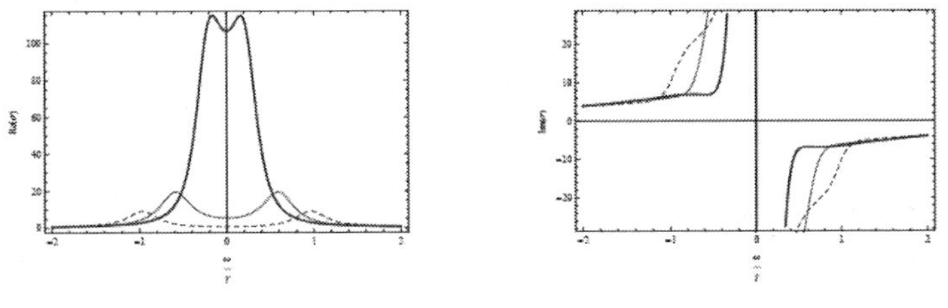

Fig. 2: Frequency response of the conductivity for $\phi = \frac{1}{r^2}$(Model-II)

3 Non-relativistic condensed matter systems with hyperscaling violation

We investigate the DC conductivities for non-relativistic condensed matter system using (3+1)-dimensional EMD (Einstein-Maxwell-Dilaton) theory with hyperscaling violation [7]. The gravity action is given by,

$$S = \int d^4x\sqrt{-g}\left(R - \frac{1}{2}(\partial\phi)^2 + V(\phi) - \frac{1}{4}\sum_{i=1}^{2} e^{\lambda_i\phi}F_i^2 - e^{\lambda_3\phi}\sum_{i=1}^{2}\partial\chi_i^2\right) \quad (7)$$

where, F_i as two different gauge fields (one for introducing Liftshitz symmetry and the other to incorporate charge in the system), $V(\phi) = -2\Lambda e^{\lambda_0\phi}$ (Liouville-like dilaton potential) and axionic fields (for momentum dissipation) are given as, $\chi_1 = \alpha x$ and $\chi_2 = \alpha y$. We obtain the following equations of motion for our holographic set-up,

$$R_{\mu\nu} = \frac{1}{2}\partial_\mu\phi\partial_\nu\phi + \frac{1}{2}e^{\lambda_3\phi}(\partial_\mu\chi_i\partial_\nu\chi_i) + \Lambda e^{\lambda_0\phi}g_{\mu\nu} + \sum_{i=1}^{2}\frac{1}{2}e^{\lambda_i\phi}(F_{i\mu\rho}F_{i\nu}^{\rho} - \frac{1}{4}F_i^2 g_{\mu\nu}) \quad (8)$$

$$\Box\phi = \frac{1}{2}\lambda_3\sum_{i=1}^{2}(\partial\chi_i) + \frac{1}{4}\sum_{i=1}^{2}\lambda_i e^{\lambda_i\phi}F_i^2 + 2\lambda_0\Lambda e^{\lambda_0\phi} \quad (9)$$

$$0 = \nabla_\mu\left(e^{\lambda_3\phi}\nabla^\mu\chi_1\right), \quad 0 = \nabla_\mu\left(e^{\lambda_3\phi}\nabla^\mu\chi_2\right) \quad (10)$$

$$0 = \nabla_\mu\left(e^{\lambda_1\phi}F_1^{\mu\nu}\right), \quad 0 = \nabla_\mu\left(e^{\lambda_2\phi}F_2^{\mu\nu}\right). \quad (11)$$

The metric ansatz is given by, $$ds^2 = r^\theta\left(-r^{2z}f(r)dt^2 + \frac{dr^2}{r^2 f(r)} + r^2(dx^2 + dy^2)\right) \quad (12)$$

where θ is the hyperscaling violation term and $z \neq 1$ is the Lifshitz scaling. The black hole factor is given by, $f(r) = 1 - \frac{m}{r^{\theta+z+2}} + \frac{Q_2^2}{2(\theta+2)(\theta+z)r^{2(\theta+z+1)}} + \frac{\alpha^2}{(\theta+2)(z-2)r^{\theta+2z}}$ where 'm' is mass, 'Q_2' is charge of black hole and 'α' is the strength of momentum relaxation. The parameters of the gravity system are fixed considering the Null energy condition as,

$$\gamma = \sqrt{(\theta+2)(\theta+2z-2)}, \qquad \lambda_0 = \frac{-\theta}{\gamma}, \qquad \lambda_1 = \frac{-(4+\theta)}{\gamma} \quad (13)$$

$$\lambda_2 = \frac{(\theta+2z-2)}{\gamma}, \qquad \lambda_3 = \frac{-\gamma}{\theta+2}, \qquad Q_1 = \sqrt{2(z-1)(\theta+z+2)} \quad (14)$$

with $\Lambda = \frac{-1}{2}(\theta + z + 1)(\theta + z + 2)$ and $\phi = \gamma \log r$. The temporal component of both the gauge fields in the bulk are given by,

$$(A_1)'_t = Q_1 r^{z-3-\lambda_1\gamma} \quad \text{and} \quad (A_2)'_t = Q_2 r^{z-3-\lambda_2\gamma}. \tag{15}$$

Hawking temperature is given by, $T = \frac{f'(r_h)}{4\pi} = \frac{(2z+1)r_h^z}{4\pi} - \frac{Q_2^2 r_h^{3z}}{8\pi(z+1)} - \frac{\alpha^2 r_h^{2z-1}}{4\pi(z+1)}$. Applying holographic formalism we introduce the linear perturbations in the gauge fields, axion field and metric components as shown,

$$\delta A_i(t,r) = \int_{-\infty}^{\infty} \frac{d\omega}{2\pi} a_i(r) e^{-i\omega t}, \; \delta\chi_i(t,r) = \int_{-\infty}^{\infty} \frac{d\omega}{2\pi} b_i(r) e^{-i\omega t}, \tag{16}$$

$$\delta g_{ti}(t,r) = \int_{-\infty}^{\infty} \frac{d\omega}{2\pi} r^{\theta+2} h_{ti}(r) e^{-i\omega t}. \tag{17}$$

Considering perturbations in both the gauge fields we obtain second-order coupled differential equations. For simplification, we consider, $\Phi = \frac{r^{5-z} f b'}{i\omega}$ and the equations of motion in the changed form is given by,

$$(r^{3-z-\theta} f a_1')' + \frac{\omega^2 a_1}{r^{z+5+\theta} f} + \frac{Q_1}{r^{z+5+\theta}}(Q_1 a_1 + Q_2 a_2 - a\Phi) = 0 \tag{18}$$

$$(r^{3z-1+\theta} f a_2')' + \frac{\omega^2 a_2}{r^{3-z-\theta} f} + \frac{Q_2}{r^{5-z+\theta}}(Q_1 a_1 + Q_2 a_2 - a\Phi) = 0 \tag{19}$$

$$(r^{3z-3} f \Phi')' + \frac{\omega^2}{r^{5-z} f}\Phi - \frac{\alpha}{r^{5-z+\theta}}(Q_1 a_1 + Q_2 a_2 - a\Phi) = 0. \tag{20}$$

Multiplying (20) with $\frac{Q_i}{\alpha}$ and adding to (18) and (19) respectively we obtain the conserved current J_i for the gauge fields,

$$J_1 = -r^{z-3-\theta} f a_1' - \frac{Q_1}{\alpha} r^{3z-3} f \Phi', \qquad J_2 = -r^{3z-1+\theta} f a_2' - \frac{Q_2}{\alpha} r^{3z-3} f \Phi'. \tag{21}$$

Next, using the Onsager relation ($J = \tau X$) we obtain the flow equations. Here, ' J' are the response of the system from external sources 'X' and τ contain transport coefficients. We evaluate τ matrix using the formalism given in [8]. We put the above coupled field equations in the given form,

$$\tau = \begin{Vmatrix} -r^{3-z-\theta} f a_1' \\ -r^{3z-1+\theta} f a_2' \\ -r^{3z-3}\phi' \end{Vmatrix} \begin{Vmatrix} i\omega a_1 \\ i\omega a_1 \\ i\omega\phi \end{Vmatrix}^{-1}. \tag{22}$$

The radial derivative is given by,

$$\tau' = \begin{Vmatrix} -r^{3-z-\theta} f a_1' \\ -r^{3z-1+\theta} f a_2' \\ -r^{3z-3}\phi' \end{Vmatrix}' \begin{Vmatrix} i\omega a_1 \\ i\omega a_1 \\ i\omega\phi \end{Vmatrix}^{-1} - i\omega\tau \begin{Vmatrix} a_1' \\ a_2' \\ \phi' \end{Vmatrix} \begin{Vmatrix} i\omega a_1 \\ i\omega a_1 \\ i\omega\phi \end{Vmatrix}^{-1}. \tag{23}$$

In the near horizon limit, we obtain the flow equation from the above equation using $f(r_h) = 0$,

$$0 = \begin{pmatrix} \frac{-i\omega}{r_h^{z+5+\theta}} & 0 & 0 \\ 0 & \frac{-i\omega}{r_h^{3-z-\theta}} & 0 \\ 0 & 0 & \frac{-i\omega}{r_h^{5-z}} \end{pmatrix} + i\omega\tau_h \begin{pmatrix} \frac{1}{r_h^{z-3-\theta}} & 0 & 0 \\ 0 & \frac{1}{r_h^{3z-1+\theta}} & 0 \\ 0 & 0 & \frac{1}{r_h^{3z-3}} \end{pmatrix} \tau_h. \tag{24}$$

We evaluate τ_h matrix and used it to obtain first order flow equations from equation (22) in the near horizon limit,

$$-r_h^{z+1} f a_1' = i\omega a_1, \qquad -r_h^{z+1} f a_2' = i\omega a_2, \qquad -r_h^{z+1} f \phi' = i\omega \phi. \tag{25}$$

To establish an explicit dependence of transport coefficients on the second gauge field we consider the first gauge field equation (18) in the near horizon limit for $\omega \to 0$ and obtain, $\Phi\alpha = q_1 a_1 + q_2 a_2$. Substituting all the terms in the current expressions and using Ohm's law as, $\sigma_{ij} = \frac{\partial J_i}{\partial E_j}$ we evaluate the DC conductivity as,

$$\sigma_{11} = \frac{1}{r_h^{4+\theta}} + \frac{q_1^2 r_h^{2z-4}}{\alpha^2}, \qquad \sigma_{12} = \frac{q_1 q_2 r_h^{2z-4}}{\alpha^2} = \sigma_{21}, \qquad \sigma_{22} = r_h^{2z-2+\theta} + \frac{q_2^2 r_h^{2z-4}}{\alpha^2}. \tag{26}$$

Also, since a_1 is the auxiliary field in the system, we can set $J_1 = 0$ [7] in the near horizon limit and obtain, $\sigma_{DC} = r_h^{2z-2+\theta}\left[1 + \frac{q_2^2}{r_h^{2+\theta}(\alpha^2 + q_1^2 r_h^{2z+\theta})}\right]$. For $\alpha \to 0$ we obtain a constant conductivity from σ_{DC}. Also, we noticed the conductivity is a combination of two different components, a charge-conjugation symmetric term and the dissipation term [9]. The explicit dependence of conductivity on Lifshitz scaling(z) and θ is observed. Here, we summarize our main findings of this work as,

a) At high temperature ($q_2 = 0$), we obtain linear resistivity ($\rho_{11} = \frac{1}{\sigma_{11}}$) for $z = 4+\theta$ and $-4 < \theta < 8/3$.
b) At low temperature ($q_2 \neq 0$), we obtain linear resistivity keeping $\theta = 0$ and $z = 4/3$.
c) For $J_1 = 0$, the temperature dependence of the DC conductivity is given by, $\sigma_{DC} \sim T^{\frac{2z-2+\theta}{z}} + \frac{q_2^2}{q_1^2} T^{\frac{-\theta-4}{z}} + \alpha^2 T^{\frac{-2}{z}}$

4 Conclusions and Discussions

In this report, we discussed the applications of holography to investigate strongly coupled systems which otherwise were inaccessible. We evaluated the frequency response of conductivity using soft-wall model for (1+1)-dimensional systems. We also obtained the expression of DC conductivities for (2+1)-dimensional non-relativistic condensed matter systems after introducing the perturbations in both the gauge fields. Temperature dependence of DC conductivity for these systems has been explored. Further, we have studied the effect of the external magnetic field on DC transport[10].

References

1. J. M. Maldacena, Adv. Theor. Math. Phys. **2**, 231 (1998).
2. S. S. Gubser, Phys. Rev. D **78**, 065034 (2008)
3. S. A. Hartnoll, et. al. Phys. Rev. Lett. **101**, 031601 (2008).
4. D. Maity et. al., Nucl. Phys. B **839**, 526 (2010).
5. J. Ren, JHEP **1011**, 055 (2010).
6. S. S. Afonin and I. V. Pusenkov, Eur. Phys. J. C **76**, no. 6, 342 (2016).
7. S. Cremonini, H. S. Liu, H. Lu and C. N. Pope, JHEP **1704**, 009 (2017).
8. Y. Tian, et.al., Phys. Rev. D **96**, no. 4, 046011 (2017).
9. M. Blake and A. Donos, Phys. Rev. Lett. **114**, no. 2, 021601 (2015).
10. N. Bhatnagar and S. Siwach, arXiv:1707.04013 [hep-th].

EPJ Web of Conferences **177**, 09003 (2018) https://doi.org/10.1051/epjconf/201817709003
AYSS-2017

Spectral Triples in Particle Physics

Arkadiusz Bochniak[1,⋆]

[1] *Institute of Physics, Jagiellonian University, prof. Stanisława Łojasiewicza 11, 30-348 Kraków, Poland*

Abstract. We give an overview of the approach to the Standard Model of Particle Physics and its extensions based on the Noncommutative Geometry. The notion of spectral triples is introduced and their applications in particle physics are presented. We revisit known results based on different approaches within Noncommutative Geometry, list problems which appeared in these methods, propose possible solutions and indicate future directions of research.

1 Introduction

The classical field-theoretical description of the Standard Model of Particle Physics is rather complicated and it is not so easy to see at first sight hidden fundamental structures behind such formalism. On the other hand we know that there are physical theories like General Relativity or classical mechanics which are described in a geometrical language, using the notion of Riemannian or symplectic manifolds, respectively. There is a natural question if there exists some purely geometrical description of the Standard Model. It turns out that it is possible but we have to extend the notion of geometry. To do that we use the idea of Noncommutative Geometry (NCG) which is based on the observation made by Gelfand and Naimark, known as the Gelfand-Naimark duality theorem, that the category of locally compact Hausdorff spaces is anti-equivalent to the category of commutative C^*-algebras. It means that the information about the topology of our space can be encoded in a properly choosen algebra. This algebra, for the case of a compact space M, is just the C^*-algebra $C(M)$ of continuous $\mathbb{C}$-valued functions on M. The idea of using functions on space instead of the space is nothing strange for physicists, because in physics we really manipulate functions defined on a space instead of using its points. This dual picture of geometry indicates its natural extension, mainly we can consider noncommutative C^*-algebras and think of them as a set of functions on *noncommutative topological space*, thus we get the notion of *noncommutative topology*. Due to the Gelfand-Naimark-Segal construction any C^*-algebra $\mathcal{A}$ is isometrically isomorphic to a C^*-algebra $B(H)$ of bounded operators on some Hilbert space H, i.e. there exists a faithful representation $\pi : \mathcal{A} \to B(H)$. Let us also mention one important fact, mainly the Serre-Swan theorem about the equivalence of the category of complex vector bundles on a compact Hausdorff space X and the category of finite projective modules over the algebra $C(X)$. The vast majority of physical objects are just sections of some bundles (e.g. vector fields, spinors, etc.) or are objects defined in terms of bundles (connections, strength tensors, etc.), therefore projective modules over noncommutative C^*-algebras are natural candidates for noncommutative vector bundles.

⋆e-mail: arkadiusz.bochniak@student.uj.edu.pl

Now, let us consider the *differential part of the theory*. The knowledge of the topology is not enough to have the whole set of information about the *geometry*. Consider a Dirac operator on closed, oriented Riemannian manifold that admits spinors, i.e. the operator given in local coordinates by $\not{D}_M = i\gamma^\mu\left(\partial_\mu + \omega_\mu\right)$, where γ^μ are γ-matrices and ω_μ can be expressed by Levi-Civita connection lifted to act on spinor-valued functions. The Dirac operator determines the metric structure [4] and is the crucial ingredient in the construction of K-homology classes. The Dirac operator $\not{D}_M$ together with $C^\infty(M)$ and a Hilbert space $L^2(M, S)$ of square-integrable spinors form a canonical example of a spectral triple - an object which we will define precisely in the next section, where we will demonstrate how to apply such a concept in case of SM and explain how to search for its possible extensions.

2 Spectral triples and the Standard Model

2.1 Spectral triples

From now on, the geometry will be understood as a *unital* spectral triple $(\mathcal{A}, H, \mathcal{D})$ defined as in [7], i.e. $\mathcal{A}$ is an unital involutive algebra (not necessary a C^*-algebra) representable as bounded operators on a separable Hilbert space H, and $\mathcal{D}$ is a selfadjoint operator on H with compact resolvent such that $a \cdot \mathrm{Dom}(\mathcal{D}) \subset \mathrm{Dom}(\mathcal{D})$ and $[\mathcal{D}, a]$ extends to a bounded operator on H, for all $a \in \mathcal{A}$. Moreover, we will always assume that our spectral triple is even and real, i.e. there exists grading operator $\gamma \in \mathrm{End}(H)$, $\gamma^* = \gamma, \gamma^2 = 1$ that commutes with all elements of the algebra and anticommutes with the Dirac operator, and there exists $J \in \mathrm{Isom}(H)$ that is antilinear and satisfies

$$J^2 = \varepsilon 1, \; J\mathcal{D} = \varepsilon \mathcal{D} J, \; J\gamma = \varepsilon'' \gamma J. \tag{1}$$

with $\varepsilon, \varepsilon', \varepsilon'' \in \{\pm 1\}$ dependent on the KO-dimension [8]. Moreover, there are two important conditions which have to be satisfied for all a and b :

- 0th order condition : $[a, JbJ^{-1}] = 0$,
- 1st order condition : $\left[[\mathcal{D}, a], JbJ^{-1}\right] = 0$.

The canonical spectral triple $\left(C^\infty(M), L^2(M, S), \not{D}_M\right)$ which was previously mentioned is a natural example of this general concept. Another class are finite spectral triples, which can be classified [15] [18].

2.2 The Standard Model

Our goal is to use the notion of spectral triples to explore the structure of the Standard Model of Particle Physics. To do that observe that the gauge group $U(1) \times SU(2) \times SU(3)$ describes (up to extra $U(1)$ term) the unitary elements of the following finite-dimensional algebra

$$\mathcal{A}_{sm} = \mathbb{C} \oplus \mathbb{H} \oplus M_3(\mathbb{C}), \tag{2}$$

where $\mathbb{H}$ is the algebra of quaternions. Therefore $\mathcal{A}_{sm}$ is a natural choice for the first ingredient of a spectral triple. The choice of a Hilbert space is dictated by the number of particles we have in the Standard Model. Simple counting [10] implies the dimensionality equal 96. Taking into account chiralities of particles and existance of antiparticles the Hilbert space can presented as a direct sum

$$H_{sm} = H_L \oplus H_R \oplus H_L^c \oplus H_R^c. \tag{3}$$

Therefore, the Dirac operator must be a 96×96-matrix that has to satisfy required conditions. Moreover, we also define the grading operator γ to be just the usual chirality operator and the real structure J which acts interchanging particle and antiparticle sectors and taking complex conjugation. The existence of mentioned compatibility conditions implies that the Dirac operator has a lot of zeros in its entries. The nonzero values are related to the Yukawa couplings, Majorana masses of neutrinos, CKM and PMNS mixing matrices [10].

To get some *numbers* from this construction, which can be experimentally tested, we consider almost-commutative spectral triples with an algebra $C^\infty(M) \otimes \mathcal{A}_{sm}$, where M plays a role of a background for our theory of particles. The Hilbert space is the tensor product of Hilbert spaces we had on each part. Similarly, as tensor products, we construct grading Γ and the real structure on the product spectral triple. Note that the form of the last one follows from the values of KO-dimensions [8]. The Dirac operator is defined by

$$\mathcal{D} = \not{D}_M \otimes 1 + \gamma_5 \otimes \mathcal{D}_{sm}, \tag{4}$$

where $\not{D}_M$ is the usual Dirac operator on spin manifold M, $\mathcal{D}_{sm}$ is the Dirac operator we had in the spectral triple for the Standard Model and γ_5 is the usual grading on M.

For a given spectral triple we have the space of 1-forms

$$\Omega^1_{\mathcal{D}}(\mathcal{A}) = \left\{ \sum_k a_k \, [\mathcal{D}, b_k] \, , \; a_k, b_k \in \mathcal{A} \right\} \tag{5}$$

and we can use them to construct *fluctuations* of the Dirac operator by taking

$$\mathcal{D}_A = \mathcal{D} + A + JAJ^{-1}. \tag{6}$$

To obtain some physical quantities we use *the spectral action*

$$S = \frac{1}{2}\langle J\xi, \mathcal{D}_A \xi\rangle + \mathrm{Tr} f\left(\frac{\mathcal{D}_A}{\Lambda}\right), \tag{7}$$

where $\xi \in \frac{1}{2}(1 + \gamma_5 \otimes \gamma)H$, f is a smooth approximation of the characteristic function of the interval $[0, 1]$ and Λ is a cutt-off parameter. Using heat kernel expansion [2] we can calculate the second part in powers of Λ^{-1}. After collecting terms in first order we get the bosonic part of action of the Standard Model on the curved background given by a manifold M, i.e. with Hilbert-Einstein terms but also with a coupling of Higgs field to the Ricci scalar. This approach gives us an effective Lagrangian at some energy scale Λ, on which all gauge coupling constants are equal. Assuming so-called Big Desert Hypothesis and using the standard tools of renormalization group we can explore the behaviour of running coupling constants and β-functions. This allows us to get values of these parameters at lower scales. Moreover, in the input we had only parameters of fermions and the bosonic fields appear from the inner fluctuations, therefore we can express the bosonic parameters in terms of fermionic one, for example we can calculate the mass of Higgs boson and the value we obtained is $\sim$ 170GeV [2] which is near the experimental one ($\sim$ 126GeV), but different. Possible explanations of the source of this difference and another problems in the Standard Model will be discussed in the next section. Let us emphasize here that the NCG reveals the origin of the spontaneous symmetry breaking in the Standard Model.

2.3 Problems

There are few problems related to the previously mentioned results. First of all, the difference in the Higgs mass suggest that we are on the right way to explain the structure of the Standard Model on the fundamental, purely geometrical level, but still something has to be improved. Moreover, the low value of the Higgs mass is related to another problem of the instability of the electroweak vacuum known as *Higgs vacuum problem*. Another important issue is the question if the assumption of *the Big Desert* is adequate. From the most actual data we know that the extrapolated values of gauge couplings do not meet at exactly one point, but they rather form a triangle, so the predicted property of SM following from NCG is only approximately satisfied. Additional problems are related to the presence of right-handed neutrinos in the model, which Majorana mass terms can cure some of the previous problems, but we encounter the issue of see-saw mechanism, the same problem as in the *classical* approach to the SM. Another problems arise from considerations of Majorana-type particles. There are also some other questions related to the choice of specific mathematical structures what we will briefly discuss in the next section.

3 Extensions of the Standard Model

Problems mentioned in the previous section suggest not only that something has to be improved, but also indirectly indicate the way how to do it [9]. The problem of the difference in Higgs mass can be cured by adding additional fermions to the theory. The vacuum instability can be repaired by the introduction of a new σ-field, which was done many years ago in the particle physics in the context of Weinberg-Salam theory. One possibility to incorporate such a field in NCG approach is to change by hand the Yukawa coupling related to right handed neutrino Majorana mass y_R into a field $y_R\sigma(x)$. As a result we cure the instability and additionally get the proper value of Higgs mass, but in NCG scalar fields should be produced from fluctuations of Dirac operator, so we have to take another spectral triple or relax some of the axioms to get the σ-field in purely geometric way. If we fix all ingredients of the spectral triple unless an algebra, then the more general form of the last constituent is [10],[9]

$$M_a(\mathbb{H}) \oplus M_{2a}(\mathbb{C}), \text{ where } a \in \mathbb{N}. \tag{8}$$

The existence of the grading implies that $a \geq 2$. For the smallest possible case, i.e. $a = 2$, one can proof [10] that requiring commutation relation between the grading and the algebra the last one is reduced to

$$\mathbb{H}_L \oplus \mathbb{H}_R \oplus M_4(\mathbb{C}). \tag{9}$$

If, in addition, we require the 1st-order condition then we get the algebra $\mathcal{A}_{sm}$ of the Standard Model. Therefore, the Standard Model is just the smallest possible noncommutative manifold. One can consider the exact form of the representation of this algebra on Hilbert space H_{sm} and try to find possible generalizations or extension of the SM. If we want to save number of fermions in our theory and do not relax any condition of spectral triple, then the only possibility, due to dimensionality argument, is to introduce mixing of spinorial and gauge indices and take larger algebra $M_2(\mathbb{H}) \oplus M_4(\mathbb{C})$ - this choice is called *Grand Symmetry model* [10]. Repeating all steps done in case of the SM one can get σ-field which follows from fluctuations of Dirac operator in a similar way as the Higgs field [11]. Moreover, in that approach we can go back into the Standard Model by imposing additional condition, but this last symmetry breaking is not given by a dynamical mechanism [9]. Nevertheless, the idea of mixing spinorial and gauge indices is important one for study possible extensions of SM. The problem of symmetry breaking from Grand Symmetry into the SM can be solved by using twisted version of Dirac operator as was proposed in [12]. Nowadays, we have more general notion of twisted spectral triples, which can be use as an alternative approach to study particle physic models by the consideration of so-called Twisted Standard Model [16]. Moreover, from the general considerations we know that in the NCG approach there is no room for $SU(5)$ or $SO(10)$ GUTs and we have some kind of *no-go theorem* for supersymmetry at least in the case of MSSM [1]. We can also ask if our axioms of noncommutative manifolds are adequate. We can, for example, relax 1st-order condition, then for the smallest noncommutative manifolds we end up with a family of Pati-Salam models with $SU(2)_L \times SU(2)_R \times SU(4)$ gauge group. In such models the unification scale is of the order 10^{16}GeV and the symmetry breaking to the SM appears at $m_R = 10^{11} - 10^{13}$GeV, depending on the details of the model [3]. It is also a scale for the see-saw mechanism. Symmetries of Dirac operator in such case provide us with the existence of few Higgs-like fields and in the simplest, so-called *composite*, Pati-Salam model we get additional field in Higgs sector which lowers the value of Higgs mass to the value consistent with experiment. We can also consider other possible changes of algebraic conditions for spectral triples, especially introduction of 2nd-order condition related to Clifford structure and the aspects of Morita equivalence for spectral triples [7],[6] or consider more deeply other concepts like leptoquark models[17] and the role of the S^0-symmetry [5],[15]. Moreover, we can also take into account possible changes in the structure of almost-commutative geometries by replacing the standard tensor product by the graded one [13]. Another approaches based on Loretzian spectral triples are also considered [14] and provide alternative framework to study such phenomena.

4 Summary

We presented a short overview of alternative approach to the description of particle physics based on the notion of spectral triples and discussed main results, issues and indications for possible extensions and future searches. We should use our knowledge from experiments to perform some kind of reverse engineering to find proper geometric description of particle physics, especially by analysing the algebraic structure of finite spectral triples and their modifications.

References

[1] W. Beenakker, T. van den Broek, W. D. van Suijlekom, *Supersymmetry and Noncommutative Geometry* (Springer, Cham, 2016) 136
[2] A. H. Chamseddine, A. Connes, M. Marcolli, Adv. Theor. Math. Phys. **11**, 991-1089 (2007)
[3] A. H. Chamseddine, A. Connes, W. D. van Suijlekom, J. High Energ. Phys. **11** (2015)
[4] A. Connes, Ergod. Theory Dyn. Syst. **9**, 207-220 (1989)
[5] A. Connes, J. Math. Phys. **36**, 6194 (1995)
[6] F. D'Andrea, L. Dąbrowski, J. Noncommut. Geom. **10**, 551–578 (2016)
[7] F. D'Andrea, L. Dąbrowski, A. Sitarz, arXiv:1703.05279v1 [math-ph]
[8] L. Dąbrowski, G. Dossena, Int. J. Geom. Methods Mod. Phys. **8**, 1833-1848 (2010)
[9] A. Devastato, J. Phys.: Conf. Ser. **634** 012008 (2015)
[10] A. Devastato, F. Lizzi, P. Martinetti, J. High Energy Phys. **01** 042 (2014)
[11] A. Devastato, F. Lizzi, P. Martinetti, Fortsch.Phys. **62** 863-868 (2014)
[12] A. Devastato, P. Martinetti, Math. Phys. Anal. Geom. **20** (2017)
[13] S. Farnsworth, J. Math. Phys. **58**, 023507 (2017)
[14] N. Franco, arXiv:1710.10959 [math-ph]
[15] T. Krajewski, J. Geom. Phys. **28**, 1-30 (1998)
[16] G. Landi, P. Martinetti, Lett. Math. Phys. **106** no.11, 1499-1530 (2016)
[17] M. Paschke, F. Scheck, A. Sitarz, Phys. Rev. D **59**, 035003 (1999)
[18] M. Paschke, A. Sitarz, J. Math. Phys. **39** 6191 (1998)

Gluing operation and form factors of local operators in $\mathcal{N} = 4$ Super Yang-Mills theory

A. E. Bolshov[1,2,a]

[1] *Department of General and Applied Physics, Moscow Institute of Physics and Technology*
[2] *Bogoliubov Laboratory of Theoretical Physics, Joint institute of Nuclear Research*

Abstract. The gluing operation is an effective way to get form factors of both local and non-local operators starting from different representations of on-shell scattering amplitudes. In this paper it is shown how it works on the example of form factors of operators from stress-tensor operator supermultiplet in Grassmannian and spinor helicity representations.

1 Introduction

$\mathcal{N} = 4$ Super Yang-Mills theory (SYM) turns out to be a very popular object for theoretical research. Mainly this is due to the development of new techniques, such as recursive methods for tree-level amplitudes, on-shell diagrams and so on (see [3] for review), allowing to obtain analytical results for the matrix elements of the S-matrix. The Grassmannian integral representation for the amplitude is of particular interest, since it makes symmetry properties of the amplitudes manifest and relates scattering amplitudes to on-shell diagrams [4]. After a while, many of the techniques, developed for scattering amplitudes, were generalized to more complicated class of quantities - form factors, which are matrix elements of the form

$$\langle 0|O(x)|1, ..., n\rangle,$$

on which our attention is focused on. In particular, it was especially interesting to obtain the Grassmannian integral representation for form factors, which was done in [1] and [2] in slightly different ways. The common thing for both these two approaches is gluing of the minimal formfactor to the corresponding amplitude, whose integral representation is known [4]. The purpose of this article is to check formulae, obtained in [1] and [2] for different representations, i.e. for the Grassmannian integral and spinor helicity representations.

2 Form factors of the Stress-Tensor Supermultiplet operator

For the purposes, claimed in the previous section, only form factors of the chiral part of the stress-tensor supermultiplet operator would be considered. With the help of the so-called harmonic superspace [5] the operator itself can be written as

$$T(x, \theta^+) = \mathrm{tr}(\phi^{++}\phi^{++}) + ... + \frac{1}{3}(\theta^+)^4\mathcal{L}, \tag{1}$$

[a]e-mail: artem.bolshov@phystech.edu

where $\theta^{+a}_{\alpha} = \theta^{A}_{\alpha} u^{+a}_{A}$, $\theta^{-a'}_{\alpha} = \theta^{A}_{\alpha} u^{-a'}_{A}$ are projections of coordinates on the superspace to the harmonic superspace with projectors u^{+a}_{A} and $u^{-a'}_{A}$. The lowest component of $T(x, \theta^{+})$ is scalar operator $\mathrm{tr}(\phi^{++}\phi^{++})$ with $\phi^{++} = \frac{1}{2}\epsilon_{ab} u^{+a}_{A} u^{+b}_{B} \phi^{AB}$. The indices a, a' and $\pm$ correspond to $SU(2) \times SU(2)' \times U(1) \subset SU(4)$. Let q and $\gamma^{-\alpha}_{a}$ be the momentum and supermomentum, carried by the operator $O(x)$ respectively. Then the super form factor is defined by

$$\mathcal{F}_{k,n} = \int d^4x d^4\theta^{+} e^{-iqx - i\theta^{+a}_{\alpha}\gamma^{-\alpha}_{a}} \langle 1, ..., n | T(x, \theta^{+}) | 0 \rangle. \tag{2}$$

For $k = 2$, which corresponds to the Maximal Helicity Violating (MHV) case, the form factor is given by the following equation:

$$\mathcal{F}_{2,n}(1, ..., n, q, \gamma^{-}) = \frac{\delta^{(4)}(P)\hat{\delta}^{(4)}(Q^{+})\hat{\delta}^{(4)}(Q^{-})}{\langle 12\rangle\langle 23\rangle ... \langle n-1n\rangle\langle n1\rangle}, \tag{3}$$

where

$$P = \sum_{i=1}^{n} \lambda_i \tilde{\lambda}_i - q, Q^{+} = \lambda_i \eta^{+}_i, Q^{-} = \lambda_i \eta^{-}_i - \gamma^{-}. \tag{4}$$

3 Gluing operation & Grassmannian integral representation

As it was pointed out in the beginning, there were some difficulties with obtaining the grassmannian integral representation for the form factor, which was explained by the fact that form factors are partially off-shell quantities. A conjecture was made, that such a representation could be obtained using known integral representation for the scattering amplitude. Graphically it can be illustrated by the figure 3. The idea of the gluing operation is very simple, i.e. to take known representation of the

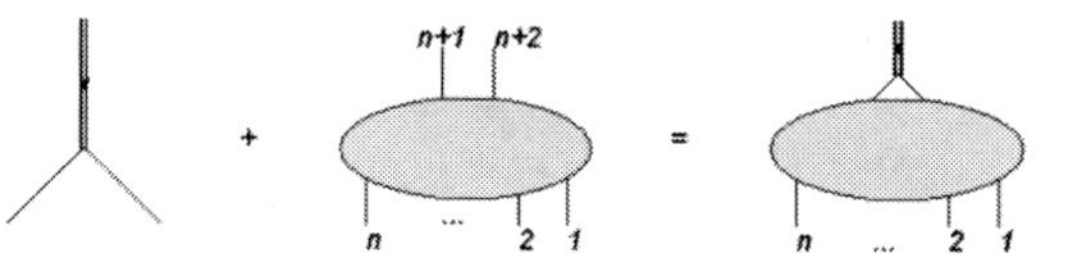

Figure 1. Graphical representation of the gluing operation

amplitude with two additional external legs and glue to it the minimal form factor, which contains delta functions, imposing proper conditions on kinematics, by integrating out these two extra degrees of freedom. Since the minimal form factor for the operator, considered here, has two external legs, one needs to glue it to $\mathcal{A}_{k,n+2}$. On mathematical language gluing reduces to the on-shell phase-space integration:

$$\mathcal{F}_{k,n} = \int \left(\prod_{i=n+1,n+2} \frac{d^2\lambda_i d^2\tilde{\lambda}_i}{Vol[GL(1)^2]} d^4\eta_i \right) \mathcal{F}_{2,2} \Big|_{\lambda_{n+1,n+2} \to -\lambda_{n+1,n+2}} \mathcal{A}_{k,n+2} + \text{other gluing positions}, \tag{5}$$

where $\mathcal{F}_{2,2} = \delta^2(\underline{\tilde{\lambda}_4})\delta^4(\underline{\eta_4})\delta^2(\underline{\tilde{\lambda}_5})\delta^4(\underline{\eta_5})$ and underscored variables impose twisted kinematics conditions (see eq. 2.21 of [1]):

$$\underline{\tilde{\lambda}_4} = \tilde{\lambda}_4 - \frac{\langle 5|q}{\langle 54\rangle} \qquad \underline{\eta^{-}_4} = \eta^{-}_4 - \frac{\langle 5|\gamma^{-}}{\langle 54\rangle} \qquad \underline{\eta^{+}_4} = \eta^{+}_4$$

$$\underline{\tilde{\lambda}_5} = \tilde{\lambda}_5 - \frac{\langle 4|q}{\langle 45\rangle} \qquad \underline{\eta^{-}_5} = \eta^{-}_5 - \frac{\langle 4|\gamma^{-}}{\langle 45\rangle} \qquad \underline{\eta^{+}_5} = \eta^{+}_5$$

After performing the integration, one arrives at the formula, valid for arbitrary k, n [1], [2]:

$$\mathcal{F}_{k,n} = \langle \xi_A \xi_B \rangle^2 \int \frac{d^{k\times(n+2)}C}{Vol[GL(k)]} \Omega_{k,n} \hat{\delta}^{2\times k}(C \cdot \underline{\underline{\tilde{\lambda}}}) \hat{\delta}^{4\times k}(C \cdot \underline{\underline{\eta}}) \hat{\delta}^{2\times(n+2-k)}(C^{\perp} \cdot \underline{\underline{\lambda}}) + \text{other gluing positions}, \quad (6)$$

where ξ_A and ξ_B are arbitrary reference spinors and

$$\Omega_{k,n} = \frac{Y(1-Y)^{-1}}{\prod_{i=1}^{n+2} M_i}, Y = \frac{(n-k+2,...n,n+1)(n+2,1,...,k-1)}{(n-k+2,...n,n+2)(n+1,1,...,k-1)}$$

and M_i are consecutive minors of the C matrix.

4 Examples of evaluation

In this section several examples of the application of the eq. (5), (6) will be provided.
The first example is 3-point MHV form factor in spinor helicity representation. Since $\mathcal{F}_{2,3}$ has 3 external legs, there are 3 gluing positions, but due to the fact, that all top-cell diagrams are equivalent in this case, all gluing positions should give the same results, so, to obtain answer for $\mathcal{F}_{2,3}$ one has to take into account only one top-cell diagram (i.e. one gluing position) since they are all equal. Substituting the expression for $\mathcal{A}_5^{\text{MHV}} = \frac{\delta^{(8)}(\sum_{i=1}^{5}\lambda_i\eta_i)}{\langle 12\rangle\langle 23\rangle\langle 34\rangle\langle 45\rangle\langle 51\rangle}$ to the integral (5) and considering gluing between legs 3 and 1, one has to integrate over $\lambda_{4,5}$, $\tilde{\lambda}_{4,5}$ and $\eta_{4,5}$. Integrations over $\tilde{\lambda}_{4,5}$ and $\eta_{4,5}$ could be done with the help of delta functions of the minimal form factor $\mathcal{F}_{2,2}$ which in turn mean that corresponding variables are just to be relabelled (see definitions of the underlined variables in the previous section). Thus, at this stage one can write:

$$\mathcal{F}_{2,3} = \int \left(\prod_{i=4,5} \frac{d^2\lambda_i}{Vol[GL(1)^2]} \right) \mathcal{A}_5^{\text{MHV}}(\lambda, \underline{\tilde{\lambda}}, \underline{\eta}).$$

To do the last interation one has to eliminate $GL(1)^2$ invariance, reparametrizing spinors as follows, introducing arbitrary reference spinors:

$$\lambda_4 = \xi_A - \beta_1 \xi_B$$
$$\lambda_5 = \xi_B - \beta_2 \xi_A.$$

This reparametrization is just change of variables, thus one can rewrite the result for $\mathcal{F}_{2,3}$ as integral over parameters $\beta_{1,2}$:

$$\mathcal{F}_{2,3} = -\langle \xi_A \xi_B \rangle^2 \frac{\delta^{(4)}(q - \sum_{i=1}^{3} p_i)\hat{\delta}^{(+4)}(Q^+)\hat{\delta}^{(-4)}(Q^-)}{\langle 12\rangle\langle 23\rangle} \cdot I$$

$$I = \frac{1}{\langle 3\xi_B\rangle\langle 1\xi_A\rangle\langle \xi_A\xi_B\rangle} \int \frac{d\beta_1 d\beta_2}{\left(\beta_1 - \frac{\langle 3\xi_A\rangle}{\langle 3\xi_B\rangle}\right)\left(\beta_2 - \frac{\langle 1\xi_B\rangle}{\langle 1\xi_A\rangle}\right)} = -\frac{1}{\langle 31\rangle\langle \xi_A\xi_B\rangle^2}.$$

The integral I is evaluated via the Cauchy residue theorem. For the case, when the minimal form factor is glued between legs i and j, one obtains for I:

$$I = -\frac{1}{\langle ij\rangle\langle \xi_A\xi_B\rangle^2}.$$

Thus the full answer reads [6]:

$$\mathcal{F}_{2,3} = \frac{\delta^{(4)}(q - \sum_{i=1}^{3} p_i)\hat{\delta}^{(+4)}(\sum_{i=1}^{3} \lambda_i \eta_i^+)\hat{\delta}^{(-4)}(\sum_{i=1}^{3} \lambda_i \eta_i^- - \gamma^-)}{\langle 12\rangle\langle 23\rangle\langle 31\rangle}. \tag{7}$$

The same answer could be derived via grassmannian integral (6). As can be easily shown, there are no integrations over grassmannian, since the number of integration variables is equal to zero in this case:

$$\#(\text{integrations}) = (k-2)(n-k)\Big|_{k=2,n=3} = 0,$$

which means that all coefficients of the C matrix are determined by the delta functions. Since

$$\hat{\delta}^6(C^\perp \cdot \underline{\underline{\lambda}}) = \prod_{i=3,4,5} \hat{\delta}^2(\underline{\underline{\lambda}}_i - c_{1i}\underline{\underline{\lambda}}_1 - c_{2i}\underline{\underline{\lambda}}_2) = \prod_{i=3,4,5} \frac{1}{\underline{\underline{\langle 12\rangle}}}\hat{\delta}\left(c_{1i} - \frac{\underline{\underline{\langle i2\rangle}}}{\underline{\underline{\langle 12\rangle}}}\right)\hat{\delta}\left(c_{2i} - \frac{\underline{\underline{\langle i1\rangle}}}{\underline{\underline{\langle 21\rangle}}}\right),$$

on the support of these delta functions one arrives at eq. (7) again, where double underscore mens twisted kinematics conditions (see eq. 3.9 of [1]).
The next example is 3-point next-to-MHV form factor. As in the previous case, all top-cell diagrams are equivalent here, so

$$\mathcal{F}_{3,3} = \frac{\langle \xi_A \xi_B\rangle^2 \delta^{(4)}(q - \sum p_i)\delta^4(\sum \lambda_i \eta_i^+)\delta^4(\sum \lambda_i \eta_i^- - \gamma^-)}{[12][23]\langle 12\rangle^4} \cdot I,$$

$$I = q^2\delta^2(\eta_3^+)\int \frac{d\beta_1 d\beta_2}{1-\beta_1\beta_2}\frac{\delta^2(q^2\eta_3^- - \langle 4|q|3]\eta_4^- - \langle 5|q|3]\eta_5^-)}{(\langle \xi_B|q|3] - \beta_1\langle \xi_A|q|3])(\langle \xi_A|q|1] - \beta_2\langle \xi_B|q|1])}.$$

Taking residues in simple poles of the integrand one gets

$$\mathcal{F}_{3,3} = \frac{q^2(\eta_3^+)^2\hat{\delta}^{(-2)}(q^2\eta_3^- + [3|q\gamma^-)}{\langle 12\rangle^4[12][23][31]}\delta^{(4)}\left(q - \sum_{i=1}^{3} p_i\right)\hat{\delta}^{(+4)}\left(\sum_{i=1}^{3} \lambda_i \eta_i^+\right)\hat{\delta}^{(-4)}\left(\sum_{i=1}^{3} \lambda_i \eta_i^- - \gamma^-\right) \tag{8}$$

Let's extract a specific component from this super form factor, namely the component at the maximal η degree ([6]), which is 12 in this case:

$$F^{max} = \frac{q^4}{[12][23][31]}. \tag{9}$$

This quantity would be necessary to check that results, derived using different approaches, agree.
In the full analogy with the first example it is possible to compute the same quantity via integral over the Grassmannian. Again, for $\mathcal{F}_{3,3}$ there are no integrations, so on the support of the

$$\hat{\delta}^6(C \cdot \underline{\underline{\tilde{\lambda}}}) = \prod_{i=1}^{3} \hat{\delta}^2(\underline{\underline{\tilde{\lambda}}}_i + c_{i4}\underline{\underline{\tilde{\lambda}}}_4 + c_{i5}\underline{\underline{\tilde{\lambda}}}_5)$$

one can solve the constraints to obtain coefficients of the C matrix:

$$c_{i4} = -\frac{\underline{\underline{[i5]}}}{\underline{\underline{[45]}}}, c_{i5} = -\frac{\underline{\underline{[i4]}}}{\underline{\underline{[54]}}}. \tag{10}$$

Substituting these values to the integrand yields:

$$\mathcal{F}_{3,3} = \frac{q^4}{[12][23][31]}\delta^{(4)}(q - \sum_{i=1}^{3} p_i)\hat{\delta}^{12}(C \cdot \underline{\underline{\eta}}). \tag{11}$$

By virtue of $\hat{\delta}^{12}(C \cdot \underline{\underline{\eta}}) = \left(\prod_{i=1}^{3}(\eta_i^+)^2\right) \times \prod_{i=1}^{3} \hat{\delta}^{(-2)}(\eta_i^- - ...)$ one can extract the component with the maximal η degree, which agrees with eq. (9).

The simplest example, containing non-trivial integrations over the Grassmannian manifold is Next-to-MHV form factor with 4 legs, or $NMHV_4$. In the case of 4 external legs there are 4 possible gluing positions, shown on the picture:

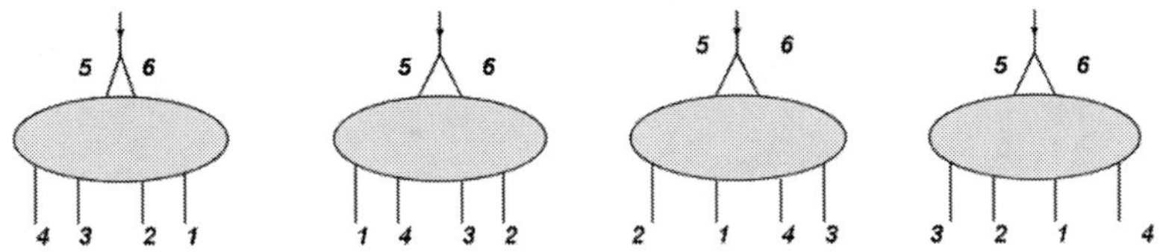

Figure 2. Gluing positions

It's worth mentioning here that a priori there is no prescription for choosing contour of integration, which gives the correct answer for arbitrary n, k. The integral over Grassmannian, corresponding to gluing between legs 1 and 4 has the form:

$$\mathcal{F}_{3,4} = \langle\xi_A\xi_B\rangle^2 \int \frac{d^{3\times 6}C}{Vol[GL(3)]}\Omega_{3,4}\hat{\delta}^{(6)}(C \cdot \underline{\underline{\tilde{\lambda}}})\hat{\delta}^{(12)}(C \cdot \underline{\underline{\eta}})\hat{\delta}^{(6)}(C^{\perp} \cdot \underline{\underline{\lambda}}). \tag{12}$$

Using the formula #(integrations) $= (k-2)(n-k)$, one notices that in this case the integral involves one non-trivial integration. The strategy of integration is described in [3], [7]: since the integral is 1-dimensional, there must exist one-parameter family of solutions $\hat{c}_{ij} = c_{ij}(\tau)$ that solve constraints, imposed by the delta functions of the integrand on the coefficients of the C-matrix. Then bosonic delta functions could be replaced by the momentum conserving delta function by localizing the integral on the solution of the constraints. Thus, the whole thing could be rewritten as follows:

$$\mathcal{F}_{3,4} = \langle\xi_A\xi_B\rangle^2\delta^{(4)}(q - \sum_{i=1}^{4} p_i) \int d\tau\hat{\Omega}_{3,4} \prod_{i=2,4,6} \hat{\delta}^{(4)}(\eta_i + \sum_{j=1,3,5} \hat{c}_{ij}\eta_j). \tag{13}$$

In the latter formula the C-matrix is gauge-fixed to be

$$C = \begin{pmatrix} c_{21} & 1 & c_{23} & 0 & c_{25} & 0 \\ c_{41} & 0 & c_{43} & 1 & c_{45} & 0 \\ c_{61} & 0 & c_{63} & 0 & c_{65} & 1 \end{pmatrix} \tag{14}$$

This for of the matrix suits for calculation residues of «even» minors ((234) and (456)), while for «odd» minors ((123) and (561)) one has to use another gauge:

$$C = \begin{pmatrix} 1 & c_{12} & 0 & c_{14} & 0 & c_{16} \\ 0 & c_{32} & 1 & c_{34} & 0 & c_{36} \\ 0 & c_{52} & 0 & c_{54} & 1 & c_{56} \end{pmatrix} \tag{15}$$

Hats here indicate τ-dependence. Let's denote (ijk) - minor of the C-matrix, composed of i-th, j-th and k-th columns of C. The integrand has poles where $(i\,i+1\,i+2) = 0$. With the help of this condition one can express c_{ij} in terms of helicity spinors, assigned to external particles (see [3] for details). As it has already been mentioned, there is no prescription in which poles one has to take residues, so it has to be figured out somehow. It turns out that in this particular case only 2 of 4 gluing positions contribute to the answer, and only in 2 of 4 poles one must take residues in each gluing position, namely

$$\mathcal{F}_{3,4} = \mathrm{Res}_{41}(123) + \mathrm{Res}_{41}(561) + \mathrm{Res}_{23}(123) + \mathrm{Res}_{23}(561). \tag{16}$$

It's worth metioning that eq. (16) agrees with eq. 3.76 of [1]. Contributing diagrams are shown on the figure 3. Here the subscript on $\mathrm{Res}_{ij}(klm)$ denotes legs, between which the minimal form factor is

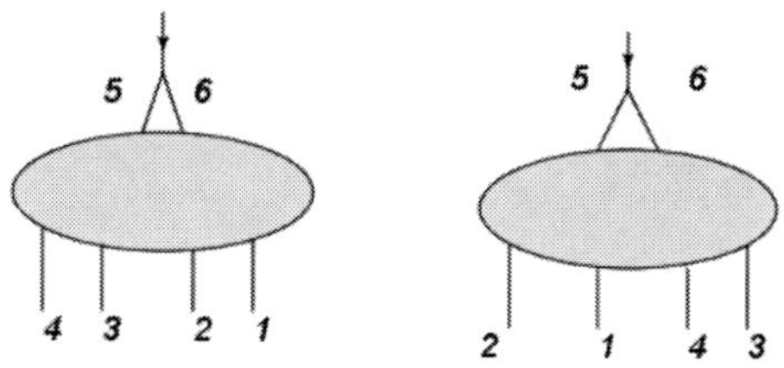

Figure 3. Diagrams contributing to $\mathcal{F}_{3,4}$

pasted in. Expressions for residues are given below:

$$\mathrm{Res}_{41}(123) = \frac{\langle\xi_A\xi_B\rangle^2\langle 13\rangle^4[4\underline{\underline{6}}]^4\hat{\delta}^{(12)}(C_{(123)}\cdot\underline{\underline{\eta}})}{\langle 12\rangle\langle 23\rangle[\underline{\underline{56}}]^2\langle 3|1+2|4]\langle 1|2+3|4]P_{123}^2}$$

$$\mathrm{Res}_{41}(561) = \frac{\langle\xi_A\xi_B\rangle^2\langle 1\underline{\underline{5}}\rangle^4[24]^4\hat{\delta}^{(12)}(C_{(561)}\cdot\underline{\underline{\eta}})}{[23][34]\langle\underline{\underline{56}}\rangle^2\langle 1|\underline{\underline{5}}+\underline{\underline{6}}|2]\langle 1|\underline{\underline{5}}+\underline{\underline{6}}|4]P_{234}^2}.$$

To obtain $\mathrm{Res}_{23}(123)$ and $\mathrm{Res}_{23}(561)$ one has to apply $s = 2$ shift to the latter formulas, i.e. relabel $1 \to 3$, $2 \to 4$, etc. $C_{(ijk)}$ denotes C-matrix evaluated at residue (ijk). Extracting $F(\phi_{12}, \phi_{12}, g^-, g^+)$ component (i.e. coefficient at $(\eta_1)^2(\eta_2)^2(\eta_3)^4$), one arrives at the following formula:

$$F(\phi_{12}, \phi_{12}, g^-, g^+) = c_{16}^2 c_{52}^2 \mathrm{Res}_{41}(561) + c_{36}^2 c_{54}^2 \mathrm{Res}_{23}(123). \tag{17}$$

Values of coefficients could be evaluated exactly as in sec. 9.3.2 of [3]. Thus, the final result for this particular component of 4-point NMHV super form factor, evaluated from Grassmannian integral, is represented by the the following formula ($q = p_1 + p_2 + p_3 + p_4$):

$$F(\phi_{12}, \phi_{12}, g^-, g^+) = \frac{1}{\langle 1|q|2]}\left[\frac{[24]^2\langle 1|q|4]}{[23][34]P_{234}^2} + \frac{\langle 13\rangle^2\langle 3|q|2]}{\langle 34\rangle\langle 41\rangle P_{134}^2}\right]. \tag{18}$$

The answer, obtained with the help of Grassmannian integral representation coincides with the formula 2.10 derived in [6] by means of BCFW recursion relations.

5 Conclusion

In recent papers [1], [2] it was shown that gluing operation could be used to derive different representations for form factors. In order to confirm their conjecture, several examples have been considered in the present article. The obtained results agree with those derived earlier. Also it's worth mentioning that presented results can be further generalized to the case of gluing non-local operators, such as Wilson line operators, to on-shell amplitudes to obtain Grassmannian integral representation for amplitudes with one leg off-shell, which will be the next step.

6 Acknowledgements

The author thanks D.I. Kazakov, A.I. Onishchenko and L.V. Bork for their support. The author is supported by RFBR grant # 17-02-00872.

References

[1] R. Frassek, D. Meidinger, D. Nandan, M. Wilhelm, JHEP **1601**, 182 (2016)
[2] L.V. Bork, A.I. Onishchenko, preprint (arXiv:1704.04758) (2017)
[3] H. Elvang, Y.-T. Huang, *Scattering Amplitudes in Gauge Theory and Gravity*, Cambridge University Press (2015)
[4] N. Arkani-Hamed et al., *Grassmannian Geometry of Scattering Amplitudes*, Cambridge University Press (2016)
[5] B. Eden, P. Heslop, G. P. Korchemsky, E. Sokatchev, Nucl.Phys. **B869**, 329-377 (2013)
[6] A. Brandhuber et al., JHEP **1110**, 046 (2011)
[7] H. Elvang et al., JHEP **1412**, 181 (2014)

Quark-hadron phase structure of QCD matter from SU(4) Polyakov linear sigma model

Abdel Magied Abdel Aal DIAB[1,2,*] and *Abdel Nasser* TAWFIK[1,2,**]

[1] *Egyptian Center for Theoretical Physics (ECTP), MTI University, 11571 Cairo, Egypt.*
[2] *World Laboratory for Cosmology And Particle Physics (WLCAPP), Cairo, Egypt*

Abstract. The SU(4) Polyakov linear sigma model (PLSM) is extended towards characterizing the chiral condensates, σ_l, σ_s and σ_c of light, strange and charm quarks, respectively and the deconfinement order-parameters ϕ and $\bar{\phi}$ at finite temperatures and densities (chemical potentials). The PLSM is considered to study the QCD equation of state in the presence of the chiral condensate of charm for different finite chemical potentials. The PLSM results are in a good agreement with the recent lattice QCD simulations. We conclude that, the charm condensate is likely not affected by the QCD phase-transition, where the corresponding critical temperature is greater than that of the light and strange quark condensates.

1 Introduction

Quantum Chromodynamics (QCD) is a fundamental theory of the strong interaction acting upon quarks and gluons. Nowadays, the experimental and theoretical studies of charm physics turns to be a very active field in hadronic physics [1]. It was shown that the PLSM model, which is a QCD-like model, describes very successfully the phenomenology of the SU(3) quark-hadron structure in thermal and dense medium and also at finite magnetic field [2–10]. At low densities or as the temperature decreases, the quark gluon plasma (QGP), where the degrees of freedom of quark flavors, N_f, are dominated, is conjectured to confine forming hadrons.

In the present work, the PLSM Lagrangian for N_f quark flavours is combined with the Polyakov loops in order to link the chiral symmetry restoration with aspects of the confinement-deconfinement phase transitions. The PLSM Lagrangian is invariant under chiral symmetry $U(N_f)_R \times U(N_f)_L$ ($N_f = 2$ in Ref. [11], $N_f = 3$ in Refs. [2, 3] and $N_f = 4$ in Ref. [12]). We utilize mean field approximation to the PLSM in order to estimate the pure mesonic potential with $N_f = 4$ quark flavours and to describe the dependence of the light, strange and charm quark chiral condensates σ_l, σ_s and σ_c respectively and the Polyakov loop order-parameters ϕ and $\bar{\phi}$ in thermal and dense QCD medium. Furthermore, we utilize the PLSM to estimate the well-studied QCD equation-of-state (EoS) and some of other thermodynamic observables. The PLSM calculations are then confirmed to recent lattice QCD simulations. The paper is organized as follows. In section (2), we briefly describe the PLSM. The results are given in section (3), while section (4) is devoted to the final conclusions.

*e-mail: a.diab@eng.mti.edu.eg
**e-mail: a.tawfik@eng.mti.edu.eg

2 SU(4) Polyakov linear sigma model

The PLSM addresses the linkage between the confinement- and deconfinement-phase, where its Lagrangian is composed as $\mathcal{L}_{\rm PLSM} = \mathcal{L}_{\rm chiral} - \mathcal{U}(\phi, \phi^*, T)$, and the chiral Lagrangian $\mathcal{L}_{\rm chiral} = \mathcal{L}_f + \mathcal{L}_m$ is constructed for N_f quark flavours. The Fermionic Lagrangian $\mathcal{L}_f$ reads,

$$\mathcal{L}_f = \sum_f \overline{q}_f \left[i\gamma^\zeta D_\zeta - g T_a(\sigma_a + i\gamma_5 \pi_a) \right] q_f, \tag{1}$$

and the mesonic Lagrangian, $\mathcal{L}_m$, is given as,

$$\begin{aligned} \mathcal{L}_m &= \mathrm{Tr}(\partial_\mu \Phi^\dagger \partial^\mu \Phi - m^2 \Phi^\dagger \Phi) - \lambda_1 [\mathrm{Tr}(\Phi^\dagger \Phi)]^2 - \lambda_2 \mathrm{Tr}(\Phi^\dagger \Phi)^2 + \mathrm{Tr}[H(\Phi + \Phi^\dagger)] \\ &+ c[\mathrm{Det}(\Phi) + \mathrm{Det}(\Phi^\dagger)]. \end{aligned} \tag{2}$$

The Polyakov-loop potential, $\mathcal{U}(\phi, \bar{\phi}, T)$, introduces color degrees-of-freedom and gluon dynamics with the expectation values of the Polyakov-loop variables ϕ and $\bar{\phi}$. There are various functional forms of the Polyakov-loop potential. In the present work, we use the logarithmic form,

$$\frac{\mathcal{U}_{log}(\phi, \bar{\phi}, T)}{T^4} = -\frac{1}{2} a(T)\, \bar{\phi}\phi + b(T)\, \ln\left[1 - 6\bar{\phi}\phi + 4(\bar{\phi}^3 + \phi^3) - 3(\bar{\phi}\phi)^2\right], \tag{3}$$

where the thermal dependent coefficients, $a(T) = a_0 + a_1(T_0/T) + a_2(T_0/T)^2$ and $b(T) = b_3(T_0/T)^3$, the constants a_0, a_1, a_2 and b_3 are fitted to the pure gauge lattice QCD data in [13].

With basis of light, strange and charm quarks, the purely mesonic potential can be driven from the mesonic Lagrangian, $\mathcal{L}_m$. The orthogonal basis transformation from the original ones (σ_0, σ_8 and σ_{15}) to the light, the strange and the charm quark flavor basis; i.e. σ_l, σ_s and σ_c, respectively, can be expressed as

$$\sigma_l = \frac{\sigma_0}{\sqrt{2}} + \frac{\sigma_8}{\sqrt{3}} + \frac{\sigma_{15}}{\sqrt{6}}, \qquad \sigma_s = \frac{\sigma_0}{2} - \sqrt{\frac{2}{3}}\sigma_8 + \frac{\sigma_{15}}{2\sqrt{3}}, \qquad \sigma_c = \frac{\sigma_0}{2} - \frac{\sqrt{3}}{2}\sigma_{15}. \tag{4}$$

Furthermore, the purely mesonic potential can be obtained as,

$$\begin{aligned} U(\sigma_l, \sigma_s, \sigma_c) &= -h_l \sigma_l - h_s \sigma_s - h_c \sigma_c + \frac{m^2\,(\sigma_l^2 + \sigma_s^2 + \sigma_c^2)}{2} - \frac{c\,\sigma_l^2 \sigma_s \sigma_c}{4} + \frac{\lambda_1\, \sigma_l^2 \sigma_s^2}{2} \\ &+ \frac{\lambda_1 \sigma_l^2 \sigma_c^2}{2} + \frac{\lambda_1 \sigma_s^2 \sigma_c^2}{2} + \frac{(2\lambda_1 + \lambda_2)\sigma_l^4}{8} + \frac{(\lambda_1 + \lambda_2)\sigma_s^4}{4} + \frac{(\lambda_1 + \lambda_2)\sigma_c^4}{4}. \end{aligned} \tag{5}$$

In mean-field approximation [14], all fields are treated as averages in space ($\vec{x}$) and imaginary time (τ). The exchanges between particles and antiparticles shall be expressed by the grand-canonical partition function (Z) which can be constructed from the thermodynamic potential density as

$$\Omega(T, \mu) = -T\, \frac{\ln Z}{V} = \Omega_{\bar{q}q}(T, \mu) + \mathcal{U}_{\rm log}(\phi, \bar{\phi}, T) + U(\sigma_l, \sigma_s, \sigma_c). \tag{6}$$

The potential of the antiquark-quark contribution reads [15],

$$\begin{aligned} \Omega_{\bar{q}q}(T, \mu) &= -2T \sum_{f=l,s,c} \int_0^\infty \frac{d^3\vec{P}}{(2\pi)^3} \left\{ \ln\left[1 + 3\left(\phi + \bar{\phi} e^{-\frac{E_f-\mu}{T}}\right) \times e^{-\frac{E_f-\mu}{T}} + e^{-3\frac{E_f-\mu}{T}}\right] \right. \\ &\left. + \ln\left[1 + 3\left(\bar{\phi} + \phi e^{-\frac{E_f+\mu}{T}}\right) \times e^{-\frac{E_f+\mu}{T}} + e^{-3\frac{E_f+\mu}{T}}\right] \right\}. \end{aligned} \tag{7}$$

The subscripts l, s, and c refer to light, strange and charm quarks, respectively. The energy-momentum dispersion relation is given as $E_f = (\vec{P}^2 + m_f^2)^{1/2}$, with m_f being the flavor mass of light, strange, and charm quark coupled to g [16]; $m_l = g\sigma_l/2$, $m_s = g\sigma_s/\sqrt{2}$ [17], and $m_c = g\sigma_c/\sqrt{2}$. In the present work, the Yukawa coupling g is fixed at $g \sim 6.5$.

In order to evaluate the dependence of the PLSM chiral condensates σ_l, σ_s, and σ_c and the deconfinement order-parameters ϕ and $\bar{\phi}$ as a function temperature (T) and baryon chemical potential μ, the real part of thermodynamic potential, Re(Ω) of Eq. (6), should be minimized at the saddle point

$$\frac{\partial\Omega}{\partial\sigma_l} = \frac{\partial\Omega}{\partial\sigma_s} = \frac{\partial\Omega}{\partial\sigma_c} = \frac{\partial\Omega}{\partial\phi} = \left.\frac{\partial\Omega}{\partial\bar{\phi}}\right|_{min} = 0, \tag{8}$$

3 Results

In the following we present the order parameters of the chiral symmetry restoration and the deconfinement phase transitions at finite temperatures and finite baryon chemical potentials. Furthermore, we explore the thermal dependence of some thermodynamic quantities. These results are compared with recent lattice QCD simulations. Fig. 1 introduces the thermal dependence of the chiral condensates for light, strange and charm quark flavors, σ_l, σ_s and σ_c, respectively [left panel (a)], while the order parameters of the deconfinement phase transitions ϕ and $\bar{\phi}$ are shown in right panel (b). The PLSM order parameters are given at different values of the baryon chemical potentials $\mu = 0$ (solid curves) 100 (dashed curves) and 300 MeV (dotted curves), respectively. The quark chiral-condensates are normalized with respect to their vacuum values, $\sigma_{l_0} = 92.4$ MeV, $\sigma_{s_0} = 94.5$ MeV and $\sigma_{c_0} = 293.87$ MeV.

Left panel of Fig. 1 illustrates the effect of the baryon chemical potential on the chiral condensates. Whereas the baryon chemical potential increases, the strange and the light condensates move backwards, i.e. the chiral critical temperatures decrease. For the charm condensate, the effect of the baryon chemical potential is absent (not appearing in the temperature range depicted here). Furthermore, the chiral critical temperatures can be obtained from the chiral susceptibility. We find that the charm condensate remains almost temperature independent until the temperature approaches 0.6 – 0.7 GeV. It seems that the corresponding critical temperature is much greater than that of the strange and light ones.

Right panel of Fig. 1 shows the order-parameters of the Polyakov-loops ϕ and $\bar{\phi}$ as functions of temperatures. At $\mu = 0$ MeV, we find that $\phi = \bar{\phi}$, while the Polyakov loop variables are different at finite baryon chemical potentials. We find that with increasing the baryon chemical potentials, the order-parameters of Polyakov-loops ϕ moves backwards but $\bar{\phi}$ moves towards, i.e. the deconfinement critical temperatures of ϕ decrease but that of $\bar{\phi}$ increase with the increase in baryon chemical potentials.

It is believed that the thermodynamic quantities are sensitive to the change accompanying the confinement-deconfinement phase transitions. Some of the thermodynamic quantities such as pressure, entropy density, energy and interaction measure, are given as $p = -\Omega(T,\mu)$, $s = \partial p/\partial T$, $\epsilon = -p + Ts$ and $(\epsilon - 3p)/T^4$, respectively. In the present work, we focus on comparing our PLSM results on the thermal dependence of normalized pressure and of interaction measure with recent lattice QCD simulation [18].

Fig. 2 shows the thermal dependence of the normalized pressure [left panel (a)] and the interaction measure [right panel (b)] at vanishing baryon chemical potential. The PLSM calculations are confronted to recent lattice QCD simulations [18]. It is obvious that the normalized pressure increases with increasing temperatures and saturates at high temperatures. We observe that the pressure doesn't exceed the Stefan-Boltzmann (SB) limit at $N_f = 4$, i.e. massless quarks and gluons,

$$\frac{P_{SB}}{T^4} = \frac{\pi^2}{45}(N_c^2 - 1) + N_c N_f \left[\frac{7\pi^2}{180} + \frac{1}{6}\left(\frac{\mu}{T}\right)^2 + O\left(\frac{\mu}{T}\right)^4\right], \tag{9}$$

https://doi.org/10.1051/epjconf/201817709005

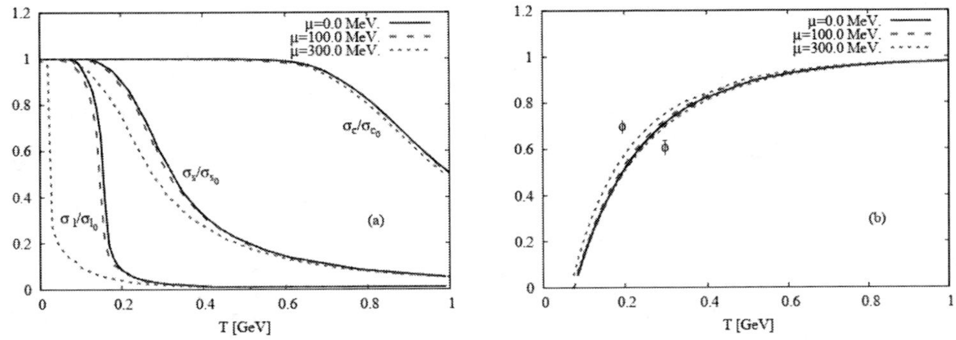

Figure 1. In dependence on temperatures, the normalized chiral condensates σ_l, σ_s and σ_c normalized with respect to their vacuum values, $\sigma_{l_0} = 92.4$ MeV, $\sigma_{s_0} = 94.5$ MeV and $\sigma_{c_0} = 293.87$ MeV, respectively [left panel (a)] and the order-parameters of Polyakov-loops ϕ and $\bar{\phi}$ [right panel (b)] are given at different values of the baryon chemical potentials $\mu = 0$, 100 and 300 MeV (solid, dashed and dotted curves, respectively).

where N_c are the color dof, the first term refers to the contributions of gluons and the second one to the quarks. The value of thermodynamic pressure is low in confined phase and begins to rise in the deconfined phase, when temperature increases. In all calculations such as the lattice QCD simulations [18] and even other non-perturbative approaches such as [19], the normalized pressure stays below the SB limit. In out calculations as well. At vanishing chemical potential, the pressure in dependence on temperature behaves smoothly through the smooth crossover. At temperatures as much as two times the critical value, the pressure reaches approximately 80% of the SB limit. In both hadronic (low temperatures) and tQGP (high temperatures) regions, a good agreement with recent lattice QCD calculations [18] is obtained.

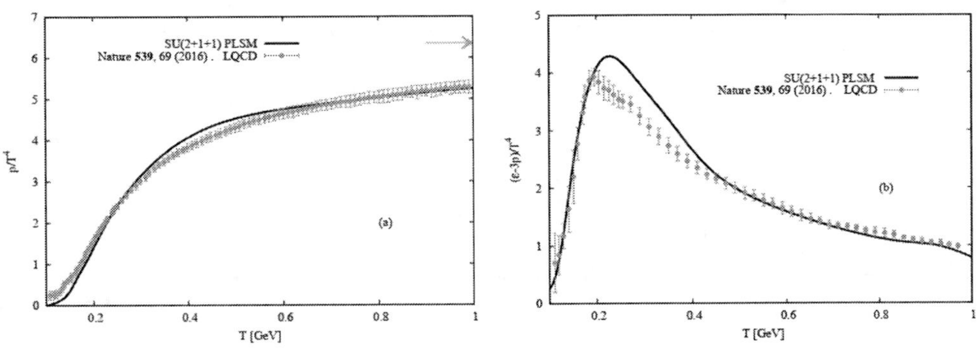

Figure 2. The normalized pressure [left panel (a)] and the the interaction measure [right panel (b)] are depicted in dependence on temperature. The PLSM calculations (solid curves) are compared with recent lattice QCD simulations (symbols) [18].

The interaction measure or the trace anomaly of the energy-momentum tensor $\mathcal{T}^{\mu\nu}$,

$$\Delta = \frac{\mathcal{T}^{\mu\mu}}{T^4} = \frac{\epsilon - 3p}{T^4} = T\frac{\partial}{\partial T}(p/T^4), \tag{10}$$

is a dimensionless thermodynamic quantity depicted in right panel of Fig. 2 at vanishing baryon chemical potential and compared with recent lattice QCD calculations. We observe that the interaction measure changes smoothly around T_c, where the interacting system is conjectured to experience crossover. The thermal dependence of the interaction measure appears to agree well with the recent lattice QCD calculations [18], in both hadronic (low temperatures) and QGP (high temperatures) phases. In the broken phase, it shows a peak around T_c, but a rapid decrease is seen in the restored phase.

4 Conclusion

A systematic study for the chiral phase-structure of heavy mesons and their decay widths in thermal and dense medium is very closely connected with recent theoretical, computational and experimental results. Thus, we present an extension to the PLSM to $N_f = 4$ quarks flavors. Accordingly, the extra degrees-of-freedom modifies the PLSM thermodynamic potential $\Omega(T,\mu)$. We present the thermal dependence of the light, strange and charm quark chiral condensates, σ_l, σ_s and σ_c, respectively and the deconfinement order-parameters (ϕ and $\bar{\phi}$) at different baryon chemical potentials. For the charm condensate, we notice that its sensitivity to the baryon chemical potentials and temperatures is very weak until $0.6 - 0.7$ GeV. We also observe a fair agreement with the recent lattice QCD calculations for the thermodynamic pressure and the interaction measure.

References

[1] N. Brambilla, A. Pineda, J. Soto and A. Vairo, Rev. Mod. Phys. **77**, 1423 (2005).
[2] A. Tawfik, N. Magdy, and A. Diab, Phys. Rev. C **89**, 055210 (2014).
[3] Abdel Nasser Tawfik, Abdel Magied Diab, Phys. Rev. C **91**, 015204 (2015).
[4] Abdel Nasser Tawfik, Abdel Magied Diab, and M.T. Hussein, *"Quark-hadron phase structure and QCD equations of state in vanishing and finite magnetic field"*, 1604.08174 [hep-lat].
[5] Abdel Nasser Tawfik, Abdel Magied Diab, and M. T. Hussein, Int. J. Adv. Res. Phys. Sci. **3**, 4 (2016).
[6] Abdel Nasser Tawfik, Abdel Magied Diab, and M. T. Hussein, Int. J. Mod. Phys. A **31**, 1650175 (2016).
[7] Abdel Nasser Tawfik, Abdel Magied Diab, N. Ezzelarab, and A. G. Shalaby, Adv. High Energy Phys. **2016**, 1381479 (2016).
[8] A. Tawfik and N. Magdy, J. Phys. G **42**, 015004 (2015).
[9] A. Tawfik and N. Magdy, Phys. Rev. C **91**, 015206 (2015).
[10] A. Tawfik and N. Magdy, Phys. Rev. C **90**, 015204 (2014).
[11] D. Parganlija, F. Giacosa and D. H. Rischke, Phys. Rev. D **82**, 054024 (2010).
[12] Abdel Magied Diab, Azar I. Ahmadov, Abdel Nasser Tawfik, Eiman Abou El Dahab, PoS **ICHEP2016**, 634 (2016).
[13] S. Roessner, C. Ratti, and W. Weise, Phys. Rev. D **75**, 034007 (2007).
[14] J. I. Kapusta and C. Gale, *"Finite-temperature field theory: Principles and applications"*, (Cambridge University Press, UK, 2006).
[15] K. Fukushima, Phys. Rev. D **77**, 114028 (2008).
[16] O. Scavenius, A. Mocsy, I. N. Mishustin, and D. H. Rischke, Phys. Rev. C **64**, 045202 (2001).
[17] P. Kovacs and Z. Szep, Phys. Rev. D **75**, 025015 (2007).
[18] S. Borsanyi, *et. al.* "*Lattice QCD for cosmology*", Nature **539**, 69 (2016).
[19] B-J Schaefer, M. Wagner, J. Wambach Phys. Rev. D **81**, 074013 (2010).

Tunneling of Two Interacting Fermions

Ilyas Ishmukhamedov[1,2,3,*] and *Altay* Ishmukhamedov[2,4,**]

[1] *Bogoliubov Laboratory of Theoretical Physics, Joint Institute for Nuclear Research, 141980 Dubna, Moscow Region, Russian Federation*
[2] *Al-Farabi Kazakh National University, 050040 Almaty, Republic of Kazakhstan*
[3] *Institute of Nuclear Physics, the Ministry of Energy of the Republic of Kazakhstan, 050032 Almaty, Republic of Kazakhstan*
[4] *K.I. Satpaev Institute of Geological Sciences, 050010 Almaty, Republic of Kazakhstan*

Abstract. We consider two interacting atoms subject to a one-dimensional anharmonic trap and magnetic field gradient. This system has been recently investigated by the Heidelberg group in the experiment on two ^{6}Li atoms. In the present paper the tunneling of two cold ^{6}Li atoms, initially prepared in the center-of-mass and relative motion excited state, is explored and full time-dependent simulation of the tunneling dynamics is performed. The dynamics is analyzed for the interatomic coupling strength ranging from strong attraction to strong repulsion.

1 Introduction

Description of quantum tunneling, which, although, was discovered about a century ago, still remains challenging, especially when one considers the tunneling of correlated systems. In the context of ultracold physics, there are only a few works that completely take into account the pair correlation impact in the problem of a decay of a quantum system [1–8]. In this work we present an efficient method [1–4] which completely takes into account the interatomic interaction and also an external field. This situation corresponds to two cold atoms, subject to an anharmonic optical trap and magnetic field gradient, which have been investigated in the recent experiment of the Heidelberg group on two interacting ^{6}Li atoms [9, 10] and theoretically studied in [5, 11, 12]. The atoms tunnels through the potential barrier of the trap to open space. Here we focus on the excited state branch which has not been considered previously. Due to the tight confinement in the transverse direction the atomic dynamics could be analyzed by solving the one-dimensional Schrödinger equation [13], which we solve by means of a split-operator technique.

2 Model

Tunneling dynamics of the system considered in this work is modeled by the time-dependent Schrödinger equation

$$i\hbar\frac{\partial\psi(z_1,z_2,t)}{\partial t} = H(z_1,z_2)\psi(z_1,z_2,t), \tag{1}$$

*e-mail: i.ishmukhamedov@mail.ru
**e-mail: ishmukhamedov.altay@gmail.com

where the two-body Hamiltonian reads

$$H(z_1, z_2) = H_1(z_1) + H_2(z_2) + V^{(aa)}(z_1 - z_2). \tag{2}$$

Here

$$H_j(z_j) = -\frac{\hbar^2}{2m}\frac{\partial^2}{\partial z_j^2} + V^{(at)}(z_j), \qquad j = 1, 2. \tag{3}$$

are the single-particle Hamiltonians, in which $V^{(at)}$ describes the atom-trap interaction and m is the atomic mass.

The atom-trap potential is taken from [5, 9, 10] which represents an optical trap and magnetic field gradient (figure 1):

$$V^{(at)}(z) = pV_0\left[1 - \frac{1}{(z/z_R)^2 + 1}\right] - \mu_B C_{|j\rangle} z, \tag{4}$$

where μ_B is the Bohr magneton, $z_R = 8.548\ell$ ($\ell = \sqrt{\hbar/m\omega}$) is the Rayleigh range, $V_0 = 56.16\hbar\omega$ is the maximum depth of the optical trap, p is the optical trap depth, $C_{|j\rangle}$ depends on external magnetic-field strength, magnetic-field gradient and the hyperfine state of the atoms, $\omega = 2\pi \times 1234$ Hz. These values represents realistic trap parameters used in [5, 9, 10].

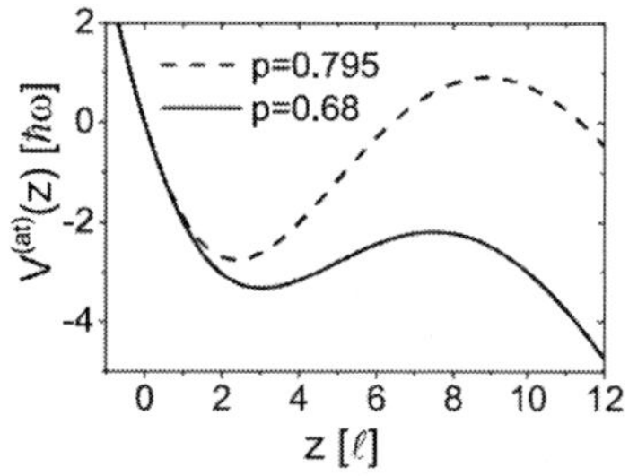

Figure 1. Plot of the trapping potential $V^{(at)}(z)$: $p = 0.795$ (dashed line) [represents the trap at $t = 0$] and $p = 0.68$ (solid line) [represents the trap at $t > 0$]

For the interatomic interaction potential $V^{(aa)}(z_1 - z_2)$ we choose the Gaussian shape [5, 14]

$$V^{(aa)}(z_1 - z_2) = -V_0 \exp\left\{-\frac{(z_1 - z_2)^2}{2r_0^2}\right\}, \tag{5}$$

where V_0 and r_0 - are the depth and range of the interaction (5). The value $r_0 = 0.1\ell$ [5, 14] - well represents the short-range interaction, which atoms experiences at low energies. It is more common to represent the interatomic interaction in terms of the 1D contact coupling strength g, which incorporates V_0 and r_0 via the 1D scattering length, a, as $g = -2\hbar^2/(ma)$. The 1D scattering length is computed by solving the scattering problem of the two atoms in the absence of the trapping potential [1, 14].

In order to integrate (1) we use the split-operator method, which is based on ideas [15] and has been developed in the works [2–4] in application to confined ultracold atom-atom collisions in waveguide-like traps:

$$\psi(z_1, z_2, t + \Delta t) = \exp\left\{-i\frac{\Delta t}{2\hbar}V^{(aa)}(z_1 - z_2)\right\}\exp\left\{-\frac{i\Delta t H_1(z_1)}{\hbar}\right\}\exp\left\{-\frac{i\Delta t H_2(z_2)}{\hbar}\right\}$$
$$\times \exp\left\{-i\frac{\Delta t}{2\hbar}V^{(aa)}(z_1 - z_2)\right\}\psi(z_1, z_2, t) \tag{6}$$

The action of the operators $\exp\left\{-i\Delta t H_j(z_j)\right\}$ is approximated by the Crank-Nicolson scheme, which maintains the accuracy order of the split-operator scheme (6) to $O(\Delta t^3)$. The partial derivatives in (3) are approximated by the sixth-order finite-differences.

To model the outgoing wave we introduce into the original Hamiltonian (2) a complex absorbing potential (CAP) $iW(z_j)$ [16], which absorbs the wave-packet at the edges of the simulation grid:

$$W(z_j) = w_c(|z_j| - z_c)^2\theta(|z_j| - z_c), \qquad j = 1, 2, \tag{7}$$

where $\theta(z)$ is the Heaviside step function. We find that the parameter values $w_c = -1\hbar\omega\ell^{-2}$ and $z_c = 25\ell$, lie in the domain where final results do not change significantly under the variation of these parameters.

3 Results

To prepare the initial state we fix the parameters $p = 0.795$ and $C_{|j\rangle} = 1894.18$ G/m (for both atoms). In this case tunneling is highly suppressed, due to the deep trap depth (figure 1) and the system is in a metastable state. In order to completely suppress tunneling and calculate eigenenergies and eigenfunctions we put a hard wall at $z = 11\ell$. This procedure models the experimental realization of the initially trapped atoms [5]. To compute the spectrum we use the method from [17]. The spectrum in such a trap is shown in figure 2. We focus on the lower excited state branch. Both excited state branches correspond to doubly excited states with respect to relative and center-of-mass motions [1]. The nodal patterns of the wave functions of these states are mixed when reaching the noninteracting limit, i.e. $g = 0$. This mixing occurs due to the rotational symmetry breaking, which is caused by anharmonic terms of the potential (4). Similar effect has been observed in [1, 18, 19].

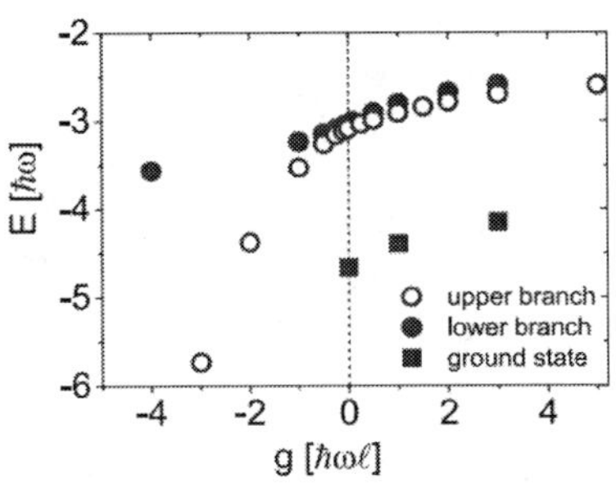

Figure 2. Energy spectrum of the "closed" trapping potential (4) using $p = 0.795$ and putting a hard wall at $z = 11\ell$. Upper and lower excited state branches correspond to doubly excited states with respect to relative and center-of-mass motions [1].

Now, with the obtained initial solutions we proceed to calculation of the evolution of $\psi(z_1, z_2, t)$ from (6). At $t > 0$ we take $p = 0.68$ and assume $C_{|1\rangle} = 1896$ G/m and $C_{|2\rangle} = 1894$ G/m. These trap values lower the potential barrier and provide the tunneling time scales to be experimentally detectable [5, 9, 10].

The plot (figure 3(a)) shows the mean number (experimentally measurable quantity [9, 10]) of trapped particles as a function of time calculated as

$$\bar{N}(t) = 2P_2(t) + P_1(t), \tag{8}$$

where $P_2(t)$ is the probability to find the two particles in the box $z_j \in [-2, 13]$ [in ℓ units]. The probability to find one particle in the trap, $P_1(t)$, is calculated by integrating the density $|\psi(z_1, z_2, t)|^2$ over the regions $z_i \in [13, 25]$ and $z_j \in [-2, 12]$ [in ℓ units] [5] (figure 3(b)). The simulation grid ends at $z_j = 40\ell$.

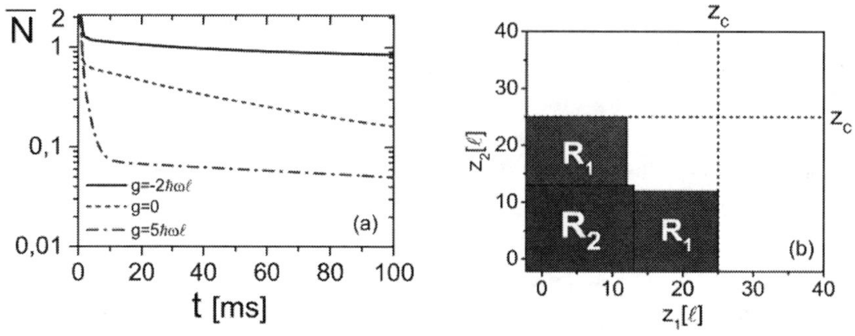

Figure 3. (Color online) (a) Mean number of trapped atoms $\bar{N}$ as a function of time on a logarithmic scale. (b) Configuration space for the two-atom system: R_2 is the region over which we integrate the density $|\psi(z_1, z_2, t)|^2$ to find $P_2(t)$; R_1 are the regions over which we integrate the density $|\psi(z_1, z_2, t)|^2$ to find $P_1(t)$.

Another interesting feature of the system tunneling is the behavior of the probability current $|\mathbf{j}(z_1, z_2, t)|$ (figure 4)

$$j_k(z_1, z_2, t) = \frac{\hbar}{2mi}\left(\psi^* \frac{\partial \psi}{\partial z_k} - \psi \frac{\partial \psi^*}{\partial z_k}\right), \qquad k = 1, 2. \tag{9}$$

From figure 4 we see that at strong attraction, $g = -2\hbar\omega\ell$, the tunneling direction dominates along the axis $z_1 = z_2$ and, with increasing g, tunneling goes along the axes z_1 and z_2.

From figure 4 one can notice how the strong attraction, $g = -2\hbar\omega\ell$, modifies the tunneling direction. At this coupling strength, from figure 3(a), we see that the mean atom number decreases, within the considered time domain, approximately to the value 0.85, which implies that one particle remains in the trap during the tunneling. This suggests that this particle has such a small energy that its tunneling time scale is much slower than that of the first particle. Tunneling in other cases, i.e. $g = 0$ and $g = 5\hbar\omega\ell$, goes in such a way that the two particles leave the trap independently (figure 4).

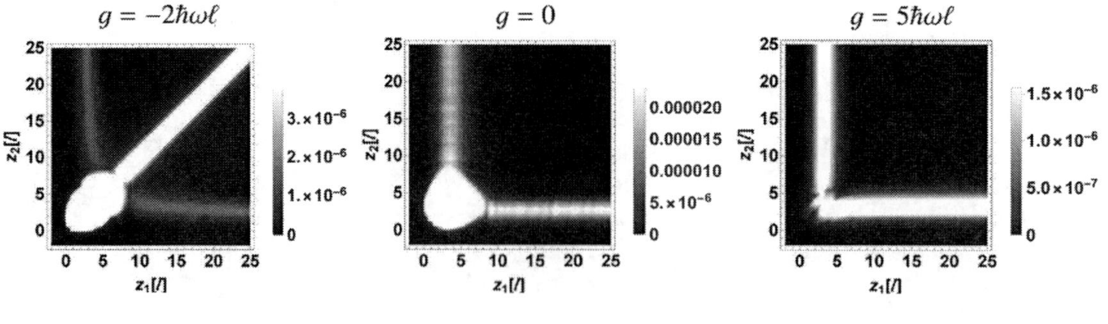

Figure 4. (Color online) Probability current $|\mathbf{j}(z_1, z_2, t)|$ of the lower excited state branch for the coupling strength values $g/(\hbar\omega\ell) = -2, 0, 5$ computed at $t = 100$ ms. The values of the probability current shown in the legends are in ω/ℓ units (note the different scale for all the graphs).

4 Conclusion

In this paper tunneling of two interacting ^{6}Li atoms from the relative and center-of-mass motion excited state is computed. This investigation is related to the recent experiments of the Heidelberg group [9, 10] and extends the analysis made in [5], where this excited state branch has not been considered. A decay of the mean atom number for strong attraction ($g = -2\hbar\omega\ell$), no interaction ($g = 0$) and strong repulsion ($g = 5\hbar\omega\ell$) cases is demonstrated. Probability current shows different tunneling directions when varying g. The present study could be related to the investigation of pairing in BEC (Bardeen, Cooper, and Schrieffer) to Bose-Einstein crossover and to topics of shell structure evolution in the context of nuclear physics [10].

Our method performs full time-dependent simulation of atom tunneling by using split-operator and Crank-Nicolson schemes. The method could be extended to more complicated problems with more degrees of freedom.

Acknowledgement

The work was financially supported by the Grant of the Plenipotentiary Representative of the Republic of Kazakhstan to JINR.

References

[1] I.S. Ishmukhamedov and V.S. Melezhik, Phys. Rev. A **95**, 062701 (2017).

[2] V.S. Melezhik, *Atoms and Molecules in Strong External Fields* (Plenum, New-York and London, 1998) pp.89-94.

[3] V.S. Melezhik, J.I. Kim and P. Schmelcher, Phys. Rev. A **76**, 053611 (2007).

[4] V.S. Melezhik, EPJ Web Conf. **108**, 01008 (2016).

[5] S.E. Gharashi and D. Blume, Phys. Rev. A **92**, 033629 (2015).

[6] A. U. J. Lode, A. I. Streltsov, O. E. Alon, H.-D. Meyer, and L. S. Cederbaum, J. Phys. B **42**, 044018 (2009).

[7] P. M. Krassovitskiy and F. M. Pen'kov, J. Phys. B: At. Mol. Opt. Phys. **47**, 225210 (2014).

[8] V. O. Nesterenko, A. N. Novikov, and E. Suraud, Laser Phys. **24**, 125501 (2014).

[9] G. Zürn, F. Serwane, T. Lompe, A. N. Wenz, M. G. Ries, J. E. Bohn, and S. Jochim, Physical Review Letters, **108**, 075303, (2012).

[10] G.Zürn, A.N.Wenz, S.Murmann, A.Bergschneider, T.Lompe, and S. Jochim, Physical Review Letters, **111**, 175302, (2013).

[11] M. Rontani, Phys. Rev. A **108**, 115302 (2012).

[12] M. Rontani, Phys. Rev. A **88**, 043633 (2013).

[13] Z. Idziaszek and T. Calarco, Phys. Rev. A, **71**, 050701, (2005)

[14] I.S. Ishmukhamedov, D.S. Valiolda, and S.A. Zhaugasheva, Phys. Part. Nuclei Lett. **11**: 238 (2014).

[15] G.I. Marchuk, *Methods of Numerical Mathematics*, (Springer-Verlag, New York,1975) Sec. 4.3.3

[16] U. V. Riss and H.-D. Meyer, J. Phys. B **26**, 4503 (1993).

[17] I.S. Ishmukhamedov, D.T. Aznabayev, and S.A. Zhaugasheva, Phys. Part. Nuclei Lett. **12**: 680 (2015).

[18] S. Sala and A. Saenz, Phys. Rev. A **94**, 022713 (2016).

[19] D. W. Noid, M. L. Koszykowski, and R. A. Marcus, J. Chem. Phys. **78**, 4018 (1983).

Study of the $B_c \to J/\psi + D_q^{(*)}$ decays in covariant confined quark model

Aidos Issadykov[1,2,⋆]

[1] *Joint Institute for Nuclear Research,141980 Dubna, Russia*
[2] *The Institute of Nuclear Physics,Ministry of Energy of the Republic of Kazakhstan, 050032 Almaty, Kazakhstan*

Abstract. In this work we study the $B_c \to J/\psi + D_q^{(*)}$ decays. It was defined ratios ($R_{D_s^*/D_s}$, $R_{D^*/D}$, R_{D^*/D_s^*} and R_{D/D_s}) of nonleptonic branching ratios of B_c meson, which will be hopefully tested on LHC experiments. We compare the obtained results with available experimental data and with the results from other theoretical approaches.

Introduction

Recently the ATLAS Collaboration reported on the measurement of the various branching fractions of the decays $B_c^+ \to J/\psi D_s^+$ and $B_c^+ \to J/\psi D_s^{*+}$ [1]. The first observations of these decays have been performed by the LHCb Collaboration [2]. In view of these developments, we decided to calculate ratios of branching fractions within the covariant confined quark model(CCQM).

The decay properties of the above processes were studied in various theoretical approaches [3–12].

Effective Hamiltonian and matrix element

The effective Hamiltonian describing the B_c nonleptonic decays into charmonium and $D(D_s)$ meson is given by (see, Ref. [13])

$$
\begin{aligned}
\mathcal{H}_{\rm eff} &= -\frac{G_F}{\sqrt{2}} V_{cb} V_{cq}^{\dagger} \sum_{i=1}^{6} C_i O_i, \\
O_1 &= (\bar{c}_{a_1} b_{a_2})_{V-A} (\bar{q}_{a_2} c_{a_1})_{V-A}, \qquad O_2 = (\bar{c}_{a_1} b_{a_1})_{V-A}, (\bar{q}_{a_2} c_{a_2})_{V-A}, \\
O_3 &= (\bar{q}_{a_1} b_{a_1})_{V-A} (\bar{c}_{a_2} c_{a_2})_{V-A}, \qquad O_4 = (\bar{q}_{a_1} b_{a_2})_{V-A} (\bar{c}_{a_2} c_{a_1})_{V-A}, \\
O_5 &= (\bar{q}_{a_1} b_{a_1})_{V-A} (\bar{c}_{a_2} c_{a_2})_{V+A}, \qquad O_6 = (\bar{q}_{a_1} b_{a_2})_{V-A} (\bar{c}_{a_2} c_{a_1})_{V+A},
\end{aligned} \tag{1}
$$

where the subscript $V-A$ refers to the usual left–chiral current $O^\mu = \gamma^\mu(1-\gamma^5)$ and $V+A$ to the usual right–chiral one $O_+^\mu = \gamma^\mu(1+\gamma^5)$. The a_i denote the color indices. The quark q stands for either s or d. The numerical values of the Wilson coefficients are taken as in Ref.[14]. They are listed in Table 1. Since in NNLO the numerical values of the C_5 and C_6 coefficients are less to the order than the C_3

⋆e-mail: issadykov@jinr.ru

Table 1. Values of the Wilson coefficients in NNLO.

C_1	C_2	C_3	C_4	C_5	C_6
−0.2632	1.0111	−0.0055	−0.0806	0.0004	0.0009

and C_4 coefficients respectively, we drop the contribution from those operators.

By using the Fierz transformation one can check that $O_3 = O_1$ and $O_4 = O_2$. Then the calculation of the matrix elements describing the nonleptonic decays of the B_c meson into charmonium and $D(D_s)$ meson is straightforward. The combinations of the Wilson coefficients appear as $a_1 = C_2+C_4+\xi\,(C_1+C_3)$ and $a_2 = C_1 + C_3 + \xi\,(C_2 + C_4)$ with $\xi = 1/N_c$. In the numerical calculations we set the color-suppressed parameter ξ to zero. Then the Wilson coefficients are equal to

$$a_1 = C_2 + C_4 = 0.93\,, \qquad \text{and} \qquad a_2 = C_1 + C_3 = -0.27 \tag{2}$$

which should be compared with the old ones $a_1 = 1.14$ and $a_2 = -0.20$ used in paper [3].

Invariant and helicity amplitudes

The invariant form factors for the semileptonic B_c decay into the hadron with spin $S = 0, 1$ are defined by

$$\begin{aligned} \mathcal{M}^{\mu}_{S=0} &= P^{\mu}\,F_{+}(q^2) + q^{\mu}\,F_{-}(q^2), &(3)\\ \mathcal{M}^{\mu}_{S=1} &= \frac{1}{m_1+m_2}\,\epsilon^{\dagger}_{\nu}\,\Big\{-g^{\mu\nu}\,Pq\,A_0(q^2) + P^{\mu}\,P^{\nu}\,A_{+}(q^2) + q^{\mu}\,P^{\nu}\,A_{-}(q^2) \\ &\qquad + i\,\varepsilon^{\mu\nu\alpha\beta}\,P_{\alpha}\,q_{\beta}\,V(q^2)\Big\}, &(4)\end{aligned}$$

where $P = p_1 + p_2$ and $q = p_1 - p_2$. Here p_1 is the momentum of the ingoing meson with a mass m_1 (B_c) and p_2 is the momentum of the outgoing meson with a mass m_2. It is convenient to express all physical observables through the helicity form factors H_m. The helicity form factors H_m can be written in terms of the invariant form factors in the following way [15]:
Spin S=0:

$$H_t = \frac{1}{\sqrt{q^2}}\,\Big\{(m_1^2-m_2^2)\,F_{+} + q^2\,F_{-}\Big\}\,, \qquad H_{\pm} = 0\,, \qquad H_0 = \frac{2\,m_1\,|\mathbf{p_2}|}{\sqrt{q^2}}\,F_{+}\,. \tag{5}$$

Spin S=1:

$$\begin{aligned} H_t &= \frac{1}{m_1+m_2}\,\frac{m_1\,|\mathbf{p_2}|}{m_2\,\sqrt{q^2}}\,\Big\{(m_1^2-m_2^2)\,(A_{+}-A_0) + q^2 A_{-}\Big\}\,,\\ H_{\pm} &= \frac{1}{m_1+m_2}\,\Big\{-(m_1^2-m_2^2)\,A_0 \pm 2\,m_1\,|\mathbf{p_2}|\,V\Big\}\,, &(6)\\ H_0 &= \frac{1}{m_1+m_2}\,\frac{1}{2\,m_2\,\sqrt{q^2}}\,\Big\{-(m_1^2-m_2^2)\,(m_1^2-m_2^2-q^2)\,A_0 + 4\,m_1^2\,|\mathbf{p_2}|^2\,A_{+}\Big\}\,.\end{aligned}$$

Here $|\mathbf{p_2}| = \lambda^{1/2}(m_1^2, m_2^2, q^2)/(2\,m_1)$ is the momentum of the outgoing meson in the B_c rest frame.

The nonleptonic B_c decay widths in terms of the helicity amplitudes are given by

$$\Gamma(B_c^+ \to J/\psi D_q^+) = N_W \left\{-a_1 f_{D_q} m_{D_q} H_t^{B_c\to J/\psi}(m_{D_q}^2) + a_2 f_{J/\psi} m_{J/\psi} H_0^{B_c\to D_q}(m_{J/\psi}^2)\right\}^2 ,$$

$$\Gamma(B_c^+ \to J/\psi D_q^{*+}) = N_W \sum_{i=0,\pm} \left\{a_1 f_{D^*} m_{D_q^*} H_i^{B_c\to J/\psi}(m_{D_q^*}^2) + a_2 f_{J/\psi} m_{J/\psi} H_i^{B_c\to D_q^*}(m_{J/\psi}^2)\right\}^2 ,$$

where we use the short notation

$$N_W \equiv \frac{G_F^2}{16\pi} \frac{|\mathbf{p}_2|}{m_1^2} |V_{cb} V_{cq}^\dagger|^2 .$$

Numerical results

All necessary details and values of the calculations of the leptonic decay constants and hadronic form factors may be found in recent publications [14, 16]. In Table 2 we show widths of the B_c meson for general values of the Wilson coefficients a_1 and a_2 obtained in this work and in other works.

Table 2. Exclusive nonleptonic decay widths of the B_c meson in units of 10^{-15} GeV for general values of the Wilson coefficients a_1 and a_2.

	$B_c^+ \to J/\psi D_s^+$	$B_c^+ \to J/\psi D_s^{*+}$	$B_c^+ \to J/\psi D^+$	$B_c^+ \to J/\psi D^{*+}$
CCQM	$(-1.84\,a_1 + 2.47\,a_2)^2$	$(3.26\,a_1 + 2.74\,a_2)^2$	$(-0.34\,a_1 + 0.36\,a_2)^2$	$(0.66\,a_1 + 0.51\,a_2)^2$
[3]	$(2.19\,a_1 + 1.32\,a_2)^2$	$(3.69\,a_1 + 2.35\,a_2)^2$	$(0.462\,a_1 + 0.277\,a_2)^2$	$(0.785\,a_1 + 0.460\,a_2)^2$
[4]	$(1.02\,a_1 + 1.95\,a_2)^2$		$(0.177\,a_1 + 0.442\,a_2)^2$	
[5]	$(0.67\,a_1 + 0.23\,a_2)^2$	$(1.1\,a_1 + 1.04\,a_2)^2$	$(0.13\,a_1 + 0.047\,a_2)^2$	$(0.20\,a_1 + 0.23\,a_2)^2$
[17]	$(1.65\,a_1 + 2.92\,a_2)^2$	$(3.31\,a_1 + 3.89\,a_2)^2$	$(0.30\,a_1 + 0.44\,a_2)^2$	$(0.48\,a_1 + 0.80\,a_2)^2$
[18]	$(1.62\,a_1 + 1.72\,a_2)^2$	$(3.13\,a_1 + 3.67\,a_2)^2$	$(0.372\,a_1 + 0.338\,a_2)^2$	$(0.686\,a_1 + 0.732\,a_2)^2$

In Table 3 we show the values of branching fractions obtained in this work for two different set of the Wilson coefficients ($a_1 = 0.93$, $a_2 = -0.27$ and $a_1 = 1.14$, $a_2 = -0.20$) and compare them with other theoretical approaches.

Table 3. Branching ratios (in %) of nonleptonic B_c decays obtained in this work for two different set of the Wilson coefficients.

Mode	$a_1 = +0.93$ $a_2 = -0.27$	$a_1 = +1.14$ $a_2 = -0.20$	[3]	[5]	[6]	[9]	[10]	[17]	[18]
$B_c \to J/\psi D_s^+$	0.10	0.22	0.34	0.34	0.17	0.14	0.81	0.12	0.15
$B_c \to J/\psi D_s^{*+}$	0.41	0.78	0.75	0.59	0.67	0.41	2.05	0.55	0.62
$B_c \to J/\psi D^+$	0.0035	0.0074	0.013	0.013	0.009	0.0055	0.028	0.0044	0.009
$B_c \to J/\psi D^{*+}$	0.017	0.031	0.023	0.019	0.028	0.018	0.067	0.010	0.028

Table 4. Comparison of the results for the ratios of nonleptonic branching fractions of B_c meson

$R_{D_s^*/D_s}$	$R_{D^*/D}$	R_{D^*/D_s^*}	R_{D/D_s}	Ref.
$2.8^{+1.2}_{-0.9}$				ATLAS [1]
2.37 ± 0.57				LHCb [2]
4.1 ± 0.80	4.86 ± 0.97	0.041 ± 0.008	0.035 ± 0.007	CCQM
2.9	1.77	0.031	0.038	RCQM [3]
1.7	1.46	0.032	0.038	QCD PM [5]
3.9	3.11	0.042	0.053	QCD SR [6]
3.01 ± 1.23	3.27	0.044	0.039	LFQM [9]
$2.54^{+0.07}_{-0.21}$	2.39	0.033	0.035	pQCD [10]

Finally, we compare our results for ratios of nonleptonic branching fractions of B_c meson with available experimental data and the results obtained in other approaches in Table 4, where

$$R_{D_{1q}^{(*)}/D_{2q}^{(*)}} = \frac{B(B_c^+ \to J/\psi D_{1q}^{(*)+})}{B(B_c^+ \to J/\psi D_{2q}^{(*)+})}\,. \tag{7}$$

Acknowledgment

Author A. Issadykov is grateful for the support by the JINR, grant number 17-302-03.

References

[1] G. Aad *et al.* [ATLAS Collaboration], Eur. Phys. J. C **76** (2016) no.1, 4 [arXiv:1507.07099 [hep-ex]].
[2] R. Aaij *et al.* [LHCb Collaboration], Phys. Rev. D **87** (2013) no.11, 112012 Addendum: [Phys. Rev. D **89** (2014) no.1, 019901] [arXiv:1304.4530 [hep-ex]].
[3] M. A. Ivanov, J. G. Körner and P. Santorelli, Phys. Rev. D **73** (2006) 054024 [hep-ph/0602050].
[4] C. H. Chang and Y. Q. Chen, Phys. Rev. D **49**, 3399 (1994).
[5] P. Colangelo and F. De Fazio, Phys. Rev. D **61**, 034012 (2000) [hep-ph/9909423].
[6] V. V. Kiselev, hep-ph/0211021.
[7] D. Ebert, R. N. Faustov and V. O. Galkin, Phys. Rev. D **68**, 094020 (2003) [hep-ph/0306306].
[8] A. K. Likhoded and A. V. Luchinsky, Phys. Rev. D **81**, 014015 (2010) [arXiv:0910.3089 [hep-ph]].
[9] H. W. Ke, T. Liu and X. Q. Li, Phys. Rev. D **89**, no. 1, 017501 (2014) [arXiv:1307.5925 [hep-ph]].
[10] Z. Rui and Z. T. Zou, Phys. Rev. D **90**, no. 11, 114030 (2014) [arXiv:1407.5550 [hep-ph]].
[11] S. Kar, P. C. Dash, M. Priyadarsini, S. Naimuddin and N. Barik, Phys. Rev. D **88**, no. 9, 094014 (2013).

[12] R. Zhu, Y. Ma, X. L. Han and Z. J. Xiao, Phys. Rev. D **95**, no. 9, 094012 (2017) [arXiv:1703.03875 [hep-ph]].

[13] G. Buchalla, A. J. Buras, M. E. Lautenbacher, Rev. Mod. Phys. **68**, 1125-1144 (1996). [hep-ph/9512380].

[14] S. Dubnicka, A. Z. Dubnickova, A. Issadykov, M. A. Ivanov and A. Liptaj, Phys. Rev. D **96** (2017) no.7, 076017 doi:10.1103/PhysRevD.96.076017 [arXiv:1708.09607 [hep-ph]].

[15] M. A. Ivanov, J. G. Körner and P. Santorelli, Phys. Rev. D **63** (2001) 074010 [hep-ph/0007169].

[16] A. Issadykov, M. A. Ivanov and G. Nurbakova, EPJ Web Conf. **158** (2017) 03002. doi:10.1051/epjconf/201715803002

[17] A. Y. Anisimov, P. Y. Kulikov, I. M. Narodetsky and K. A. Ter-Martirosian, Phys. Atom. Nucl. **62** (1999) 1739 [Yad. Fiz. **62** (1999) 1868] [hep-ph/9809249].

[18] A. Abd El-Hady, J. H. Munoz and J. P. Vary, Phys. Rev. D **62** (2000) 014019 doi:10.1103/PhysRevD.62.014019 [hep-ph/9909406].

EPJ Web of Conferences **177**, 09008 (2018) https://doi.org/10.1051/epjconf/201817709008
AYSS-2017

Magnetoelectric Effect in Lorentz-violating Extensions of the Standard Model

Oleg Kharlanov[1,*] and *Ivan* Degtev[2]

[1] *Lomonosov Moscow State University, Faculty of Physics, 1/2 Leninskie Gory, 119991 Moscow, Russia*
[2] *Lebedev Physical Institute of the Russian Academy of Sciences, 53 Leninskiy Prospekt, 119991 Moscow, Russia*

Abstract. We demonstrate that an atom placed in a constant background interacting with the electronic axial-vector current exhibits a new type of parity-breaking transverse magnetoelectric polarizability. Within beyond-Standard-Model (BSM) frameworks, such as the Lorentz-violating Standard Model Extension, such constant backgrounds are considered to be condensates of Planck-scale fields, thus, precision measurements of the magnetoelectric polarizability is able to constrain non-trivial BSM physics. Indeed, we demonstrate that the contribution to this effect within the Standard Model due to weak interaction is strongly suppressed, the effect virtually being purely exotic. We calculate magnetoelectric polarizability for a simple atom and discuss its observability.

1 Introduction

Lorentz invariance has been an underlying principle of virtually all modern physical theories since their emergence, including quantum field theory and high-energy physics. With the advent of the string theory and other attempts to find a consistent description of quantum gravity, however, it turned out that this symmetry could be spontaneously broken on an energy scale close to the Planck energy [1]. Moreover, a possibility for dynamical Lorentz violation by radiative corrections was also pointed out in field theories, such as the Axion–Wess–Zumino (AWZ) model [2]. Finally, quite recently, Petr Hořava presented an example of a field theory, whose non-Lorentzian behavior in the high-energy regime results in power-counting renormalizability, whereas in the low-energy regime, the conventional $SO(1,3)$ symmetry group is restored and the theory reproduces General Relativity [3]. These discoveries have triggered interest in precision tests of Lorentz invariance one could perform at attainable energies, as a window into beyond-Standard-Model (BSM) physics.

In order to implement the strategy of constraining BSM physics with precision tests of Lorentz invariance, the Standard Model Extension (SME) framework was proposed [4]. This framework assumes that high-energy particles present in the hypothetical fundamental theory have masses that are too large to be observable at any of the present experiments, while signatures of Lorentz violation could be found in condensates of Planck-scale fields whose nonzero values led to the corresponding symmetry breaking. Thus, SME is assumed to be a low-energy mean-field theory with a number of order parameters interacting with conventional fields featuring in the Standard Model via a set of extra

[*] e-mail: kharlanov@physics.msu.ru

terms in the SME Lagrangian. Note that no additional dynamical fields are present in SME, compared with the Standard Model. Diversity of exotic terms in SME is usually limited by a number of 'natural requirements', such as energy-momentum conservation, $SU(3)_c \times SU(2)_I \times U(1)_Y$ gauge invariance, and anomaly cancellation; as a result, the number of allowed Lorentz-violating terms can even be made finite (for details, see Ref. [4]).

After the electroweak symmetry breaking, SME leads to extended electrodynamics, whose Lagrangian reads

$$\mathcal{L}_{\rm EQED} = -\frac{1}{4}F_{\mu\nu}^2 + \frac{1}{2}\epsilon_{\mu\nu\alpha\beta}k_{AF}^{\mu}A^{\nu}F^{\alpha\beta} - \frac{1}{4}k_F^{\mu\nu\alpha\beta}F_{\mu\nu}F_{\alpha\beta} + \bar{\psi}(\mathrm{i}\Gamma^{\mu}(\partial_\mu + \mathrm{i}eA_\mu) - M)\psi, \tag{1}$$

$$\Gamma^{\mu} = \gamma^{\mu} + c^{\nu\mu}\gamma_\nu - d^{\nu\mu}\gamma_\nu\gamma_5 + e^{\mu} + \mathrm{i}f^{\mu}\gamma_5 + \frac{1}{2}g^{\alpha\beta\mu}\sigma_{\alpha\beta}, \quad M = m + a^{\mu}\gamma_\mu - b^{\mu}\gamma_\mu\gamma_5 + \frac{1}{2}H^{\mu\nu}\sigma_{\mu\nu}, \tag{2}$$

where m and e are the electron mass and charge, respectively, and $k_{AF}, k_F, a, b, \ldots, H$ are constant (pseudo)tensor couplings (condensates) with certain symmetry properties; $\gamma_\mu, \gamma_5, \sigma_{\mu\nu}$ are conventional 4×4 Dirac matrices and the electromagnetic field strength $F_{\mu\nu} = \partial_\mu A_\nu - \partial_\nu A_\mu$. Note that the same 'Lorentz-violating' terms can also describe physical situations arising within the conventional Standard Model, e.g., the axial-vector b^{μ} term can arise in intensive neutrino fluxes due to forward neutrino-electron scattering [5]. This term can also reflect the presence of a nonzero background torsion [6]. In the present paper, we study the effect of the b^{μ} term, taking

$$\mathcal{L} = -\frac{1}{4}F_{\mu\nu}^2 + \bar{\psi}(\gamma^{\mu}(\mathrm{i}\partial_\mu - eA_\mu + b_\mu\gamma_5) - m)\psi, \tag{3}$$

and consider the magnetoelectric response of a hydrogen atom, i.e., induction of an electric dipole moment of the atom by a weak external magnetic field, and vice versa. Electric and magnetic field strengths being a 3-vector and a 3-pseudovector, respectively, the corresponding polarizability $\kappa_{\rm ME}$ should be parity-violating, thus, one can expect its exotic nature, e.g., from the timelike component of the axial condensate b^{μ}. Indeed, the ground state of a hydrogen atom is characterized by the only (pseudo)vector $\boldsymbol{S}$ of the electron spin, in the case of a moving atom one also has the velocity $\boldsymbol{V}$ of the nucleus, which is a polar vector. Now, to construct a pseudovector $\boldsymbol{\mu}$ of the magnetic dipole moment induced by a (polar) vector electric field $\boldsymbol{E}$ in the $b^0 = 0$ case, one can only choose $\boldsymbol{\mu} \propto [\boldsymbol{V} \times \boldsymbol{E}]$ (trilinear expressions in $\boldsymbol{S}, \boldsymbol{V}, \boldsymbol{E}$ are T-even and cannot contribute to the T-odd magnetic moment), but such a contribution is odd in the velocity and its net effect should therefore be suppressed for systems of atoms with symmetric velocity distributions. In the $b^0 \neq 0$ case, however, another possibility, namely, $b^0[\boldsymbol{E} \times \boldsymbol{S}]$, opens up, which is nonzero for atoms at rest. In search of such an effect, we thus take $b^{\mu} = (b, \boldsymbol{0})$ in the rest frame of the nucleus and assume b to be small. Note that similar effects, having to do with mixing of electric and magnetic observables, have been discussed recently to appear in core-collapse supernovae, Weyl semimetals, and heavy-ion collisions [7–9].

The paper is organized as follows: in Sec. 2, we introduce the setup of the problem in question and evaluate the magnetoelectric polarizability of an atom; Sec. 3 discusses Standard-Model contributions to the ME effect due to weak interaction; the concluding Sec. 4 summarizes the results obtained.

2 Magnetoelectric effect in a hydrogen atom in the b condensate

The electron in a hydrogen atom is described by the following 1-particle Hamiltonian, which follows from (3) (we adopt units with $\hbar = c = 1$ hereinafter),

$$\hat{H} = \boldsymbol{\alpha}(\hat{\boldsymbol{p}} - e\boldsymbol{A}) + eA^0 + m\gamma^0 + b\gamma_5, \qquad \boldsymbol{\alpha} \equiv \gamma^0\boldsymbol{\gamma}, \ \hat{\boldsymbol{p}} \equiv -\mathrm{i}\boldsymbol{\nabla}. \tag{4}$$

We will work in the nonrelativistic (Pauli) approximation, in which the 2×2 Hamiltonian reads [10]

$$\hat{h}_b = \frac{(\hat{\boldsymbol{p}} - e\boldsymbol{A} - b\boldsymbol{\sigma})^2}{2m} + eA^0(x) - \frac{e}{2m}\boldsymbol{\sigma}\boldsymbol{B}, \tag{5}$$

where $\boldsymbol{\sigma}$ are the Pauli matrices, $A^\mu(x) = (\varphi_{\rm nucl}(r) - \boldsymbol{E} \cdot \boldsymbol{x}, \boldsymbol{A})$, $\varphi_{\rm nucl}(r)$, $\boldsymbol{E}$, and $\boldsymbol{B} = \boldsymbol{\nabla} \times \boldsymbol{A}$ being the potential of the nucleus and the external electric and magnetic fields, respectively. To evaluate the magnetic moment induced by $\boldsymbol{E}$ in the ground state of the atom

$$\boldsymbol{\mu}(\boldsymbol{E}) = -\frac{\partial \langle \psi_{\boldsymbol{E},\boldsymbol{B}} | \hat{h}_b(\boldsymbol{E}, \boldsymbol{B}) | \psi_{\boldsymbol{E},\boldsymbol{B}} \rangle}{\partial \boldsymbol{B}} = -\langle \psi_{\boldsymbol{E},\boldsymbol{0}} | \left. \frac{\partial \hat{h}_b(\boldsymbol{E}, \boldsymbol{B})}{\partial \boldsymbol{B}} \right|_{\boldsymbol{B}=\boldsymbol{0}} | \psi_{\boldsymbol{E},\boldsymbol{0}} \rangle, \tag{6}$$

we will need the ground-state wave function $\psi_{\boldsymbol{E},\boldsymbol{0}}$ of $\hat{h}_b$ in a weak electric field. The latter can be found by virtue of a unitary transformation that eliminates b from the Hamiltonian (5) [10]

$$\hat{h}_b(\boldsymbol{E}, \boldsymbol{0}) = e^{ib\boldsymbol{\sigma}\hat{\boldsymbol{x}}} \hat{h}_0(\boldsymbol{E}, \boldsymbol{0}) e^{-ib\boldsymbol{\sigma}\hat{\boldsymbol{x}}} + O(b^2), \quad |\psi_{\boldsymbol{E},\boldsymbol{0}}\rangle \approx e^{ib\boldsymbol{\sigma}\hat{\boldsymbol{x}}} |\psi^{(0)}_{\boldsymbol{E},\boldsymbol{0}}\rangle, \tag{7}$$

where $\psi^{(0)}_{\boldsymbol{E},\boldsymbol{0}}$ is the wave function in the conventional, $b = 0$ case. Plugging this transformation into the definition (6) of the magnetic moment and subsequent perturbative expansion in $\boldsymbol{E} \to 0$ yield

$$\begin{aligned} \boldsymbol{\mu} &\approx -\mu_{\rm B} \langle \psi^{(0)}_{\boldsymbol{E},\boldsymbol{0}} | (\hat{\boldsymbol{l}} + \boldsymbol{\sigma} + 2b[\boldsymbol{\sigma}\boldsymbol{x}]) | \psi^{(0)}_{\boldsymbol{E},\boldsymbol{0}} \rangle \\ &= \boldsymbol{\mu}_0 - \left(2eb\mu_{\rm B} \sum_{\nu \neq {\rm gnd}} \frac{\langle \nu | \boldsymbol{E} \cdot \boldsymbol{x} | {\rm gnd} \rangle \, \langle {\rm gnd} | [\boldsymbol{\sigma}\boldsymbol{x}] | \nu \rangle}{E^{(0)}_\nu - E^{(0)}_{\rm gnd}} + {\rm c.c.} \right) + O(E^2), \end{aligned} \tag{8}$$

where $\mu_{\rm B} = -e/2m > 0$ is the Bohr magneton, $\hat{\boldsymbol{l}} \equiv \hat{\boldsymbol{x}} \times \hat{\boldsymbol{p}}$, and $\boldsymbol{\mu}_0$ is the magnetic moment of the ground state $|{\rm gnd}\rangle$ for $\boldsymbol{E}, \boldsymbol{B}, b = 0$; the excited states $|\nu\rangle$ have the energies $E^{(0)}_\nu$ inversely proportional to the principal quantum number n squared.

Note now that the above expression is featuring matrix elements of the coordinate operator $\boldsymbol{x}$, which leads to selection rules, namely, nonvanishing contributions to the sum over ν come from the $n{\rm p}_{x,y,z}$ states ($n = 2, 3, 4, \ldots$) having the same spin polarization as the ground state. That is, one can use the real spherical harmonics for these states, $\psi^{(0)}_{n{\rm p}_{x,y,z}}(\boldsymbol{x}) \propto R_{n{\rm p}}(|\boldsymbol{x}|) \frac{\boldsymbol{e}_{x,y,z} \cdot \boldsymbol{x}}{|\boldsymbol{x}|}$, $\boldsymbol{e}_{x,y,z}$ being the three Cartesian basis vectors, thus arriving at a sum over n and the angular momentum projection. Quite surprisingly, this sum can be evaluated analytically [11], leading to the magnetic momentum shift induced by the external electric field

$$\delta\boldsymbol{\mu} \equiv \boldsymbol{\mu}(\boldsymbol{E}) - \boldsymbol{\mu}_0 \approx \frac{72ebr_{\rm B}^2}{\rm Ry} [\boldsymbol{E} \times \boldsymbol{S}] \, \mu_{\rm B}, \tag{9}$$

where $r_{\rm B} = 1/me^2$ is the Bohr's radius and ${\rm Ry} = e^2/2r_{\rm B} \approx 13.6\,{\rm eV}$. Note that the effect we have calculated is linear in the condensate, unlike many other effects associated with it, including the energy shifts [10]. Moreover, technically, the above expression does not include the contribution of the delocalized states with $E > 0$, however, their account changes the numerical factor in (9) by no more than 5% (note that continuum states should be used with care, since their wave functions are delocalized and sensitive to the environment of the atom). Moreover, a dual effect, i.e., a nonzero electric dipole induced by the external magnetic field, also exists and exhibits the same structure of the expression $\boldsymbol{d} \propto \frac{ebr_{\rm B}^2}{\rm Ry} [\boldsymbol{B} \times \boldsymbol{S}] \, er_{\rm B}$. Both effects are transverse, since the induced dipole moments are perpendicular to the external field, for example, the magnetoelectric polarizability tensor describing $\delta\boldsymbol{\mu}(\boldsymbol{E})$ reads

$$(\kappa_{\rm ME})_{ij} = \frac{72ebr_{\rm B}^2}{\rm Ry} \epsilon_{ijk} S_k, \qquad \delta\boldsymbol{\mu} \equiv \kappa_{\rm ME} \cdot \boldsymbol{E}. \tag{10}$$

It is worth examining, however, whether any Standard-Model effect can produce nontrivial magnetoelectric responses similar to Eq. (9) that could mask the Lorentz-violating contributions to the effect. In the above, we have identified the source of the nontrivial magnetoelectric polarizability in the P parity violation. Therefore, in search for the non-exotic ME effect, one is led towards the study of weak interactions in a hydrogen atom.

3 Magnetoelectric effect from the weak interaction of the electron and the nucleus

Weak interactions in atomic systems have been studied since 1970s [12, 13] using precision optical resonant techniques, which has revealed a tiny but observable effect of the electron-nucleus interaction. For a spin-0 nucleus, this interaction can be written in a four-fermion fashion, leading to the following additional term in the electron Hamiltonian

$$\hat{H} \supset -G\gamma_5\delta^3(\boldsymbol{x}), \tag{11}$$

where the G factor is proportional to the Fermi constant and the weak charge of the nucleus [14]. Formally, the above term looks like an interaction of the electron with the b condensate localized at the origin. In the nonrelativistic regime, such a term results in a correction to the Pauli Hamiltonian, namely, now, instead of Eq. (5), one has

$$\hat{h}_G = \frac{(\hat{\boldsymbol{p}} - e\boldsymbol{A})^2}{2m} + eA^0(\boldsymbol{x}) + \frac{G}{2m}\{\boldsymbol{\sigma}\cdot\hat{\boldsymbol{p}}, \delta^3(\hat{\boldsymbol{x}})\}, \tag{12}$$

where braces denote the anticommutator. Let us now analyze whether this, competing mechanism of P violation in the Standard Model is able to generate a nontrivial magnetoelectric response $\delta\boldsymbol{\mu}$ to a weak external field $\boldsymbol{E}$. Now we cannot use the transformation (7) to get rid of the parity-violating G term, thus, we treat both the G term and the external electric field term $-e\boldsymbol{E}\cdot\boldsymbol{x}$ as perturbations. By extracting the terms that are linear in both $\boldsymbol{E}$ and G in the magnetic moment (6), we arrive at

$$\delta\boldsymbol{\mu} \approx -\frac{eG\mu_{\mathrm{B}}}{2m}\sum_{\nu,\lambda\neq\mathrm{gnd}} \frac{\langle\mathrm{gnd}|\{\boldsymbol{\sigma}\cdot\hat{\boldsymbol{p}}, \delta^3(\boldsymbol{x})\}|\nu\rangle\,\langle\nu|(\boldsymbol{l}+\boldsymbol{\sigma})|\lambda\rangle\langle\lambda|\boldsymbol{E}\cdot\boldsymbol{x}|\mathrm{gnd}\rangle}{(E^{(0)}_\nu - E^{(0)}_{\mathrm{gnd}})(E^{(0)}_\lambda - E^{(0)}_{\mathrm{gnd}})} + \mathrm{c.c.} \tag{13}$$

The matrix element of $\boldsymbol{x}$ selects only the p states among all possible λ's; action of $\boldsymbol{l}+\boldsymbol{\sigma}$ does not change the total momentum, so ν states should also be p states. Now, one can express these states in terms of real spherical harmonics, and a careful analysis of the structure of the scalar products reveals that the contributions coming from the $\boldsymbol{l}$ and the $\boldsymbol{\sigma}$ operators in the central term in the numerator cancel each other, leading to

$$\delta\boldsymbol{\mu} = 0. \tag{14}$$

In principle, the above result is a consequence of the localization of the weak interaction term at the origin (because of the Dirac delta function), where the wave functions of p states vanish. Also, the anomalous magnetic moment of the electron transforms $\boldsymbol{l}+\boldsymbol{\sigma}$ into an imbalanced combination with nonequal coefficients and the cancelation (14) does not take place. In any case, we conclude that the suppression of the magnetoelectric effect in the Standard Model is quite strong.

4 Conclusion

The effect we have studied in the present paper has a number of features that distinguish it favorably as a basis for precision tests of Lorentz invariance. Firstly, the corresponding effects in the Standard

Model cancel quite surprisingly in the leading order (to the best of our knowledge, this does not happen due to any obvious symmetry of the system). Secondly, the effect is linear in the small Lorentz-violating coupling, unlike other effects the latter one generates, which are at least quadratic. On the other hand, the magnetoelectric polarizability is nonzero for spin-polarized atoms, so certain polarization techniques should be used to avoid averaging it to zero. For the latter reason, the Lorentz-induced magnetoelectric effect might show up in magnetic materials with nonzero mean magnetization.

References

[1] V. A. Kostelecký and S. Samuel, Phys. Rev. D **40**, 1886 (1989).
[2] A. A. Andrianov, R. Soldati, and L. Sorbo, Phys. Rev. D **59**, 025002 (1999).
[3] P. Hořava, Phys. Rev. D **79**, 084008 (2009).
[4] D. Colladay and V. A. Kostelecký, Phys. Rev. D **58**, 116002 (1998).
[5] A. I. Studenikin, Phys. Atom. Nucl. **67**, 993 (2004).
[6] I. L. Shapiro, Phys. Rept. **357**, 113 (2002).
[7] A. F. Bubnov, N. V. Gubina, and V. Ch. Zhukovsky, Phys. Rev. D **96**, 016011 (2017).
[8] M. Dvornikov, JCAP **05(2015)**, 037 (2015).
[9] K. Fukushima, D. E. Kharzeev, and H. J. Warringa, Phys. Rev. D **78**, 074033 (2008).
[10] O .G. Kharlanov and V. Ch. Zhukovsky, J. Math. Phys. **48**, 092503 (2007).
[11] H. Bethe and E. E. Salpeter, *Quantum Mechanics of One- and Two-Electron Atoms* (Springer-Verlag, Berlin, 1957).
[12] M. A. Bouchiat and C. C. Bouchiat, Phys. Lett. B **48**, 111 (1974).
[13] V. N. Novikov and I. B. Khriplovich, JETP Lett. **22**(3), 74 (1975).
[14] F. Curtis-Michel, Phys. Rev. **138B**, 408 (1965).

Confinement, Chiral Symmetry Breaking and it's Restoration using Dual QCD Formalism

Garima Punetha[a] and H. C. Chandola

Center of Advanced Study, Department of Physics, Kumaun University Nainital-263001, India

Abstract. Utilizing the dual QCD model in term of magnetic symmetry structure of non-Abelian gauge theories, the dynamical chiral-symmetry breaking using Schwinger-Dyson equation has been investigated. A close relation among the color confinement and chiral-symmetry breaking has been observed and demonstrated by computing dynamical parameters. The recovery of the chiral symmetry has also been discussed at finite temperature through the variation of quark mass function and quark condensate which gradually decreases with temperature and vanishes suddenly near the critical temperature.

1 Introduction

Color confinement and chiral-symmetry breaking are two outstanding non-perturbative phenomena in quantum chromodynamics (QCD) and their relation have been studied as one of the important difficult problems in theoretical particle physics. In this direction, the dual superconductor picture for the Yang-Mills theory [1–5] is an attractive hypothesis for explaining quark confinement and for this mechanism to work, the existence of chromomagnetic monopole and their condensation play a leading role to cause the dual Meissner effect. In QCD, chiral symmetry is approximately realized at the Lagrangian level and is broken by the strong interaction at zero temperature. On the other hand, at finite temperature, the broken symmetries are restored. As a theoretical approach to understand the phase structure of QCD, the present paper, therefore, deals to present a gauge independent field decomposition formalism in which the QCD potential splits into the non-topological abelian part and the topological monopole part to establish the color confinement and chiral symmetry breaking using Schwinger-Dyson equation (SDE) method.

2 Dual QCD Formulation and Confinement

For the case of the simplest non-abelian gauge group, $G = SU(2)$ [6–12] with its little group H as $U(1)$, the multiplet $\hat{m}$ may thus be viewed to identify the homotopy of the mapping $\pi_2(S^2)$ in the following form, $\hat{m} : S^2_R \rightarrow S^2 = SU(2)/U(1)$, where, S^2_R is the two dimensional sphere of the three dimensional space and S^2 is the group coset space completely fixed by $\hat{m}$. The typical gauge potential and field strength may then identified as

$$\mathbf{W}_\mu = A_\mu \hat{m} - g^{-1}(\hat{m} \times \partial_\mu \hat{m}) \text{ and } \mathbf{G}_{\mu\nu} = \partial_\mu \mathbf{W}_\nu - \partial_\nu \mathbf{W}_\mu + g(\mathbf{W}_\mu \times \mathbf{W}_\nu) = (F_{\mu\nu} + B^{(d)}_{\mu\nu})\hat{m}, \quad (1)$$

[a]e-mail: garimapunetha@gmail.com

where, $A_\mu \equiv \hat{m} \cdot \mathbf{W}_\mu$ is the Abelian component of $\mathbf{W}_\mu$, while the second part on the right hand side is completely determined by the magnetic symmetry and is of topological origin. The topological structure of the theory may be then brought into dynamics explicitly by performing the separation of gauge fields in the magnetic gauge obtained by rotating $\hat{m}$ to the third direction in isospace such that, $\hat{m} \xrightarrow{U} \hat{\xi}_3 = U\hat{m} = (0\ 0\ 1)^T$ along with the parameterisation $\hat{m} = (\sin\alpha\cos\beta,\ \sin\alpha\sin\beta,\ \cos\alpha)^T$. The parameterisation with the choice $U = \exp(-\alpha t_2 - \beta t_3)$ then leads to the expression of the potential and the corresponding field strength as

$$W_\mu \xrightarrow{U} W_\mu^{'} = (A_\mu + B_\mu)\hat{\xi}_3 \text{ and } G_{\mu\nu} \xrightarrow{U} G_{\mu\nu} = (F_{\mu\nu} + B_{\mu\nu}^{(d)})\hat{\xi}_3, \tag{2}$$

where, the part $B_{\mu\nu}^{(d)}$ may now be identified as $B_{\mu\nu}^{(d)} = -g^{-1}\sin\alpha\,(\partial_\mu\alpha\,\partial_\nu\beta - \partial_\nu\alpha\,\partial_\mu\beta) = \partial_\mu B_\nu - \partial_\nu B_\mu$. In order to explore the typical dual QCD vacuum properties and the dynamics of the system, the Lagrangian in the quenched approximation may be expressed as [6]

$$\mathcal{L}_d^{(m)} = -\frac{1}{4}B_{\mu\nu}^2 + \left|\left[\partial_\mu + i\frac{4\pi}{g}B_\mu^{(d)}\right]\phi\right|^2 - V(\phi^*\phi). \tag{3}$$

Using $V(\phi^*\phi) = 3\lambda\alpha_s^{-2}(\phi^*\phi - \phi_0^2)^2$ potential the Lagrangian given by equation (3) generate the dynamical symmetry breaking and leads to the following field equations given as below [10]

$$(\partial_\mu - i\frac{4\pi}{g}B_\mu^{(d)})(\partial_\mu + i\frac{4\pi}{g}B_\mu^{(d)})\phi + 96\lambda\alpha_s^{-2}(\phi^*\phi - \phi_0^2)\phi = 0,$$

$$\partial^\nu B_{\mu\nu} + \frac{i4\pi}{g}(\phi\partial_\mu\phi^* - \phi^*\partial_\mu\phi) - \frac{32\pi^2}{g^2}B_\mu^{(d)}\phi\phi^* = 0. \tag{4}$$

Utilizing the asymptotic solutions of the associated dual QCD fields, the energy per unit length of the resulting flux tube configuration may be derived as [10]

$$k = 2\pi\int_0^\infty \rho d\rho\Big[\frac{n^2g^2}{32\pi^2\rho^2}(F_\lambda^{'}(\rho))^2 + \frac{n^2g^2}{16\pi^2\rho^2}F_\lambda^2(\rho)\chi^2(\rho) + (\chi^{'}(\rho))^2 + \frac{48\lambda\pi^2}{g^4}(\phi^*\phi - \phi_0^2)^2\Big]. \tag{5}$$

In terms of a set of more convenient parameters, $r = 2\sqrt{3\lambda}\alpha_s^{-1}\phi_0\rho$, $F_\lambda(r) = (4\pi\alpha_s^{-1})^{\frac{1}{2}}\rho B(\rho)$, $H_\lambda(r) = \phi_0^{-1}\chi(\rho)$, eq. (5) may be reduced to a simple form as [10]

$$k = \gamma\phi_0^2,\ \text{with},\ \gamma_\lambda = 2\pi\int_0^\infty rdr\Big[\frac{6\lambda}{g^2}\frac{(F_\lambda^{'})^2}{r^2} + \frac{(n+F_\lambda)^2}{r^2}H_\lambda^2 + (H_\lambda^{'})^2 + \frac{(H_\lambda^2-1)^2}{4}\Big]. \tag{6}$$

Using the numerical computation of equation (6) for γ_λ, we obtain the characteristics mass scales, the scalar (m_ϕ) and vector (m_B) glueball masses for some typical values of coupling in infrared sector of QCD and are summarized in Table 1 [7, 10].

3 Chiral Symmetry Breaking using Dual QCD Formalism

In the low energy sector the dual gluon propagator leads to a linear confinement potential and the chiral symmetry associated with the quarks is broken dynamically. The SDE for quark propagator is expressed as

$$S^{-1}(p) = Z_2(i\gamma.p + m_b) + Z_1\int\frac{d^4q}{(2\pi)^4}g^2\gamma^\mu\frac{\lambda^a}{2}S(q)\Gamma^{\nu b} \times D_{\mu\nu}^{ab}(p-q), \tag{7}$$

Table 1. The masses of vector and scalar glueball as functions of α_s using Quartic potential.

α_s	γ	ϕ_0	$m_B(GeV)$	$m_\phi(GeV)$	$\lambda^{(d)}_{QCD}(GeV^{-1})$	$\xi^{(d)}_{QCD}(GeV^{-1})$	$\kappa^{(d)}_{QCD}$
0.12	8.30	0.143	2.11	4.20	0.473	0.238	2
0.22	7.89	0.149	1.51	2.2	0.662	0.450	1.5
0.47	6.28	0.167	1.21	1.22	0.826	0.819	1
0.96	5.40	0.181	0.929	0.655	1.07	1.53	0.7

where $S(p)$ is the full quark propagator, $D^{ab}_{\mu\nu}$ is the dressed gluon propagator, $\Gamma^{\nu b}$ is the dressed quark-gluon vertex, m_b is the bare mass of the quark, Z_1 is the quark-gluon vertex renormalization and Z_2 is the quark wave function renormalization. In solving SDE, a system of two coupled integral equation for $A(p^2)$ and $M(p^2)$ are obtained and expressed as

$$[A(p^2)-1]p^2 = \frac{4}{3}g^2\beta\int\frac{d^4q}{(2\pi)^4}\frac{A(q^2)}{q^2A^2(q^2)+B^2(q^2)}\Big[pq+\frac{q(p-q)(p-q)p}{(p-q)^2}\Big], \tag{8}$$

$$M(p^2) = 4g^2\beta\int\frac{d^4q}{(2pi)^4}\delta^4(p-q)\frac{B(q^2)}{q^2A^2(q^2)+B^2(q^2)}, \tag{9}$$

where the gluon propagator [13, 14] is given as $D^{ab}_{\mu\nu}(q) = \delta^{ab}(\delta_{\mu\nu} - \frac{q_\mu q_\nu}{q^2})\beta\delta^4(q)$. The equation for $M(p^2)$ gives the momentum dependent quark mass function generated dynamically and indicates the breaking of the chiral symmetry in the dual QCD vacuum. The solution of the above coupled integral equations determined in [15–18] showing the validity in the infrared region of QCD expressed as

$$M(p^2) = \sqrt{\frac{2m_B^2\alpha_s}{\pi} - \frac{4p^2}{9}} \quad p^2 < \frac{9}{4}\Big(\frac{2m_B^2\alpha_s}{\pi}\Big) \text{ and } M(p^2) = 0 \quad p^2 > \frac{9}{4}\Big(\frac{2m_B^2\alpha_s}{\pi}\Big), \tag{10}$$

The non-trivial solution of the SDE is denoted by the solid curve in Figure 1 through the variation of $M(p^2)$ with momentum scale (p) for different values of α_s. The above equation reduces to the square of dynamical quark mass as $M^2(0) = \frac{2m_B^2\alpha_s}{\pi}$ which consists of two parameters m_B and α_s in the infrared region. The relationship between quark condensate ($<q\bar{q}>$) and $M(p^2)$ for dual QCD is expressed as [19]

$$<q\bar{q}> = -\frac{2}{4\pi^2}\int dp^2\frac{p^2M(p^2)}{\frac{4}{9}p^2+M^2(p^2)} = -\frac{27}{40\pi^2}\Big(\frac{2m_B^2\alpha_s}{\pi}\Big)^{3/2}. \tag{11}$$

The corresponding plot of dynamical quark mass ($M(0)$) and $<q\bar{q}>$ as a function of α_s have been shown is Figure 1 and their numerical values are estimated in Table 2 for various coupling. The

Table 2. The dynamical quark mass $M(0)$ and quark condensate $<q\bar{q}>$ as functions of α_s.

α_s	$m_B(GeV)$	$M(0)(GeV)$	$<q\bar{q}>(GeV^3)$
0.12	2.11	0.583	0.0135
0.22	1.51	0.565	0.0123
0.47	1.21	0.662	0.0198
0.96	0.929	0.726	0.0262

non-zero value of $<q\bar{q}>$ in condensed QCD vacuum appears strong for the large value of m_B,

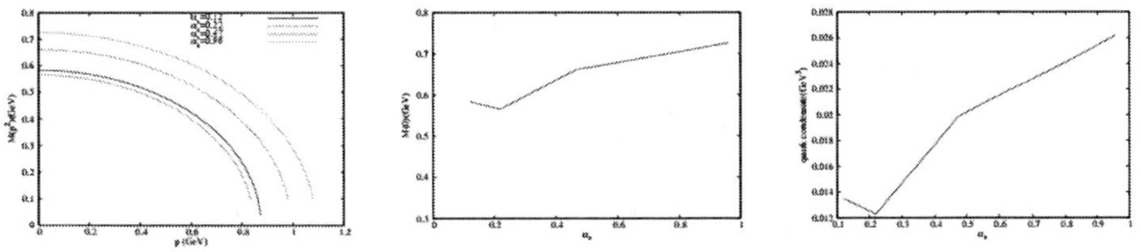

Figure 1. (Color online.) The variation $M(p^2)$ with p, $M(0)$ and $< q\bar{q} >$ with α_s for dual QCD vacuum.

and therefore a large amount of dynamical symmetry breaking seems to be brought on by the QCD-monopole condensation, which is responsible for the color confinement.

Further we study the manifestation of chiral symmetry at finite temperature in the dual QCD vacuum by making the following replacement in the solution of the SDE (10) of the dynamically generated quark-mass function. The solution at finite-temperature SDE in the limit $T \to 0$ is exactly reduced to the SDE at $T = 0$. The quark mass function at finite temperature ($M_T(p^2)$) is expressed as

$$M_T(p^2) = \sqrt{\frac{2(m_B^{(T)})^2\alpha_s}{\pi} - \frac{4p^2}{9}}, \tag{12}$$

where $m_B^{(T)}$ is the thermal glueball masses which reduces monotonously near the critical temperature [20]. The variation of $M_T(p^2)$ at finite temperature is shown in Figure 2 for dual QCD vacuum and demonstrate a rapid decrease with temperature. The quark condensate at finite temperature ($< q\bar{q} >_T$) using the $M_T(p^2)$ at finite temperature for dual QCD vacuum is obtained as

$$< q\bar{q} >_T = -\frac{27}{40\pi^2}\left(\frac{2(m_B^{(T)})^2\alpha_s}{\pi}\right)^{3/2}. \tag{13}$$

The ratio of $< \bar{q}q >_T$ to that of $< \bar{q}q >_{T=0}$ has been plotted as a function of temperature for dual QCD vacuum at $\alpha_s = 0.47$ coupling in Figure 2 which gradually decreases with temperature and vanishes suddenly near the critical temperature, indicating the recovery of the chiral symmetry.

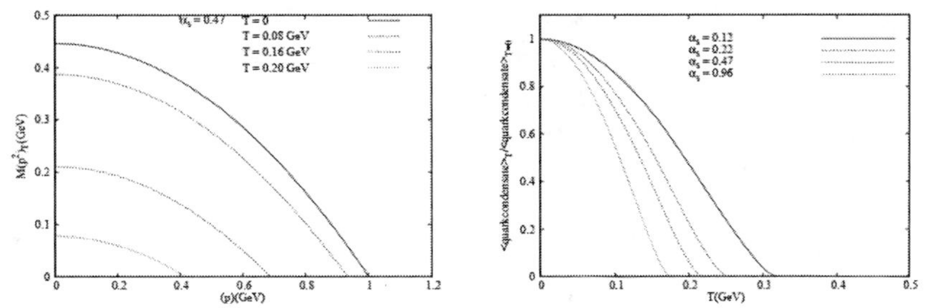

Figure 2. (Color online.) The variation of $M(p^2)$ with p at different T for $\alpha_s = 0.47$ coupling and the ratio of $< \bar{q}q >_T / < \bar{q}q >_{T=0}$ as a function of T for dual QCD vacuum at different coupling.

4 Results and Conclusions

The mechanism of quark confinement has been studied in the context of a gauge theory of non-Abelian monopoles which describes an interesting dual dynamics between the color isocharge and the topological charge of a non-Abelian symmetry and allow an extra internal symmetry called magnetic symmetry which preserves the natural topological structure of the full gauge symmetry. In addition to the color confinement, the breaking of the chiral symmetry is also studied using Schwinger-Dyson techniques, characterized by large effective quark mass and quark condensate. The dependency of the quark mass function on the strong coupling constant and dual gluon mass has been established by solving the SDE for dual QCD and has been shown to lead to an intimate connection between confinement and dynamical chiral symmetry breaking in dual QCD vacuum. It has been shown that the strength of the dynamical chiral symmetry breaking increases in the deep infrared sector in dual QCD where the coupling are very intense. The recovery of the chiral symmetry breaking has also been discussed at finite temperature through the variation of the quark mass function and ratio of quark condensate at finite temperature to that at zero temperature which gradually decreases with temperature and vanishes suddenly near the critical temperature.

5 Acknowledgments

Garima Punetha, is thankful to the organizers of AYSS 2017 and University Grant Commission, New Delhi, India, under the UGC-RFSMS fellowship for financial assistance. Thanks are also due to referees for their valuable comments and suggestions.

References

[1] Y. Nambu, Phys. Rev. **D10**, (1974) 4262.
[2] G. 't Hooft, *High Energy Physics edited by A. Zichichi* (Editorice Compositori, Bologna, 1975).
[3] S. Mandelstam, Phys. Rep. **23**, 245 (1976).
[4] C. N. Yang and R. L. Mills, Phys. Rev. **96**, 191 (1954).
[5] R. Utiyama, Phys. Rev. **101**, 1597 (1956).
[6] Y. M. Cho, Phys. Rev. **D 21**, 1080 (1980).
[7] Y. M. Cho, Phys. Rev. **D 23**, 2415 (1981).
[8] N. Cundy, Y. M. Cho, W. Lee, J. Leem, Phys. Lett. **B 729**, 192 (2014).
[9] H. C. Pandey and H. C. Chandola, Phys. Lett. **B 476**, 193 (2000).
[10] H. C. Chandola, Garima Punetha, H. Dehnen, Nucl. Phys. **A 945**, 226 (2016).
[11] Garima Punetha and H. C. Chandola, *Springer Proceedings in Physics edited by Bhuyan Bipul*, (Springer International Publishing, Switzerland, 2016) p. 653, Vol. 174.
[12] Garima Punetha and H. C. Chandola, EPL **116**, 11001 (2016).
[13] H. J. Munczek and A. M. Nemirovsky, Phys. Rev. **D 28**, 181 (1983).
[14] D. W. Mckay and H. J. Munczek, Phys. Rev. **D 55**, 2455 (1997).
[15] P. C. Tandy, Prog. Part. Nucl. Phys. **39**, 117 (1997).
[16] C. D. Roberts and A. G. Williams, Prog. Part. Nucl. Phys. **33**, 477 (1994).
[17] A. C. Aguilar, A. A. Natale, and R. Rosenfeld, Phys. Rev. **D 62**, 094014 (2000).
[18] Alok Kumar, R. Parthasarathy, Phys. Lett. **B 595**, 373 (2004).
[19] I. A. Shushpanov, A.V. Smilga, Phys. Lett. **B 402**, 351 (1997).
[20] H. C. Chandola and D. Yadav, Nucl. Phys. **A 829**, 151 (2009).

EPJ Web of Conferences **177**, 09010 (2018) https://doi.org/10.1051/epjconf/201817709010
AYSS-2017

The impact of the tensor interaction on the β-delayed neutron emission of the neutron-rich Ni isotopes

E. O. Sushenok[1,⋆], *A. P.* Severyukhin[1,2], *N. N.* Arsenyev[1], and *I. N.* Borzov[3,1]

[1] *Bogoliubov Laboratory of Theoretical Physics, Joint Institute for Nuclear Research, 141980 Dubna, Moscow region, Russia*
[2] *Dubna State University, 141982 Dubna, Moscow region, Russia*
[3] *National Research Centre "Kurchatov Institute", 123182 Moscow, Russia*

Abstract. The neutron emission of the β-decay of ^{74,76,78,80}Ni are studied within the quasi-particle random phase approximation with the Skyrme interaction. The coupling between one- and two-phonon terms in the wave functions of the low-energy 1^+ states of the daughter nuclei is taken into account. It is shown that the strength decrease of the neutron-proton tensor interaction leads to the increase of the half-life and the neutron-emission probability.

Study of the β-decay properties of the nuclei near the doubly magic isotope ^{78}Ni plays the crucial role in the modeling of the astrophysical r-process [1]. The multi-neutron emission is a multistep process consisting of the β-decay of the parent nucleus (N, Z) which results in feeding the excited states of the daughter nucleus (N - 1, Z + 1) followed by the multi-neutron emissions to the ground state or γ-deexcitation to the ground state of the product nucleus (N - 1 - X, Z + 1) [2]. The microscopic study of the β-decay half-lives and the β-delayed neutron-emission probabilities of the nuclei with high N/Z ratio asymmetry makes it possible to reconstruct the β-strength function [2, 3]. Taking into account the tensor interaction has a substantial effect on the Gamow–Teller (GT) transitions within the β-decay window [4]. A redistribution of the GT strength has been found due to the impact of the tensor correlations and $2p$-$2h$ fragmentation of the GT transitions [5].

One of the successful tools for studying charge-exchange nuclear modes is the quasiparticle random phase approximation (QRPA) with the self-consistent mean-field derived from a Skyrme energy-density functional (EDF), see e.g., [6–9]. These QRPA calculations enable one to describe the properties of the ground state and excited charge-exchange states using the same EDF. As proposed in Ref. [10], the study of the multineutron emission, following the β-decay, requires taken into account the phonon-phonon coupling (PPC). The finite rank separable approximation (FRSA) [11–13] for the residual interaction allows one to perform the calculations in large configurational spaces.

In this paper, we investigate the impact of the tensor correlations on the probability of the neutron emission. As the parameter set in the particle-hole channel, we use the Skyrme EDFs TIJ [14], which take into account the tensor terms. Indices I and J refer to the proton-neutron (β) and like-particle (α) coupling constants that define the contribution to the spin-orbit potential from the $\mathbf{J}^2$ terms:

$$U_{SO}^{(q)} = \frac{W_0}{2r}\left(2\frac{d\rho_q}{dr} + \frac{d\rho_{q'}}{dr}\right) + \left(\alpha\frac{J_q}{r} + \beta\frac{J_{q'}}{r}\right), \quad (1)$$

⋆e-mail: sushenok@theor.jinr.ru

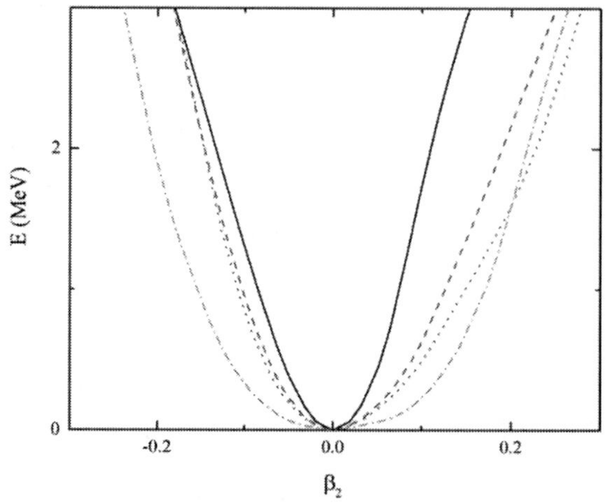

Figure 1. (Color online) Deformation energy curve as a function of the mass quadrupole deformation. The curves of ^{74}Ni, ^{76}Ni, ^{78}Ni and ^{80}Ni are denoted by the dotted, dashed, solid, and dash-dotted lines respectively.

$$\alpha = 60(J-2) \quad \text{MeV fm}^5 \tag{2}$$

$$\beta = 60(I-2) \quad \text{MeV fm}^5 \tag{3}$$

where ρ_q, J_q ($q = n, p$) are the nucleon density and the spin-orbit current respectively. The parameter W_0 describes the strength of the spin-orbit force.

The starting point is the HF-BCS calculation of the ground state properties of the even-even parent nucleus (N,Z) within a spherical symmetry assumption. The continuous part of the single-particle spectrum is discretized by diagonalizing the HF hamiltonian on a harmonic oscillator basis. The pairing correlations are generated by a zero-range force with a strength $V_0 = -270$ MeV fm^3. This value of the pairing strength has been fitted to reproduce the odd-even mass difference in the studied region of nuclei [15, 16].

$$\begin{aligned} V^{(pp)}_{T=1}(\vec{r}_1,\vec{r}_2) &= V_0\left(\frac{1-P_\sigma}{2}\right)\left(1-\eta\frac{\rho(r_1)}{\rho_0}\right)\delta(\vec{r}_1-\vec{r}_2), \\ V^{(pp)}_{T=0}(\vec{r}_1,\vec{r}_2) &= fV_0\left(\frac{1+P_\sigma}{2}\right)\left(1-\eta\frac{\rho(r_1)}{\rho_0}\right)\delta(\vec{r}_1-\vec{r}_2), \end{aligned} \tag{4}$$

where P_σ is the spin-exchange operator and ρ_0 is the nuclear matter density. We use $f = 1$ i.e. the realization of the SU(4) symmetry [17]. In this work, we use volume pairing forces ($\eta = 0$). The binding energies of the daughter $B(N-1, Z+1)$ and final nuclei $B(N-1-X, Z+1)$ are calculated with the blocking effect for unpaired nucleons [18].

Keeping spherical symmetry might not be that bad for ^{74,76,78,80}Ni. In fig. 1, we show the HF-BCS energy curve obtained with a constraint on the mass quadrupole moment Q_2 as a function of the dimensionless quadrupole deformation β_2:

$$\beta_2 = \sqrt{\frac{5}{16\pi}\frac{4\pi Q_2}{3R^2A}}, \tag{5}$$

where the radius constant is given by $R = 1.2A^{1/3}$ fm. The deformation energies are obtained using the EV8 code [19] that solves the HF-BCS equations for the same EDF via a discretization of the quasiparticle wave function on a three-dimensional Cartesian mesh.

The correct description of the Q_β values for the parent nuclei and the neutron separation energies (S_n, S_{2n}) for the daughter nuclei is the important ingredient for the reliable prediction of the neutron-emission probability. The Q_β-value and the neutron separation energies are given by

$$Q_\beta = \Delta M_{n-H} + B(N-1, Z+1) - B(N, Z), \tag{6}$$

$$S_{xn} = B(N-1, Z+1) - B(N-1-X, Z+1). \tag{7}$$

$\Delta M_{n-H} = 0.782$ MeV is the mass difference between the neutron and the hydrogen atom. Skyrme EDFs T43 and T45 has been selected to reproduce the experimental Q_β and S_n values (see fig. 2), and enough positive value of the spin-isospin Landau parameter ($G_0^{'} \geq 0.1$) [20]. The parameter set T43 corresponds to the relatively strong ($\beta/\alpha = 2.0$) neutron-proton tensor correlations and set T45 corresponds to the weak ($\beta/\alpha = 0.7$) neutron-proton tensor correlations with respect to the like-particle tensor part. The results of the HF-BCS calculations with the EDFs T43 and T45 are in a reasonable agreement with the experimental data [21, 22] (see Ref. [23] for more details).

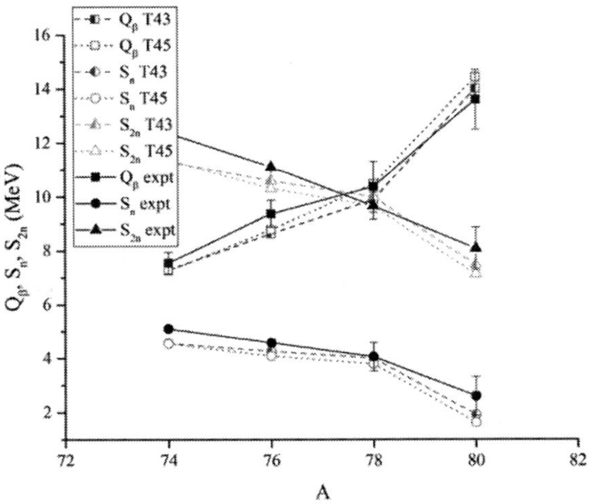

Figure 2. (Color online) The β-decay energy Q_β of ^{74,76,78,80}Ni and one- and two-neutron separation energies of Cu isotopes with the same mass number, calculated with Skyrme forces T43 and T45. Experimental data are taken from the Ref. [21, 22].

To build the QRPA equations on the basis of HF-BCS quasiparticle states of the parent nucleus is the standard procedure [24]. Using the FRSA model, the QRPA eigenvalues (E_k) are obtained as the roots of the secular equation [11–13], and we carry out QRPA calculations in very large two-quasiparticle spaces. To take into account the PPC effects we construct the wave functions from a linear combination of one-phonon and two-phonon configurations [5]

$$\Psi_\nu(JM) = \left(\sum_i R_i(J\nu) Q^+_{JMi} + \sum_{\lambda_1 i_1 \lambda_2 i_2} P^{\lambda_1 i_1}_{\lambda_2 i_2}(J\nu) \left[Q^+_{\lambda_1\mu_1 i_1} \bar{Q}^+_{\lambda_2\mu_2 i_2} \right]_{JM} \right) |0\rangle , \tag{8}$$

where λ denotes the total angular momentum and μ is its z-projection in the laboratory system. The ground state of the parent nucleus (N, Z) is the QRPA phonon vacuum $|0\rangle$. The wave functions $Q^+_{\lambda\mu i}|0\rangle$ of the one-phonon excited states of the daughter nucleus (N - 1, Z + 1) are described as linear combinations of 2qp configurations; $\bar{Q}^+_{\lambda\mu i}|0\rangle$ is a one-phonon excitation of the parent nucleus

Table 1. The half-lives and the neutron-emission probabilities for the β-decay ^{74,76,78,80}Ni calculated with Skyrme interactions T43 and T45. Experimental data are taken from the Ref. [21].

A	T43			T45			Experiment
	$T_{1/2}$, 10^{-3}s	P_{1n}, %	P_{2n}, %	$T_{1/2}$, 10^{-3}s	P_{1n}, %	P_{2n}, %	$T_{1/2}$, 10^{-3}s
74	40	1	0	281	3	0	507.7±4.6
76	19	9	0	162	11	0	234.6±2.7
78	10	12	0	115	100	0	122.2±5.1
80	4	80	19	40	0	94	$23.9^{+26.0}_{-17.2}$

(N, Z). All one- and two-phonon configurations with the excitation energy of the daughter nucleus up to 19 MeV are included. Simultaneous taking into account the effects from the tensor force and the phonon-phonon coupling allows us not to use any quenching factors [25].

In the allowed GT approximation, the β^--decay half-life is expressed by summing the probabilities (in units of $G_A^2/4\pi$) of the energetically allowed transitions ($E_k^{\rm GT} \leq Q_\beta$) weighted with the integrated Fermi function

$$T_{1/2}^{-1} = D^{-1}\left(\frac{G_A}{G_V}\right)^2 \sum_k f_0(Z+1, A, E_k^{\rm GT})B(GT)_k, \tag{9}$$

$$E_k^{\rm GT} = Q_\beta - E_{1_k^+}, \tag{10}$$

where G_A/G_V=1.25 and D=6147 s [26]. $E_{1_k^+}$ denotes the excitation energy of the 1_k^+ state of the daughter nucleus. As proposed in Ref. [6], this energy can be estimated by the following expression:

$$E_{1_k^+} \approx E_k - E_{2qp,lowest}. \tag{11}$$

The difference in the characteristic time scales of the β-decay and subsequent neutron emission processes justifies an assumption of their statistical independence. As proposed in Ref. [27], the P_{xn} probability of the βxn emission accompanying the β-decay to the excited states in the daughter nucleus can be expressed as

$$P_{xn} = T_{1/2}D^{-1}\left(\frac{G_A}{G_V}\right)^2 \sum_{k'} f_0(Z+1, A, E_{k'}^{\rm GT})B(GT)_{k'},$$

where the GT transition energy ($E_{k'}^{\rm GT}$) is located within the neutron emission window $Q_{\beta xn} \equiv Q_\beta - S_{xn}$.

The impact of the tensor interaction on the half-lives and the probabilities of β-delayed neutron emission is shown in the table. The half-life description of $^{74-80}$Ni, obtained with the T45 EDF, are in a reasonable agreement with the experimental data [21]. The strong neutron-proton tensor force in the case of T43 set leads to the reduction of the β-decay half-lives for all studied Ni isotopes. In case of the β-decay of ^{76}Ni the neutron-emission probabilities, calculated with T43 and T45 EDFs, are in a reasonable agreement with the experimental data, $P_{tot} = 14.0 \pm 3.6\%$ [21]. The $N = 50$ shell-closure is reflected in the tensor-force effect, as is shown in the table.

The neutron emission of the β-decay of ^{74,76,78,80}Ni are studied within the quasiparticle random phase approximation with the Skyrme interaction taking into account the tensor terms. The coupling between one- and two-phonon terms in the wave functions of the low-energy 1^+ states of the daughter nuclei is taken into account. It is shown that the reduction of the neutron-proton tensor interaction leads to the increase of the half-life and the neutron-emission probability. The results of calculations with the Skyrme interaction T45 are in a reasonable agreement with available experimental data.

Acknowledgments

This work is supported by the Russian Science Foundation (Grant No. RSF-16-12-10161).

References

[1] M. Madurga, S. V. Paulauskas, R. Grzywacz, D. Miller, D. W. Bardayan, J. C. Batchelder, N. T. Brewer, J. A. Cizewski, A. Fijałkowska, C. J. Gross, M. E. Howard, S. V. Ilyushkin, B. Manning, M. Matoš, A. J. Mendez II, K. Miernik, S. W. Padgett, W. A. Peters, B. C. Rasco, A. Ratkiewicz, K. P. Rykaczewski, D. W. Stracener, E. H. Wang, M. Wolińska-Cichocka, E. F. Zganjar, Phys. Rev. Lett. **117**, 092502 (2016)

[2] I. N. Borzov, Phys. Rev. C **71**, 065801 (2005)

[3] I. N. Borzov, EPJ Web of Conf. **107**, 06002 (2016)

[4] F. Minato, C. L. Bai, Phys. Rev. Lett. **110**, 122501 (2013); **116**, 089902(E) (2016).

[5] A. P. Severyukhin, V. V. Voronov, I. N. Borzov, N. N. Arsenyev, Nguyen Van Giai, Phys. Rev. C **90**, 044320 (2014)

[6] J. Engel, M. Bender, J. Dobaczewski, W. Nazarewicz, R. Surman, Phys. Rev. C **60**, 014302 (1999)

[7] M. Bender, J. Dobaczewski, J. Engel, W. Nazarewicz, Phys. Rev. C **65**, 054322 (2002)

[8] S. Fracasso, G. Colò, Phys. Rev. C **76**, 044307 (2007)

[9] C. L. Bai, H. Sagawa, H. Q. Zhang, X. Z. Zhang, G. Coló, F. R. Xu, Phys. Lett. B **675**, 28 (2009)

[10] A. P. Severyukhin, N. N. Arsenyev, I. N. Borzov, E. O. Sushenok, Phys. Rev. C **95**, 034314 (2017)

[11] Nguyen Van Giai, Ch. Stoyanov, V. V. Voronov, Phys. Rev. C **57**, 1204 (1998)

[12] A. P. Severyukhin, V. V. Voronov, Nguyen Van Giai, Prog. Theor. Phys. **128**, 489 (2012)

[13] A. P. Severyukhin, H. Sagawa, Prog. Theor. Exp. Phys. **2013**, 103D03 (2013)

[14] T. Lesinski, M. Bender, K. Bennaceur, T. Duguet, J. Meyer, Phys. Rev. C **76**, 014312 (2007)

[15] A. P. Severyukhin, V. V. Voronov, Nguyen Van Giai, Phys. Rev. C **77**, 024322 (2008)

[16] A. P. Severyukhin, N. N. Arsenyev, N. Pietralla, Phys. Rev. C **86**, 024311 (2012)

[17] Yu. V. Gaponov, Yu. S. Lyutostansky, Sov. Phys. Part. Nucl. **12**, 1324 (1984)

[18] V. G. Soloviev, Kgl. Dan. Vid. Selsk. Mat. Fys. Skr. **1**, 238 (1961)

[19] P. Bonche, H. Flocard, P. H. Heenen, Comput. Phys. Commun. **171**, 49 (2005)

[20] A. P. Severyukhin, E. O. Sushenok, Phys. Atom. Nucl. **78**, 680 (2015)

[21] M. Birch, B. Singh, I. Dillmann, D. Abriola, T. D. Johnson, E. A. McCutchan, A. A. Sonzogni, Nuclear Data Sheets **128**, 131 (2015)

[22] M. Wang , G. Audi, A. H. Wapstra, F. G. Kondev, M. MacCormick, X. Xu, B. Pfeiffer, Chin. Phys. C **36**(12), 1603 (2012)

[23] E. O. Sushenok, A. P. Severyukhin, N. N. Arsenyev, I. N. Borzov, Phys. Atom. Nucl. **81**, 15-22 (2018)

[24] J. Terasaki, J. Engel, M. Bender, J. Dobaczewski, W. Nazarewicz, M. Stoitsov, Phys. Rev. C **71**, 034310 (2005)

[25] G. F. Bertsch, I. Hamamoto, Phys. Rev. C **26**, 1323 (1982)

[26] J. Suhonen, *From Nucleons to Nucleus* (Springer-Verlag, Berlin, 2007)

[27] A. C. Pappas, T. Sverdrup, Nucl. Phys. A **188**, 48 (1972)

AUTHOR INDEX

AUTHOR INDEX

EDP Sciences – Web of Conferences
17, Avenue du Hoggar
Parc d'Activité de Courtabouf
BP 112
F-91944 Les Ulis Cedex A
France

ISBN 978-1-5108-6195-4

CURRAN ASSOCIATES INC.
proceedings.com